NANOTECHNOLOGY AND NANOMATERIAL APPLICATIONS IN FOOD, HEALTH, AND BIOMEDICAL SCIENCES

Innovations in Agricultural and Biological Engineering

NANOTECHNOLOGY AND NANOMATERIAL APPLICATIONS IN FOOD, HEALTH, AND BIOMEDICAL SCIENCES

Edited by

Deepak Kumar Verma, PhD
Megh R. Goyal, PhD, PE
Hafiz Ansar Rasul Suleria, PhD

AAP | APPLE ACADEMIC PRESS

Apple Academic Press Inc.
3333 Mistwell Crescent
Oakville, ON L6L 0A2
Canada USA

Apple Academic Press Inc.
1265 Goldenrod Circle NE
Palm Bay, Florida 32905
USA

Exclusive worldwide distribution by CRC Press, a member of Taylor & Francis Group

International Standard Book Number-13: 978-1-77188-764-9 (Hardcover)
International Standard Book Number-13: 978-0-42942-566-0 (eBook)

Library and Archives Canada Cataloguing in Publication

Title: Nanotechnology and nanomaterial applications in food, health, and biomedical sciences / edited by Deepak Kumar Verma, PhD, Megh R. Goyal, PhD, PE, Hafiz Ansar Rasul Suleria, PhD.

Names: Verma, Deepak Kumar, 1986- editor. | Goyal, Megh Raj, editor. | Suleria, Hafiz, editor.

Series: Innovations in agricultural and biological engineering.

Description: Series statement: Innovations in agricultural and biological engineering | Includes bibliographical references and index.

Identifiers: Canadiana (print) 20190111240 | Canadiana (ebook) 20190111372 | ISBN 9781771887649 (hardcover) | ISBN 9780429425660 (PDF)

Subjects: LCSH: Nanotechnology. | LCSH: Nanostructured materials. | LCSH: Biotechnology industries. | LCSH: Food industry and trade.

Classification: LCC T174.7 .N3453 2019 | DDC 620/.5—dc23

CIP data on file with US Library of Congress

Apple Academic Press also publishes its books in a variety of electronic formats. Some content that appears in print may not be available in electronic format. For information about Apple Academic Press products, visit our website at **www.appleacademicpress.com** and the CRC Press website at **www.crcpress.com**

THE BOOK SERIES: INNOVATIONS IN AGRICULTURAL AND BIOLOGICAL ENGINEERING, APPLE ACADEMIC PRESS INC.

Under the book series titled *Innovations in Agricultural and Biological Engineering*, Apple Academic Press Inc. is publishing subsequent volumes in the specialty areas defined by the American Society of Agricultural and Biological Engineers (<asabe.org>) over a span of 8–10 years. Academic Press Inc. wants to be principal source of books in agricultural biological engineering. We seek book proposals from readers in areas of their expertise.

The mission of this series is to provide knowledge and techniques for agricultural and biological engineers (ABEs). The series offers high-quality reference and academic content in agricultural and biological engineering (ABE) that is accessible to academicians, researchers, scientists, university faculty and university-level students, and professionals around the world.

Agricultural and biological engineers ensure that the world has the necessities of life, including safe and plentiful food, clean air and water, renewable fuel and energy, safe working conditions, and a healthy environment by employing knowledge and expertise of the sciences, both pure and applied, and engineering principles. Biological engineering applies engineering practices to problems and opportunities presented by living things and the natural environment in agriculture.

ABE embraces a variety of the following specialty areas (asabe.org): aquacultural engineering, biological engineering, energy, farm machinery and power engineering, food and process engineering, forest engineering, information & electrical technologies engineering, natural resources, nursery and greenhouse engineering, safety and health, and structures and environment.

For this book series, we welcome chapters on the following specialty areas (but not limited to):

1. Academia-to-industry-to-end-user loop in agricultural engineering
2. Agricultural mechanization
3. Aquaculture engineering

4. Biological engineering in agriculture
5. Biotechnology applications in agricultural engineering
6. Energy source engineering
7. Food and bioprocess engineering
8. Forest engineering
9. Hill land agriculture
10. Human factors in engineering
11. Information and electrical technologies
12. Irrigation and drainage engineering
13. Nanotechnology applications in agricultural engineering
14. Natural resources engineering
15. Nursery and greenhouse engineering
16. Potential of phytochemicals from agricultural and wild plants for human health
17. Power systems and machinery design
18. GPS and remote sensing potential in agricultural engineering
19. Robot engineering in agriculture
20. Simulation and computer modeling
21. Smart engineering applications in agriculture
22. Soil and water engineering
23. Structures and environment engineering
24. Waste management and recycling
25. Any other focus area

For more information on this series, readers may contact:

Megh R. Goyal, PhD, PE
Book Series Senior Editor-in-Chief
Innovations in Agricultural and
Biological Engineering
E-mail: goyalmegh@gmail.com

BOOKS ON AGRICULTURAL & BIOLOGICAL ENGINEERING BY APPLE ACADEMIC PRESS, INC.

Management of Drip/Trickle or Micro Irrigation
Megh R. Goyal, PhD, PE, Senior Editor-in-Chief

Evapotranspiration: Principles and Applications for Water Management
Megh R. Goyal, PhD, PE and Eric W. Harmsen, Editors

Book Series: Research Advances in Sustainable Micro Irrigation
Senior Editor-in-Chief: Megh R. Goyal, PhD, PE

Volume 1: Sustainable Micro Irrigation: Principles and Practices
Volume 2: Sustainable Practices in Surface and Subsurface Micro Irrigation
Volume 3: Sustainable Micro Irrigation Management for Trees and Vines
Volume 4: Management, Performance, and Applications of Micro Irrigation Systems
Volume 5: Applications of Furrow and Micro Irrigation in Arid and Semi-Arid Regions
Volume 6: Best Management Practices for Drip Irrigated Crops
Volume 7: Closed Circuit Micro Irrigation Design: Theory and Applications
Volume 8: Wastewater Management for Irrigation: Principles and Practices
Volume 9: Water and Fertigation Management in Micro Irrigation
Volume 10: Innovation in Micro Irrigation Technology

Book Series: Innovations and Challenges in Micro Irrigation
Senior Editor-in-Chief: Megh R. Goyal, PhD, PE

- Micro Irrigation Engineering for Horticultural Crops: Policy Options, Scheduling and Design
- Micro Irrigation Management: Technological Advances and Their Applications
- Micro Irrigation Scheduling and Practices
- Performance Evaluation of Micro Irrigation Management: Principles and Practices

- Potential of Solar Energy and Emerging Technologies in Sustainable Micro Irrigation
- Principles and Management of Clogging in Micro Irrigation
- Sustainable Micro Irrigation Design Systems for Agricultural Crops: Methods and Practices
- Engineering Interventions in Sustainable Trickle Irrigation: Water Requirements, Uniformity, Fertigation, and Crop Performance
- Management Strategies for Water Use Efficiency and Micro Irrigated Crops: Principles, Practices, and Performance

Book Series: Innovations in Agricultural & Biological Engineering
Senior Editor-in-Chief: Megh R. Goyal, PhD, PE

- Dairy Engineering: Advanced Technologies and Their Applications
- Developing Technologies in Food Science: Status, Applications, and Challenges
- Engineering Interventions in Agricultural Processing
- Engineering Practices for Agricultural Production and Water Conservation: An Inter-disciplinary Approach
- Emerging Technologies in Agricultural Engineering
- Flood Assessment: Modeling and Parameterization
- Food Engineering: Emerging Issues, Modeling, and Applications
- Food Process Engineering: Emerging Trends in Research and Their Applications
- Food Technology: Applied Research and Production Techniques
- Modeling Methods and Practices in Soil and Water Engineering
- Processing Technologies for Milk and Dairy Products: Methods Application and Energy Usage
- Soil and Water Engineering: Principles and Applications of Modeling
- Soil Salinity Management in Agriculture: Technological Advances and Applications
- Technological Interventions in the Processing of Fruits and Vegetables
- Technological Interventions in Management of Irrigated Agriculture
- Engineering Interventions in Foods and Plants
- Technological Interventions in Dairy Science: Innovative Approaches in Processing, Preservation, and Analysis of Milk Products

- Novel Dairy Processing Technologies: Techniques, Management, and Energy Conservation
- Sustainable Biological Systems for Agriculture: Emerging Issues in Nanotechnology, Biofertilizers, Wastewater, and Farm Machines
- State-of-the-Art Technologies in Food Science: Human Health, Emerging Issues and Specialty Topics
- Scientific and Technical Terms in Bioengineering and Biological Engineering
- Engineering Practices for Management of Soil Salinity: Agricultural, Physiological, and Adaptive Approaches
- Processing of Fruits and Vegetables: From Farm to Fork
- Technological Processes for Marine Foods, from Water to Fork: Bioactive Compounds, Industrial Applications, and Genomics
- Engineering Practices for Milk Products: Dairyceuticals, Novel Technologies, and Quality
- Nanotechnology and Nanomaterial Applications in Food, Health, and Biomedical Sciences
- Nanotechnology Applications in Dairy Science: Packaging, Processing, and Preservation

ABOUT THE EDITOR
DEEPAK KUMAR VERMA

Deepak Kumar Verma is an agricultural science professional and is currently a PhD Research Scholar in the specialization of food processing engineering in the Agricultural and Food Engineering Department at the Indian Institute of Technology, Kharagpur (WB), India. In 2012, he received a DST-INSPIRE Fellowship for PhD study from the Department of Science & Technology (DST), Ministry of Science and Technology, Government of India. Mr. Verma is currently working on the research project "Isolation and Characterization of Aroma Volatile and Flavoring Compounds from Aromatic and Non-Aromatic Rice Cultivars of India." His previous research work included "Physico-Chemical and Cooking Characteristics of Azad Basmati (CSAR 839-3): A Newly Evolved Variety of Basmati Rice (*Oryza sativa* L.)." He earned his BSc degree in agricultural science from the Faculty of Agriculture at Gorakhpur University, Gorakhpur, and his MSc in Agricultural Biochemistry in 2011. He also received an award from the Department of Agricultural Biochemistry, Chandra Shekhar Azad University of Agriculture and Technology, Kanpur, India. Apart from his area of specialization in plant biochemistry, he has also built a sound background in plant physiology, microbiology, plant pathology, genetics and plant breeding, plant biotechnology and genetic engineering, seed science and technology, and food science and technology. In addition, he is member of several professional bodies, and his activities and accomplishments include conferences, seminars, workshops, training, and the publication of research articles, books, and book chapters.

ABOUT THE SENIOR EDITOR-IN-CHIEF MEGH R. GOYAL

Megh R. Goyal, PhD

Retired Professor in Agricultural and Biomedical Engineering, University of Puerto Rico, Mayaguez Campus; Senior Acquisitions Editor, Biomedical Engineering and Agricultural Science, Apple Academic Press, Inc.

Megh R. Goyal, PhD, PE, is a Retired Professor in Agricultural and Biomedical Engineering from the General Engineering Department in the College of Engineering at the University of Puerto Rico–Mayaguez Campus; and Senior Acquisitions Editor and Senior Technical Editor-in-Chief in Agriculture and Biomedical Engineering for Apple Academic Press, Inc. He has worked as a Soil Conservation Inspector and as a Research Assistant at Haryana Agricultural University and Ohio State University.

During his professional career of 49 years, Dr. Goyal has received many prestigious awards and honors. He was the first agricultural engineer to receive the professional license in Agricultural Engineering in 1986 from the College of Engineers and Surveyors of Puerto Rico. In 2005, he was proclaimed as "Father of Irrigation Engineering in Puerto Rico for the Twentieth Century" by the American Society of Agricultural and Biological Engineers (ASABE), Puerto Rico Section, for his pioneering work on micro irrigation, evapotranspiration, agroclimatology, and soil and water engineering. The Water Technology Centre of Tamil Nadu Agricultural University in Coimbatore, India, recognized Dr. Goyal as one of the experts "who rendered meritorious service for the development of micro irrigation sector in India" by bestowing the Award of Outstanding Contribution in Micro Irrigation. This award was presented to Dr. Goyal during the inaugural session of the National Congress on "New Challenges and Advances in Sustainable Micro Irrigation" on March 1, 2017, held at Tamil Nadu Agricultural University. Dr. Goyal also received the Netafim Award for Advancements in Microirrigation: 2018 from the American Society of Agricultural Engineers at the ASABE International Meeting in August 2018.

A prolific author and editor, he has written more than 200 journal articles and textbooks and has edited over 65 books. He is the editor of three book series published by Apple Academic Press: Innovations in Agricultural & Biological Engineering, Innovations and Challenges in Micro Irrigation, and Research Advances in Sustainable Micro Irrigation. He is also instrumental in the development of the new book series Innovations in Plant Science for Better Health: From Soil to Fork.

Dr. Goyal received his BSc degree in engineering from Punjab Agricultural University, Ludhiana, India; his MSc and PhD degrees from Ohio State University, Columbus; and his Master of Divinity degree from Puerto Rico Evangelical Seminary, Hato Rey, Puerto Rico, USA.

ABOUT THE EDITOR HAFIZ SULERIA

Hafiz Ansar Rasul Suleria, PhD

Hafiz Ansar Rasul Suleria, PhD, is currently working to the Alfred Deakin Research Fellow at Deakin University, Melbourne, Australia. He is also an Honorary Fellow in the Diamantina Institute, Faculty of Medicine, The University of Queensland, Australia.

Recently he worked as Postdoc Research Fellow in the Department of Food, Nutrition, Dietetic and Health at Kansas State University, USA.

Previously, he has been awarded an International Postgraduate Research Scholarship (IPRS) and Australian Postgraduate Award (APA) for his PhD research at the UQ School of Medicine, the Translational Research Institute (TRI) in collaboration with the Commonwealth and Scientific and Industrial Research Organization (CSIRO, Australia).

Before joining the UQ, he worked as a Lecturer in the Department of Food Sciences, Government College University Faisalabad, Pakistan. He also worked as a Research Associate in the PAK-US Joint Project funded by the Higher Education Commission, Pakistan, and Department of State, USA, with the collaboration of the University of Massachusetts, USA, and National Institute of Food Science and Technology, University of Agriculture Faisalabad, Pakistan.

He has a significant research focus on food nutrition, particularly in the screening of bioactive molecules—isolation, purification, and characterization using various cutting-edge techniques from different plant, marine, and animal source, and *in vitro, in vivo* bioactivities, cell culture, and animal modeling. He has also done a reasonable amount of work on functional foods and nutraceutical, food and function, and alternative medicine.

Dr. Suleria has published more than 60 peer-reviewed scientific papers in different reputed/impacted journals. He is also in collaboration with more than ten universities where he is working as a co-supervisor/special member for PhD and postgraduate students and is also involved in joint publications, projects, and grants. He is Editor-in-Chief for the book series on Innovations in Plant Science for Better Health: From Soil to Fork, published by AAP. Readers may contact him at: hafiz.suleria@uqconnect.edu.au.

CONTENTS

CONTRIBUTORS

Munawar Abbas, MS
Research Associate, Institute of Home and Food Sciences, Department of Food Science, Nutrition and Home Economics, Government College University, Faisalabad, Pakistan.
E-mail: foodian2007@gmail.com

Muhammad Afzaal, MSc
PhD Research Scholar and Lecturer, Institute of Home and Food Sciences, Department of Food Science, Nutrition and Home Economics, Government College University, Faisalabad, Pakistan. E-mail: muhammadafzaal@gcuf.edu.pk

Nazir Ahmad, PhD
Assistant Professor, Institute of Home and Food Sciences, Government College University, Faisalabad, Faisalabad, Pakistan. E-mail: nazirahmad83@gmail.com

Sara Antunes, MSc
Graduate Student, Faculty of Pharmacy, University of Coimbra, Azinhaga de Santa Comba, Pólo da Saúde (III), 3000-548 Coimbra, Portugal. E-mail: sara.mg.antunes@gmail.com

Huma Bader-Ul-Ain, MS
Research Associate, Institute of Home and Food Sciences, Department of Food Science, Nutrition and Home Economics, Government College University, Faisalabad, Pakistan.
E-mail: humahums@yahoo.com

H. Bava Bakrudeen, PhD
Postdoctoral Researcher, Biotechnology Department, Indian Institute of Technology Madras, Chennai 600036, Tamil Nadu, India. E-mail: deen.pharmacy@gmail.com

Ajit Behera, PhD
Assistant Professor, Department of Metallurgical and Materials Engineering, National Institute of Technology, Rourkela 769008, Odisha, India.
E-mail: ajit.behera88@gmail.com

Prapaporn Boonme, PhD
Associate Professor, Department of Pharmaceutical Technology, Faculty of Pharmaceutical Sciences, Prince of Songkla University, Songkhla, Thailand
NANOTEC - PSU Center of Excellence on Drug Delivery System, Songkhla, Thailand.
E-mail: prapaporn.b@psu.ac.th

K. Chandraraj, PhD
Associate Professor, Department of Biotechnology, Indian Institute of Technology Madras, Chennai 600036, Tamil Nadu, India. E-mail: kcraj@iitm.ac.in

Joseph Adetunji Elegbede, BTech
Research Scholar, Laboratory of Industrial Microbiology and Nanobiotechnology, Nanotechnology Research Group (NANO+), Department of Pure and Applied Biology, Ladoke Akintola University of Technology, P. M. B. 4000, Ogbomoso, Nigeria.
E-mail: josephadetunji18@gmail.com

Lavanya Goodla, PhD
Postdoctoral Research Associate, Key Laboratory of Regenerative Biology, South China Institute for Stem Cell Biology and Regenerative Medicine, Guangzhou Institutes of Biomedicine and Health, Chinese Academy of Sciences, Guangzhou 510530, China.
E-mail: lavanya_goodla@gibh.ac.cn; lavanyag47@gmail.com

Megh R. Goyal, PhD, PE
Retired Faculty, Agricultural and Biomedical Engineering, College of Engineering, University of Puerto Rico – Mayaguez Campus
Senior Acquisitions Editor and Senior Technical Editor-in-Chief in Agricultural and Biomedical Engineering for Apple Academic Press Inc.; PO Box 86, Rincon – PR – 006770086, USA.
E-mail: goyalmegh@gmail.com

Huey-Min Hwang, PhD
Professor of Biology, Department of Biology, Jackson State University, Jackson, Mississippi 39217, USA. E-mail: huey-min.hwang@jsums.edu

J. Prasana Manikanda Kartik, MTech
PhD Research Scholar, Biotechnology Department, Indian Institute of Technology Madras, Chennai 600036, Tamil Nadu, India. E-mail: prasanamanikandakartik@gmail.com

Agbaje Lateef, PhD
Professor of Microbiology, Laboratory of Industrial Microbiology and Nanobiotechnology, Nanotechnology Research Group (NANO+), Department of Pure and Applied Biology, Ladoke Akintola University of Technology, PMB 4000, Ogbomoso, Nigeria.
E-mail: alateef@lautech.edu.ng; agbaje72@yahoo.com

Manjunath Manubolu, PhD
Postdoctoral Research Associate, Aquatic Ecology Laboratory, Department of Evolution, Ecology and Organismal Biology, 230 Reserach Center, The Ohio State University, Columbus, OH 43212, USA.
E-mail: manubolu.1@osu.edu; drmanubolubiochem@gmail.com

Soumya Sanjeeb Mohapatra, PhD
Assistant Professor, Department of Chemical Engineering, National Institute of Technology, Rourkela 769008, Odisha, India. E-mail: mohapatras@nitrkl.ac.in

V. Monika, MTech
PhD Research Scholar, Biotechnology Department, Indian Institute of Technology Madras, Chennai 600036, Tamil Nadu, India. E-mail: monika555meister@gmail.com

Kavitha Pathakoti, PhD
Postdoctoral Research Associate, Department of Biology, Jackson State University, Jackson, Mississippi 39217, USA.
E-mail: kavitha.pathakoti@jsums.edu; kavithapathakoti@gmail.com

Irina Pereira, MSc
Graduate Student, Faculty of Pharmacy, University of Coimbra, Azinhaga de Santa Comba, Pólo da Saúde (III), 3000-548 Coimbra, Portugal
Department of Biology and Environment, School of Life and Environmental Sciences, (ECVA, UTAD), University of Trás-os-Montes and Alto Douro, P.O. Box 1013; 5001-801 Vila Real, Portugal.
E-mail: irina.pereira@live.com.pt

Y. Antony Prabhu, MPhil
PhD Research Scholar, Biochemistry Department, Bharathiar University, Coimbatore 641046, Tamil Nadu, India. E-mail: antonybiochemist@gmail.com

B. Ramiya, MSc
PhD Research Scholar, Biotechnology Department, Indian Institute of Technology Madras, Chennai 600036, Tamil Nadu, India. E-mail: ramee.bask@gmail.com

Swati Rana, MTech
PhD Research Scholar, School of Biotechnology, Gauatm Buddha University, Gautam Budh Nagar, Greater Noida 201310, Uttar Pradesh, India. E-mail: 1995swatirana@gmail.com

Farhan Saeed, PhD
Assistant Professor, Institute of Home and Food Sciences, Department of Food Science, Nutrition and Home Economics, Government College University, Faisalabad, Pakistan.
E-mail: F.saeed@gcuf.edu.pk; far1552@yahoo.com

Ana C. Santos, PharmD
Graduate Student, Faculty of Pharmacy, University of Coimbra, Azinhaga de Santa Comba, Pólo da Saúde (III), 3000-548 Coimbra, Portugal
Institute for Innovation and Health Research, Group Genetics of Cognitive Dysfunction, Institute for Molecular and Cell Biology, Rua do Campo Alegre, 823, 4150-180 Porto, Portugal.
E-mail: ana.cl.santos@gmail.com

Amélia M. Silva, PhD
Assistant Professor, Department of Biology and Environment, School of Life and Environmental Sciences, (ECVA, UTAD), University of Trás-os-Montes and Alto Douro, P.O. Box 1013; 5001-801 Vila-Real, Portugal
Centre for the Research and Technology of Agro-Environmental and Biological Sciences, University of Trás-os-Montes and Alto Douro (CITAB-UTAD), Vila-Real, Portugal.
E-mail: amsilva@utad.pt

Barkha Singhal, PhD
Assistant Professor, School of Biotechnology, Gautam Buddha University, Greater Noida, Gautam Budh Nagar 201310, Uttar Pradesh, India.
E-mail: barkha@gbu.ac.in; gupta.barkha@gmail.com

Eliana B. Souto, PhD
Assistant Professor (Habilitation), Faculty of Pharmacy of University of Coimbra, Azinhaga de Santa Comba, Pólo da Saúde (III), 3000-548 Coimbra, Portugal
REQUIMTE/LAQV, Group of Pharmaceutical Technology, Faculty of Pharmacy, University of Coimbra, Portugal. E-mail: ebsouto@ff.uc.pt; ebsouto@ebsouto.pt

Hafiz Ansar Rasul Suleria
PhD, McKenzie Fellow, Food Nutrition Department of Agriculture and Food Systems, The University of Melbourne, Level 1, 142 Royal Parade, Parkville Victoria, 3010 Australia; Mobile: +61- 470439670; E-mail: hafiz.suleria@unimelb.edu.au

S. P. Suriyaraj, PhD
Post Doctoral Researcher, Biotechnology Department, Indian Institute of Technology Madras, Chennai 600036, Tamil Nadu, India. E-mail: jarayirus.nano@gmail.com

S. Vaidevi, MTech
PhD Research Scholar, Pharmaceutical Technology Department, Anna University, BIT Campus, Tiruchirappalli 620024, India. E-mail: vaidevipavi@gmail.com

Francisco J. Veiga, PhD
Professor, Faculty of Pharmacy, University of Coimbra, Azinhaga de Santa Comba, Pólo da Saúde (III), 3000-548 Coimbra, Portugal
REQUIMTE/LAQV, Group of Pharmaceutical Technology, Faculty of Pharmacy, University of Coimbra, Portugal. E-mail: fveiga@ff.uc.pt

Deepak Kumar Verma
PhD Research Scholar, Agricultural and Food Engineering Department, Indian Institute of Technology, Kharagpur 721302, West Bengal, India.
E-mail: deepak.verma@agfe.iitkgp.ernet.in; rajadkv@rediffmail.com

ABBREVIATIONS

AAS	atomic absorption spectroscopy
AChE	acetylcholinsterase
AFM	atomic force microscopy
AgNPs	silver nanoparticles
AIDS	acquired immune deficiency syndrome
AI-FeNPs	*Azadiracta indica* Iron Nanoparticles
aPTT	activated partial thromboplastin time
ASV	anodic stripping voltammetric
ATPS	aqueous two-phase systems
AuNPs	gold nanoparticles
BEEM	ballistic electron emission microscopy
BSA	bovine serum albumin
C-AFM	conductive atomic force microscopy
CBE	cocoa bean extract
CFM	chemical atomic force microscopy
CKD	chronic kidney disease
CMNC	ceramic matrix nanocomposites
CNFs	carbon nanofibers
CNT	carbon nanotube
COD	chemical oxygen demand
CPHE	cocoa pod husk extract
CROs	cathode ray oscilloscopes
CRT	cathode rays tube
Cryo-TEM	cryo-transmission electron microscopy
CT	computed tomography
CVD	chemical vapor deposition
DDW	deuterium-depleted water
DLS	dynamic light scattering
DMEM	Dulbecco's modified Eagle's medium
DSC	differential scanning calorimetry
DTA	differential thermal analysis
ECM	extracellular matrix

EDS	energy dispersive spectroscopy
EDXS	energy dispersive X-ray spectroscopy
EGFR	epidermal growth factor receptor
ETA	elemental trace analysis
ETEM	environmental transmission electron microscope
FCC	face-centered cubic
FE-SEM	field emission scanning electron microscope
FMM	force modulation microscopy
FRAP	ferric reducing antioxidant power
FTIR	Fourier-transform infrared
FTO	fluorine-doped tin oxide
FWHM	full width at half maximum
GPS	geosynchronous positioning system
GtO	graphite oxide
HBsAg	hepatitis B surface antigen
HIV	human immunodeficiency virus
HMDS	hexamethyldisilazane
HOPG	highly oriented pyrolytic graphite
HRTEM	high-resolution transmission electron microscope
IC	inhibitory concentration
ISCOM	immunostimulating complex
ITO	indium tin oxide
KPFM	Kelvin probe force microscopy
LC	lethal concentration
MDR	multidrug resistance
MeNPs	metallic nanoparticles
Met-NPs	metal nanoparticles
MFM	magnetic force microscopy
MIC	minimum inhibitory concentration
MMNC	metal matrix nanocomposites
MNPs	magnetic nanoparticles
MPM	magnetic probe microscopy
MPS	mononuclear phagocytes system
MRFM	magnetic resonance force microscopy
MRI	magnetic resonance imaging
MS	mass spectrometry
MSNs	mesoporous silica nanoparticles

MuF-NPs	multifunctional nanoparticles
NCs	nanocrystals
NMR	nuclear magnetic resonance
NMRS	nuclear magnetic resonance spectroscopy
NP	nanoparticle
NSCLC	nonsmall-cell lung cancer
NSOM	near-field scanning optical microscopy
O/W	oil-in-water
O-NPs	oxide nanoparticles
PBS	phosphate buffered saline
PDI	polydispersity index
PdNPs	palladium nanoparticles
PEG	polyethylene glycol
PEI	polyethylenimine
PET	positron emission tomography
PFG-NMR	pulsed field gradient nuclear magnetic resonance
PMNC	polymer matrix nanocomposites
PoM-NPs	polymeric nanoparticles
PSA	particle size analysis
PT	prothrombin time
PTFE	polytetrafluoroethylene
PTMS	photothermal microspectroscopy/microscopy
PtNPs	platinum nanoparticles
PVA	polyvinyl alcohol
PVD	physical vapor deposition
PVP	polyvinyl pyrrolidone
QD	quantum dot
RES	reticuloendothelial system
rGO	reduced graphene oxide
ROS	reactive oxygen species
SAED	selected area electron diffraction
SANS	small angle neutron scattering
SAXS	small angle X-ray scattering
SEM	scanning electron microscope
SERS	surface enhance Raman scattering
SGM	scanning gate microscopy
SHPC	soy hydrogenated phosphatidylcholine
SICM	scanning ion-conductance microscopy
siRNA	small interfering RNA

SMCs	smooth muscle cells
SPM	scanning probe microscopy
SPR	surface plasmon resonance
SPSM	spin polarized scanning tunnelling microscope
STEM	scanning transmission electron microscopy
STM	scanning tunnelling microscopy
SWNT	single-walled nanotubes
TEM	transmission electron microscopy
TGA	thermogravimetric analysis
TG-DSC	thermogravimetric differential scanning colorimeter
TNF	tumor necrosis factor
TT	thrombin time
USPIO	ultra-small superparamagnetic iron oxide
VEGF	vascular endothelial growth factor
VLPs	virus-like particles
W/O	water-in-oil
XRD	X-ray diffraction
ZOI	zone of inhibition

Preface 1 by Deepak Kumar Verma

A set of applied sciences fall under the biomedical sciences, which employ some of natural or formal and/or both sciences to develop the immense knowledge, interventions, and/or technology, with the aim to use it for the welfare of human health or healthcare. In other words, we can say that the study of human physiology, its structure and function in health and disease can be referred to as biomedical science. In the biomedical sciences, biomedical engineering, clinical epidemiology, clinical virology, genetic epidemiology, and medical microbiology are known as applied sciences and considered to be in the medical sciences whereas a group of interdisciplinary areas of study concerned with the design, action, delivery, and disposition of drugs are referred to as pharmaceutical sciences.

Today, scientists, groups of researchers, and research institutions are playing their important roles to meet many goals with their sound knowledge and understanding of different disciplines of sciences, such as medical biochemistry, microbiology, pharmacology, cell and molecular biology, anatomy, physiology, infectious diseases, and neuroscience. Among them, nanotechnology is a new and emerging science, which plays an important role in overcoming the problems and challenges of the biomedical industry, which occur during the development of devices and materials that are intended to benefit both patients as well as their healthcare. The better understanding and sound knowledge of nanotechnology combines with biology and open revolutionized research in the biomedical and pharmaceutical fields by using the novel phenomena and unique properties of materials such as physical, chemical, and biological at 10^{-9} nanometer scales with the direct application to biological targets. These small scale materials are called "nanomaterials" that are especially designed for a variety of biomedical applications including biosensors, enzyme encapsulation, neuronal nanotechnology, and many more for human health and healthcare.

The present book, titled *Nanotechnology and Nanomaterial Applications in Food, Health, and Biomedical Sciences,* meets an urgent need for students, researchers, scientists, academicians, etc.; it addresses nanotechnology, nanoparticles, and the emerging scope and potential applications for food, human health, and healthcare.

The contributions from eminent scientists and researchers have enabled us to present their work in this book volume. We hope that this book will serve as a guide for researchers and students working in the area of nanotechnology and nanomaterials sciences in the food, biomedical, and pharmaceutical fields.

—Deepak Kumar Verma, PhD
Editor

Preface 2 by Megh R. Goyal

At the 49th annual meeting of the Indian Society of Agricultural Engineers at Punjab Agricultural University during February 22–25, 2015, a group of ABEs convinced me that there is a dire need to publish book volumes on the focus areas of agricultural and biological engineering (ABE). This is how the idea was born on new book series titled, *Innovations in Agricultural and Biological Engineering*.

The contributions by the cooperating authors to this book volume have been most valuable in the compilation. Their names are mentioned in each chapter and in the list of contributors. This book would not have been written without the valuable cooperation of these investigators many of whom are renowned scientists who have worked in the field of ABE throughout their professional careers.

The goal of this book volume, *Nanotechnology and Nanomaterial Applications in Food, Health, and Biomedical Sciences,* is to guide the world science community on how nanotechnology has involved in different fields.

We thank editorial staff, and Ashish Kumar, Publisher and President at Apple Academic Press Inc., for making every effort to publish the book when the diminishing water resources are a major issue worldwide. Special thanks are due to the AAP production staff also.

I request readers to offer your constructive suggestions that may help to improve the next edition.

I express my deep admiration to my family for understanding and collaboration during the preparation of this book. As an educator, there is a piece of advice to one and all in the world: "Permit that our Almighty God, our Creator, allow us to inherit new technologies for a better life at our planet. I invite my community in agricultural engineering to contribute book chapters to the book series by getting married to my profession…". I am in total love with our profession by length, width, height, and depth. Are you?

—**Megh R. Goyal, PhD, PE**
Senior Editor-in-Chief

Preface 3 by Hafiz Ansar Rasul Suleria

Nanotechnology is a fast developing field of research with diverse applications in various sectors of the industry. Nanotechnology, with these promising new insights and innovations, is expected to have mass usage by 2020 to revolutionize many aspects of human life. This book has reviewed the research efforts and potential applications of nanotechnology in agriculture, food, biomedical, biological, and pharmaceutical applications.

Nanomaterials are increasingly used in a wide range of applications in science, engineering, technology, and medicine. Nanomaterials are already in commercial use, and the range of commercial products existing today is very broad. Currently, nanomaterials are used in various fields, such as carbon black particles for wear-resistant automobile tires; nanofibers for insulation and reinforcement of composites; iron oxide for magnetic disk drives and audio–video tapes; nano-Zinc oxides and titanium as sun shields for UV-rays, etc. Numerous techniques are employed to characterize the material, and a systematic application of one or more techniques among them leads to a complete understanding of the nanomaterial.

The growing public concern has incited research interest in the use of nanoparticles to develop nanosensors for detection of various contaminants and pathogens in the food system. The elegant amalgamation of nanotechnology with different scientific disciplines leads to refurbished impetus to the development of novel sensing platforms for enhancement in the safety of food materials. Therefore, this book also addresses the various aspects and embodiments related to the importance and types of nanosensors and their fabrication as well as comprehensive applications in food safety. Instead, nanotechnology involves enclosing a bioactive ingredient in nanoscale capsule. Nanoencapsulation holds potential for its application in nutritional additives and specialized nanopackaging.

This book addresses the number of challenges that have been overcome by the continuous inflow of new technologies and nanotechnology as a potential aid to revolutionize this ever-growing sector.

I thank Dr. Megh R. Goyal, my mentor and promoter of human health and novel technologies, for inviting me to join his team and contribute our combined input and effort for the success of the proposed book.

—Hafiz Ansar Rasul Suleria, PhD
Editor

PART I

Nanotechnology Techniques for Biomedical Applications

CHAPTER 1

GREEN SYNTHESIS OF SILVER (Ag), GOLD (Au), AND SILVER–GOLD (Ag–Au) ALLOY NANOPARTICLES: A REVIEW ON RECENT ADVANCES, TRENDS, AND BIOMEDICAL APPLICATIONS

JOSEPH ADETUNJI ELEGBEDE and AGBAJE LATEEF*

*Corresponding author. E-mail: alateef@lautech.edu.ng

ABSTRACT

The synthesis of metallic nanoparticles (MeNPs) through green approach has continued to court the interests of scientists from diverse background due to numerous beneficial reasons. The process is eco-friendly, devoid of the use of hazardous chemicals and procedures in the synthesis, which promotes their biocompatibility and low toxicity; it is economical, rapid, cost-effective, and can be accomplished under benign conditions. The abundance of several biomolecules in biological entities that can concomitantly serve as both reduction and capping agents has also fueled the growing trends in one-pot synthesis of MeNPs. Amongst the MeNPs, silver (Ag), gold (Au), and their bimetallic alloy (Ag–AuNPs) have been vividly studied owing to their novel optical, physical, chemical, photothermal, catalytic, and electrical attributes for multiple applications. Some of the important usefulness of these nanoparticles is in their deployment as antimicrobial, larvicidal, antioxidant, anticoagulant, and thrombolytic agents. These biomedical applications are envisaged to combat myriads of problems facing mankind, particularly the antimicrobial resistance phenomena, control of vector-borne diseases, deleterious activities of free radical species, and control/management of blood coagulation disorders. It is evidently clear that properties exhibited by nanoparticles aptly positioned them to be used as vital tools in the development of new generation of nanomedicals. Therefore, this review presents the contributions of green synthesized Ag, Au, and Ag–AuNPs for biomedical applications with due diligence to antimicrobial, larvicidal, antioxidant, anticoagulant, and thrombolytic activities. Until now, there is no review that summarizes the biomedical applications of Ag, Au, and Ag–AuNPs as a compendium. The review underscores the importance of these particles in the emerging disciplines of nano- and biomedicine.

1.1 INTRODUCTION

Nanotechnology is concerned with synthesis, strategy, and manipulation of structure of particles ranging from approximately 1–100 nm in size. Within this size of scope, all the properties (biological, chemical, electrical, and physical) are modified in fundamental ways of both individual atoms/molecules and their respective bulk.[10] Nanotechnology has continued to attract great interests because of its extensive applications in diverse aspects of life endeavors. It is a subject matter that cuts across various fields of science and technology.[87]

Nanobiotechnology, an offspring of nanotechnology, is a multidisciplinary discipline that combines different branches of knowledge that include nanotechnology, biotechnology, physical methodology, chemical processing, and system engineering into molecular motors, biochips, nanobiomaterials, and nanocrystals.[78] Nanobiotechnology is rapidly gaining interest by recreating and improving the mechanisms of nanomaterials in several fields such as: health care, biomedicine, food and feed, cosmetics, environment, chemical industries, catalysis, single electron transistors, drug–gene delivery, electronics, mechanics, light emitters, space industries, energy science, nonlinear optical devices, and photo-electrochemical applications and so on. Among all types of nanoparticles used for aforementioned purposes, the metallic nanoparticles (MeNPs) are most promising because of their remarkable attributes. They exhibit large surface area to volume ratio, which attracts interests of researchers, especially in the field of microbiology to combat the surge in antimicrobial resistance.[65]

Arising from the foregoing, this present chapter presents a review on the role and advancement of green synthesized silver (Ag), gold (Au), and silver–gold (Ag–Au) alloy bimetallic nanoparticles (NPs) for antimicrobial, larvicidal, antioxidant, anticoagulant, and thrombolytic applications in medicine. Till today, there is no comprehensive review that details the antimicrobial, larvicidal, antioxidant, anticoagulant and thrombolytic applications of Ag, Au, and Ag–AuNPs as a compendium.

1.2 SELECTED METALLIC NANOPARTICLES

1.2.1 SILVER NANOPARTICLES

Silver nanoparticles (AgNPs) are widely studied due to the numerous and excellent properties and applications that are ascribed to them. These include: optical, bio-imaging, catalytic, fibrinolytic, sensing, wound-healing, larvicidal, antimicrobial, antioxidant, biodesulfurization, and anticancer applications.[23,31,40,52,71,78,81,85,95,96,105,120,122,125,147,172] Recently, the potential applications have been extended to its use as antiplatelet, anticoagulant, and thrombolytic agents.[15,52,80,83,164,167] The green fabrication of nanoparticles, which precludes usage of hazardous procedures and chemicals, has contributed to the expansion of applications of nanoparticles for the production of biocompatible and eco-friendly particles using low-cost and benign approach. Furthermore, the abundance of biomolecules[5,10,79,82,87,89,102,128,168] in diverse biological entities (such as plants, microbes, agro wastes, pigments, enzymes, arthropods, and their metabolites) have also added to the growing trend in the green production of nanoparticles for varied applications as shown in Table 1.1.

1.2.2 GOLD NANOPARTICLES

Gold nanoparticles (AuNPs) have been known to exhibit considerable biocompatibility, due to the fact that gold is not readily oxidized, unlike silver. Therefore, it has the potential to be used for long-term biomedical applications as it displays low toxicity. It has been reported that conjugation of AuNPs with antibodies and proteins enhance their functionality for sensing and therapeutic functions.[66] AuNPs have been studied for different types of applications (Table 1.2) including catalytic, bioimaging, antioxidant, photothermal, anticancer, anticoagulant, fluorescent, biolabeling, biosensing, antimicrobial, and thrombolytic purposes.[9,12,24,27,30,42,69,70,124,182,185]

TABLE 1.1 Green Synthesis and Selected Biomedical Applications of Silver Nanoparticles.

Biological material	Morphology of NPs	Size (nm)	λ_{max} (nm)	Activities	Reference
Root extract of *Panax ginseng*	Spherical and monodispersed	10–30	412	Antimicrobial	[166]
Leaf extracts of *Caryota urens*	Monodispersed, spherical with presence of quasi-spherical, triangular, and pentagonal shape	1–70	437	Antimicrobial bacterial MIC: 8–64 μg/mL; fungal MIC: 32–64 μg/mL	[138]
Pongamia glabra			429	Antimicrobial bacterial MIC: 32–64 μg/mL fungal MIC: 64–128 μg/mL	
Hamelia patens			424	Antimicrobial bacterial MIC: 16–128 μg/mL fungal MIC: 64–128 μ/mL	
Thevetia peruviana			452	Antimicrobial bacterial MIC: 16–64 μg/mL fungal MIC: 64 μg/mL	
Calendula officinalis			428	Antimicrobial bacterial MIC: 16–64 μg/mL fungal MIC: 64–128 μg/mL	
Tectona grandis			434	Antimicrobial bacterial MIC: 16–128 μg/mL fungal MIC: 128 μg/mL	
Ficus petiolaris			428	Antimicrobial bacterial MIC: 64–256 μg/mL fungal MIC: 128–256 μg/mL	
Ficus busking			415	Antimicrobial bacterial MIC: 16–128 μg/mL fungal MIC: 64 μg/mL	
Juniper communis			437	Antimicrobial bacterial MIC: 16–64 μg/mL fungal MIC: 64 μg/mL	
Bauhinia purpurea			440	Antimicrobial bacterial MIC: 16–64 μg/mL fungal MIC: 32–64 μg/mL	
Bark extracts of *F. benghalensis* and *A. indica*	Spherical with rough surfaces	60	426 and 420	Antibacterial MIC: 12.5–25 μg/mL MBC: 100 μg/mL	[122]
Extracellular extract of *B. safensis* LAU 13	Spherical	5–95	419	Antimicrobial ZOI; 11–19 mm antioxidant; DPPH (IC_{50} = 15.99 μg/mL and FRAP (1.84–2.42 at 20–100 μg/mL); Larvicidal Species: *Anopheles gambiae* larvae LC_{50} = 42.19 μg/mL	[85]

TABLE 1.1 *(Continued)*

Biological material	Morphology of NPs	Size (nm)	λ_{max} (nm)	Activities	Reference
Cocoa bean extract	Spherical and fairly polydispersed	8.96–54.22	438.5	Synergistic with antibiotics (42.9–100%), paint additive (antibacterial and antifungal), antibacterial ZOI; 10–14 mm, Larvicidal species: *Anopheles gambiae* larvae LC_{50} = 44.37 µg/mL, Anticoagulant	[15]
Cocoa pod husk	Spherical and well dispersed	4–32	428.5	Synergistic with antibiotics (42.9–100%), paint additive (antibacterial and antifungal), antibacterial ZOI; 10–14 mm antioxidant DPPH (IC_{50} = 49.70 µg/mL), and FRAP (14.44–83.94 at 20–100 µg/mL), Larvicidal species: *Anopheles gambiae* larvae LC_{50} = 43.52 µg/mL	[82]
Cola nitida pod	Spherical and polydispersed in nature	12–80	431.5	Paint additive (antibacterial and antifungal), antibacterial ZOI: 12–30 mm, antioxidant: DPPH (IC_{50} = 43.98 µg/mL and FRAP (13.62–49.96 at 20–100 µg/mL)	[83]
Culture supernatant of *P. aeruginosa*	Uniformly spherical	80	430–450	Antibacterial ZOI: 8–10 mm, anticoagulant	[58]
Dolichos biflorus Linn Seed	Spherical and agglomerated	0.223	420–430	Antibacterial	[19]
Extract of *Chenopodium murale*	Nearly spherical	30–50	440	Antibacterial, antioxidant; DPPH (65.43%) and β-carotene oxidation (53.38%) at 20 mg/mL	[2]
Helicteres isora root	Spherical and monodispersed	16–95	450	Antibacterial, antioxidant; DPPH (90%), nitric oxide (80.46%), and H_2O_2 (93.31%)	[23]
Keratinase of *B. safensis* LAU 13	Spherical	5–30	409	Antibacterial ZOI; 8.6–12.5 mm at 150 µg/mL	[78]

TABLE 1.1 *(Continued)*

Biological material	Morphology of NPs	Size (nm)	λ_{max} (nm)	Activities	Reference
Pods of *P. pterocarpum*	Spherical	70–85	407	Antibacterial, anticoagulant	[141]
Root extract of Korean red ginseng	Spherical and monodispersed	10–30	410	Antibiofilm, antibacterial ZOI: 16–27 mm	[165]
Seaweed *Enteromorpha compressa*	Spherical	4–24	421	Antimicrobial bacterial ZOI: 10.5–12.0 mm Fungal ZOI; 9.2–10.2 mm	[146]
Spider cobweb	Spherical	3–50	436	Synergistic with antibiotics (improvement by 3.1–100%), paint additive (antibacterial and antifungal), Antibacterial ZOI: 10–17 mm	[86]
Cola nitida seed	Spherical	8–50	457.5	Antibacterial ZOI: 10–32 mm MIC: 50–120 μg/mL	[81]
Cola nitida seed shell		4–40	454.5	Antibacterial ZOI: 10–32 mm MIC: 50 μg/mL	
Piper longum fruit extract	Spherical and polydispersed	46	430	Antibacterial ZOI: 1.0–2.0 cm, antioxidant FRAP; DPPH (67%), superoxide (60%), nitric oxide (70%), and H_2O_2 (96%)	[151]
Culture supernatant of *Sporosarcina koreensis* DC4	Spherical	30–50	424	Antibiofilm, antibacterial MIC: 3–6.9 μg/mL	[169]
Leaves of *Panax ginseng*	Spherical and monodispersed	5–15	420	Antibiofilm, antibacterial MIC: 3 μg/mL, Anticoagulant	[167]
Roasted *Coffea arabica* seed	Spherical and ellipsoidal	20–30	459	Antibacterial ZOI: 2.7–3.1 cm, MIC: 0.28–0.54 mg/mL L	[31]
Datura stramonium	Spherical		444	Antibacterial	[47]

TABLE 1.1 *(Continued)*

Biological material	Morphology of NPs	Size (nm)	λ_{max} (nm)	Activities	Reference
Leaf and seed extract of *Embelia ribes*	Spherical	5–35	440	Antibacterial ZOI: 22–28 mm, antioxidant; DPPH (IC_{50} = 100 μg/mL), phosphomolybdenum assay (IC_{50} = 60 μg/mL)	[32]
Musa balbisiana	Spherical	200	425–475	Antibacterial ZOI: 14–16 mm	[17]
Azadirachta indica	Triangular			Antibacterial ZOI: 12–16 mm	
Ocimum tenuiflorum	Cuboidal			Antibacterial ZOI: 14 mm	
Laccase of *Lentinus edodes*	Walnut-shaped	50–100	430	Antibacterial ZOI: 11–20 mm	[96]
Leaf extract of *Synsepalum dulcificum*	Fairly spherical and polydispersed	5–22	440	Antimicrobial ZOI (bacterial): 11–24 mm at 60 μg/mL. antifungal: 73.17–100%, anticoagulant, thrombolytic	[79]
Seed extract of *Synsepalum dulcificum*		4-26	438.5		
Sesbania grandiflora leaves	Spherical and well dispersed	16	416	Antimicrobial ZOI (bacterial strains): 3–8 mm, ZOI (fungal strains): 3–7 mm	[11]
III-5 strain of *Stenotrophomonas maltophilia* and cells of D-4 strain of *Microbacterium marinilacus*	Tiny cubes	10–50	400	Antibacterial MIC of 10^{-6} dilutions = 16.6 units/μL	[112]
Paper wasp (*Polistes* sp.)	Well dispersed anisotropic structures of triangle, sphere, hexagon, rhombus, and rod	12.5–95.55	428	Antimicrobial ZOI (bacteria): 12–35 mm, antifungal: 75.61–100%, anticoagulant, thrombolytic	[80]
Plumeria alba (Frangipani) flower	Spherical	36.19	445	Antibacterial ZOI: 3–18 mm, antioxidant DPPH (IC_{50} = 100 μg/mL), FRAP, superoxide, nitric oxide (51%), and H_2O_2 (91.8%)	[105]

TABLE 1.1 *(Continued)*

Biological material	Morphology of NPs	Size (nm)	λ_{max} (nm)	Activities	Reference
Extract of leaf of *Aristolochia indica* L. (LAIL)	Spherical and polydispersed	–	–	Antimicrobial ZOI: bacterial strains (7–11 mm) and fungal strains (9–13 mm), antioxidant; DPPH (81.19%), ABTS scavenging assay (64.01%)	[159]
Crude extract of *Bergenia ciliata*	Spherical	35	425	Antimicrobial ZOI: bacterial strains (7.5–9 mm) and fungal strains (7.5–9 mm), antioxidant; DPPH (59.31%)	[134]
Leaf of *Talinum triangulare*	Spherical	13.86	417–430	Antimicrobial ZOI: 13–23 mm at 100 μg/mL antioxidant; DPPH (88%)	[39]
Leaf of *Solanum nigrum*	Cuboidal	20	428	Antibiofilm, antimicrobial ZOI: 9.6–21.5 mm antioxidant; ROS generation: 3-fold increase	[60]
Leaf of *Artocarpus altilis*	Spherical and Polydispersed	34–38	432	Antimicrobial, antioxidant; DPPH (IC_{50} = 51.17 μg/mL)	[148]
Stem of Siberian ginseng	Spherical	126	440	Antibacterial ZOI: 10.7–13.8 mm, antioxidant; DPPH (IC_{50} = 100 μg/mL)	[1]
Alternanthera sessilis Linn	Aggregated	20–30	435.04	Antibacterial, antioxidant; DPPH (IC_{50} = 300.6 μg/mL)	[123]
Caesalpinia pulcherrima flower extract	Spherical without agglomeration	12	410	Antibacterial MIC: 2.5–10 mg/mL, antioxidant; DPPH (IC_{50} = 70 μg/mL) superoxide (IC_{50} = 38.5 μg/mL) ABTS (IC_{50} = 55 μg/mL) FRAP = 8.8 Mg^{-1}	[110]
Extracellular extract of *B. safensis* LAU 13	Fairly spherical	5–95	419	Anti-candida ZOI:11–15 mm, MIC:40 μg/mL, anticoagulant, thrombolytic	[90]

TABLE 1.1 *(Continued)*

Biological material	Morphology of NPs	Size (nm)	λ_{max} (nm)	Activities	Reference
Sponge weed *Codium tomentosum*	Predominantly irregular in shape	20–40	420	Antibacterial ZOI: 17.68–20.24 mm at 150 μg /mL, Larvicidal species: *Anopheles stephensi* larvae and pupa LC_{50} = 18.1–40.7 ppm	[116]
Musa paradisiaca stem	Spherical and polydispersed	5–35	410	Antidiabetic, antiplasmodial, Larvicidal species: *Anopheles stephensi* larvae and pupa LC_{50} = 3.6–17.9 ppm	[13]
Plant extract of *Cleistanthus collinus*	Spherical	20–40	420	Antioxidant: FRAP (84.64%), DPPH (69%), hydroxyl (79%), and H_2O_2 (85.05%)	[62]
Costus pictus leaf	Spherical and polydispersed	46.7	430	Antioxidant: FRAP, DPPH, H_2O_2 superoxide (90%), nitric oxide (49.36%)	[119]
Gymnema sylvestre leaf	Spherical	33	435	Antioxidant; FRAP, DPPH, H_2O_2 (47%), Nitric oxide (82%)	[118]
Lipid of *Acutodesmus dimorphus*	Spherical and polydispersed	2–20	420	Antioxidant: DPPH (IC_{50} = 6.91 μg/mL), ABTS (IC_{50} = 14.41 μg/mL)	[25]
Pod extract of *Cola nitida*	Spherical	12–80	431.5	Antioxidant: DPPH (IC_{50} – 0.67 mg/mL at 500 ppm of AgNPs	[14]
Fruit of Shora (*Capparis petiolaris*)	Spherical and less polydispersed	10–30	423	Antioxidant: DPPH (38.98%)	[181]
Leaf extract of *Fraxinus excelsior*	Spherical and polydispersed	25–40	425	Antioxidant: DPPH (IC_{50} = 5.71 μg/mL)	[129]
Spider cobweb	Spherical and polydispersed	3–50	436	Anticoagulant, thrombolytic % lysis: 55.76–89.33%, H_2O_2 scavenging (77–99.8%)	[88]
Pod of kolanut (*Cola nitida*)		12–80	431.5		
Seed of kolanut (*Cola nitida*)		8–50	457.5		
Seed shell of kolanut (*Cola nitida*)		5–40	454.5		

TABLE 1.1 *(Continued)*

Biological material	Morphology of NPs	Size (nm)	λ_{max} (nm)	Activities	Reference
A.indica seed kernel	Spherical	20–50	450	Larvicidal species: fourth nstar larvae of *A. aegypti* and *A. stephensi* LC_{50}: 1.85 and 3.13 mg/L respectively	[180]
Arachis hypogaea peels	Spherical	35–60	450	Larvicidal species: larvae and pupae *Anopheles stephensi* LC_{50}: 3.9–8.2 ppm	[115]
Biomass of *Brevibacterium casei*	Spherical	10–50	420	Anticoagulant	[61]
Extract from *Eclipta prostrata*	Spherical	35–60	420	Larvicidal species: *C. quinquefasciatus* and *A. subpictus* LC_{50} values of 4.56 mg/L, and 5.14 mg/L, respectively.	[142]
Hymenodictyon orixense	Spherical and polydispersed	25–35	449	Larvicidal species: larvae of *A. subpictus*, *A. albopictus*, and *C. tritaeniorhynchus* LC_{50} : 17.10, 18.74, and 20.08 μg/mL, respectively	[49]
Leaf extract of *Carissa spinarum*	Spherical	40	447	Larvicidal species: *A. subpictus*, *A. albopictus*, and *C. tritaeniorhynchus* LC_{50} values of 8.37, 9.01, and 10.04 μg/mL, respectively.	[50]
Leaf extract of *Zornia diphylla*	Shapes such as triangle, spheres, decahedral and truncated triangles	30–60	418.5	Larvicidal species: larvae of *A. subpictus*, *A. albopictus*, and *C. Tritaeniorhynchus* LC_{50}: 12.53, 13.42, and 14.61 μg/mL, respectively	[51]
Leaf powder of *Diplazium esculentum* (retz.) sw.	Mostly spherical, oval, and triangular	10–45	439	Anticoagulant	[131]
Mesocarp layer extract of *Cocos nucifera*	Spherical	23	433	Larvicidal species: *Anopheles stephensi* and *C. quinquefasciatus* LC_{50}: 87.24 ± 4.75 and 49.89 ± 2.42 mg/L, respectively.	[154]

TABLE 1.1 *(Continued)*

Biological material	Morphology of NPs	Size (nm)	λ_{max} (nm)	Activities	Reference
Plant extract of *Pteridium aquilinum*	Spherical and polydispersed	35–65	420	Larvicidal species: Larvae and pupae *Anopheles stephensi* LC_{50} : 7.48–31.41 ppm	[127]
Xylan	Spherical and well dispersed	20–45	405	Fibrinolytic/thrombolytic (Fibrin plate assay)	[52]

ZOI: Zone of inhibition.

TABLE 1.2 Green Synthesis and Some Biomedical Applications of Gold Nanoparticles.

Biological Material	Morphology of NPs	Size (nm)	λ_{max} (nm)	Activities	Reference
Extracellular extract of *Bacillus safensis* LAU 13	Anisotropic, uniform Spherical	10–45	561	Antifungal % inhibition: 66.67%–75.32% at 200 μg/m, anticoagulant, thrombolytic	[124]
Red marine algae *Gracilaria corticata*	Well distributed without aggregation	45–57	540	Antibacterial, synergistic with antibiotics, antioxidant: DPPH (80.6%), FRAP (237.3 mm/mL) at 200 μg/mL	[121]
Dolichos biflorus Linn seed	Predominantly spherical with some irregular form	0.24	530	Antibacterial	[19]
Inonotus obliquus (Chaga mushroom)	Spherical, triangle, and rod shapes	11–37.7	532	Antibacterial (ZOI : 12–16 mm), Antioxidant: ABTS (max at 1 mM concentration)	[97]
Elettaria cardamomum Seeds	Spherical	18.3	527	Antibacterial, antioxidant: DPPH (62.18%), nitric oxide (64.44%), and hydroxyl (67.5%).	[143]
Nocardiopsis sp. MBRC-48	Spherical and polydispersed	11.57	530	Antimicrobial, antioxidant: DPPH (69%)	[102]
Seed extract of *Embelia ribes*	Spherical and polydispersed	10–30	500–550	Antioxidant: DPPH (IC_{50}: 20 μg/mL), phosphomolybdenum assay (IC_{50}: 40 μg/mL), antibacterial: ZOI (22–34 mm)	[32]

TABLE 1.2 *(Continued)*

Biological Material	Morphology of NPs	Size (nm)	λ_{max} (nm)	Activities	Reference
Leaf extract of *Carica papaya*	Spherical including triangle and hexagonal structures	2–20	500–600	Antibacterial: MIC (62.5–250 μg/mL)	[117]
Leaf of *Catharanthus Roseus*		3.5–9			
Leaf extract of *Carica papaya* and *Catharanthus roseus*		6–18		Antibacterial: MIC (15.6–125 μg/mL)	
Leaf of *Vetiveria Zizanioides*	Spherical and aggregated	40	538	Antifungal: ZOI (3.8–4.8 mm)	[176]
Leaf of *Cannabis sativa*					
Seed of *Abelmoschus esculentus*	Spherical	45–75	536	Antifungal: ZOI (15–18 mm)	[57]
Rind of water-melon (*Citrullus lanatus*)	Spherical	20–140	560	Synergistic with antibiotics, antibacterial: ZOI (9.23–11.58 mm), antioxidant: DPPH (24.69%), nitric oxide (25.62%), ABTS (29.42%)	[130]
Hovenia dulcis Fruit	Polydispersed, spherical, and hexagonal	15–20	536	Antibacterial: ZOI (18–19 mm), antioxidant: DPPH (59.17% at 500 μg/mL), Nitric oxide (88.75% at 500 μg/mL), H_2O_2 (48.60% at100 μg/mL), and FRAP	[18]
6-O chitosan sulphate solution	Spherical	15	530	Antibacterial, anticoagulant: aPTT and PT (doubled)	[37]
Heparin I	Spherical and monodispersed	20.26	527	Anticoagulant: PT: 13%	[70]
Heparin II	Spherical and polydispersed	40.85	523	Anticoagulant: PT: 26.7%, TT: 21.8%, aPTT: 23.2%	
Earthworm extracts	Spherical	6.13	533	Anticoagulant: increased aPTT (118.%9–134.8%)	[69]
Fruit extract of *Couroupita guianensis* Aubl.	Anisotropic structures of spherical, triangular, and hexagonal	25	530	Antioxidant: DPPH (IC_{50}: 37 μg/mL), hydroxyl (IC_{50}: 36 μg/mL), and superoxide (89.8%)	[156]

TABLE 1.2 *(Continued)*

Biological Material	Morphology of NPs	Size (nm)	λ_{max} (nm)	Activities	Reference
Stem of Siberian ginseng (*Eleutherococcus senticosus*)	Predominantly spherical	189	575	Antioxidant: DPPH (IC_{50}: 250 μg/mL)	[1]
C. tora leaf	Nearly spherical	57	538	Antioxidant: nitric oxide (70%), catalase activity (60% increase)	[4]
Costus pictus leaf	Spherical and polydispersed	37.2	530	Antioxidant: FRAP, H_2O_2, DPPH, nitric oxide (62%), and superoxide (90%)	[119]
Gymnema sylvestre Leaf	Spherical	26	536	Antioxidant: FRAP, superoxide, DPPH, nitric oxide (58%), and H_2O_2 (47%)	[118]
Lactobacillus kimchicus DCY51^T	Spherical and monodispersed	5–30	540	Antioxidant: DPPH (IC_{50}: 233.75 μg/mL)	[104]
Nerium oleander Leaf	Nearly spherical and highly dispersed	2–10	560	Antioxidant: DPPH (42%–72%)	[177]
Leaf of *A. leptotus*	Nearly spherical with few triangular shapes	13–28	557	Antioxidant: DPPH (87.33%)	[16]
Leaves of *Panax Ginseng*	Spherical and monodispersed	10–20	528	Anticoagulant	[167]
Brevibacterium Casei biomass	Spherical	10–50	540	Anticoagulant	[61]
Fibrin	Monodispersed	32	530	Fibrinolysis, detection of plasmin, and plasminogen	[59]
Biomass of *Momordica cochinchinensis*	Mostly spherical, oval, and triangular	10–80	552	Anticoagulant	[132]
Flower extract of *Couroupita guianensis*	Spherical and oval	29.2–43.8	560	Larvicidal species: larvae, pupae, and adult of *Anopheles stephensi* LC_{50}: 17.36–28.78 ppm	[173]

1.2.3 SILVER–GOLD ALLOY NANOPARTICLES (Ag–AuNPs)

Bimetallic nanoparticles have gained attention in their synthesis and applications, owing to the fact that they combine attributes of the monometallic components and by altering the molar ratios of the two metals. Unique bimetallic nanoparticles can be created with very good properties for diverse applications. Amongst such bimetallic nanoparticles of importance is Ag–AuNPs, which have been synthesized using the biological route.[89,124,155,158,162]

Silver–gold alloy nanoparticles that show a distinct surface plasmon resonance (SPR) that is intermediate of the SPR band of AuNPs and AgNPs,[109] possibly will exhibit lesser toxicity in comparison with AgNPs,[183] thereby improving biocompatibility[41] in nanomedical applications. Unlike AgNPs and AuNPs, the reports on biomedical applications of green Ag–AuNPs are scanty (Table 1.3), thereby necessitating intensive investigations on the potentials of the bimetallic material.

TABLE 1.3 Green Synthesis and Some Biomedical Applications of Silver–Gold Alloy Nanoparticles.

Biological material	Morphology of NPs	Size (nm)	Λ_{max} (nm)	Activities	Reference
Cell-free extract of *Bacillus safensis* LAU 13	Uniform spherical with irregular aggregation to form rod-like structures	13–80	545	Antifungal % inhibition: 83.33–90.78 at 200 µg/mL, anticoagulant, thrombolytic	[124]
Leaf of *Cola nitida*	Nearly spherical in shape	17–90	531	Antifungal: % inhibition (69.51–100 at 150 µg/mL), anticoagulant, thrombolytic, larvicidal	[89]
Seed of *Cola nitida*			524		
Seed shell of *Cola nitida*			497		
Pod of *Cola nitida*	Anisotropic structures of hexagon, sphere, triangle, and rod	12–91	517	species: larvae of *Anopheles gambiae* Activity: 100% mortality at 60 µg/mL with 3–72 h	
Fruit juice of pomegranate	Core-shell structure	12	479	Antioxidant: hydroxyl (76%) and nitric oxide	[72]

1.3 GREEN SYNTHESIS APPROACH FOR METALLIC NANOPARTICLES: RECENT TRENDS AND ADVANCES

The green synthesis approach is a promising branch of nanobiotechnology in which environmentally non-threatening materials in the form of whole cells, metabolites of microorganisms, or extracts derived from plants and animals are employed in the biofabrication of nanoparticles of metallic origin. Green synthesis is beneficial over physical and chemical methods because it is safe, simple, relatively reproducible, cost-effective, and often produces more stable materials. Biological method deployed to synthesize MeNPs of desirable size and morphology occupies a strategic place in nanoscience and has made enormous progress recently. The method does not utilize toxic chemicals that are often used in the synthetic procedures, with the attendant undesirable environmental consequences.[5] The quest for nanomaterials that are environmentally friendly and cost effective with desirable sizes and morphologies has led to various advances in green (biological) synthesis of MeNPs with arrays of sizes within 1–100 nm. Several biological agents including bacteria, fungi, plants, enzymes, and extracts have been employed to mediate the green synthesis of MeNPs.[5,78,80,83,86,87,89,95,96,124,189] The need to control the dimension, morphology and the contiguous surroundings of nanoparticles is of vital importance, because these factors influence several fundamental properties of MeNPs.[171] Amongst the MeNPs, AgNPs, and AuNPs are at the moment being intensively studied owing to their striking physiochemical and biological attributes, as well as prospective applications to develop novel technologies such as biosensors, optoelectronic devices, drug delivery systems, and catalysts.[100]

Over the years, the use of silver to treat medical illnesses owing to its wide-ranged antimicrobial properties has been reported.[26] Silver nanoparticles (AgNPs) are known to possess antimicrobial and anti-inflammatory properties, which are useful in preventing infections of open wounds and burns.[26,44] Gold nanoparticles (AuNPs) are recognized as a result of the distinctive and amenable physiochemical attributes, and also diverse applications in biology and medicine.[65,111] The tiny size of AuNPs, with their large surface area and elevated penetrative power is responsible for their improved action, which aids the efficient attachment of the nanoparticles to the substrates on both outer and cell membranes of organisms. AuNPs have extensive applications owing to certain precise properties like antibacterial activity[7] and its non-toxicity in human cells at low concentration. It has been used successfully in producing range of materials such as clothing, infant products, biolabeling, socks, electronic home appliances, cosmetics, surgical instruments, ocular lens, wound dressings, bone prostheses, and contraceptive devices. It has also found applications in food industries and in cancer therapy. It has also shown activities such as: anti-inflammatory, antifungal, antibacterial, antidiabetic, and antiviral properties. Among various types of bimetallic nanoparticles, Ag–AuNPs have attracted extensive development in different fields, such as catalyses, optics, biosensing, and medicine[72,158]. Due to the fact that individual metal components can be varied, with the attendant influence on the properties of bimetallic alloy nanoparticles; the fabricated alloy particles do exhibit more admirable optical, catalytic, and biological activities compared to the monometallic equivalents.[72]

Resistance of pathogenic microorganisms to antibiotics in humans has been an immense challenge in manufacturing of drugs and healthcare delivery. The worldwide prevalence of antimicrobial resistance is huge with resistant bacteria having been isolated from different sources in the environment.[6,74–76,84,91–94] Antibiotic resistance patterns of microorganisms have led to the dread as regard the frequency of appearance and recurrence of multidrug-resistant (MDR) pathogenic organisms. Therefore, improvement of or alteration in antimicrobial preparations to enhance their latent bactericidal activities is a main thrust of research in contemporary era.[73] Nanotechnology offers an excellent opportunity in transforming and developing the essential properties of metals to produce nanoparticles that have potential applications as cell labelers, biomarkers, antimicrobial agents, contrast agents for bioimaging, and drug delivery systems.[170] MeNPs (such as zinc, titanium, copper, silver, gold, and magnesium) are known to possess good bactericidal potentials. But among all, AgNPs have displayed outstanding antiviral, antifungal, and antibacterial activities as well as cidal activities on some other eukaryotic organisms.[140] Although lesser than AgNPs, AuNPs are as well known to demonstrate some degrees of antimicrobial activities with some mechanisms in place to explain the activities of the nanoparticles. For instance, it was assumed that the positive charge on the Au ion can promote antimicrobial activity of AuNPs by interacting with the negatively charged cell membrane of microorganisms. Another suggested mechanism is the fact that AuNPs can puncture bacterial cell wall to produce holes that enhance penetration of particles, as well as promoting seepage of intracellular contents and eventually cell death.[140]

Among arthropods, mosquitoes (Diptera: Culicidae) embody a foremost hazard to millions of living things throughout the world, because of being vectors of many notable pathogens and parasites, thereby serving as vehicles for the cause of diseases as filariasis, dengue, Zika, and malaria.[21,106,113] There are challenges in processes to manage the incidence of mosquitoes and associated diseases as a result of dearth of efficient protective and/or therapeutic measures that can be deployed to combat these disorders,[21] in addition to the burning demand for valuable and efficient agents that can kill mosquitoes thereby limiting the incidence of culicidae and associated diseases.[20] There are significant studies indicating that MeNPs that are fabricated using plant-based extracts could serve as promising nanoagents in the management of the incidence of mosquitoes that have both veterinary and medical consequences.[21] Presently, a rising number of plant metabolites are projected for proficient and rapid ex situ synthesis of MeNPs, with outstanding activities against *Plasmodium* parasites, as well as mosquitoes in both laboratory and field investigations.[21,48,49] However, in spite of these advances, it is still considered that limited practicable report is accessible about nanoformulates synthesized using green route with multiple activities against mosquitoes and arthropods of parasitological and public health concerns. This is viewed to constitute a major constraint in developments in the contribution of nanobiotechnology to the fields of parasitology and entomology to promote healthy living.[21,56] Several studies on mosquitocidal activities of nanoparticles have shown tremendous actions against larvae, pupa and adult insects with very low LC_{50}.[33,51,103,113,114,144,145,150,153,175] Some field experiments have also established the potency of plant-mediated nanoparticles against mosquitoes. The ability of nanoparticles to act against mosquito larvae could be due to the infiltration and penetration of the exoskeleton of the larvae by nanoparticles as a result of their small sizes. Moreover, once the nanoparticles enter, they can disrupt the functions of proteins and DNA by attaching to sulfur and phosphorus respectively, to ensure speedy destruction of intracellular constituents as well as denaturation of enzymes, which would ultimately lead to death of the larvae[22]

Studies on abilities of biosynthesized MeNPs to scavenge free radicals give an initiative about the association and activity of these nanoparticles with the biomolecules that are present in living organisms. Antioxidants generally exercise fundamental functions in the performance of biological entities by mopping lethal free radicals from the system. Measurements of antioxidants activity are still to a large extent limited to biological moieties,[159] although, some studies on antioxidant activities of MeNPs have been previously reported.[35,62,99,151,160] Researchers have proposed that antioxidants check oxidative damages that are borne out of pressures arising from oxidation reactions in some important components of cells such as lipids, proteins, and DNA thereby minimizing menace of age-related illnesses. Among the several methods of assessing antioxidant activities, in-vitro assays are simple, convenient, cost-effective, and reproducible.[99] The 2, 2-diphenyl-1-picrylhydrazyl (DPPH) is largely used to establish the free-radical scavenging activity of a compound or a plant extract in an easy and rapid manner.[121]

DPPH is a stable nitrogen-centered free radical which accepts hydrogen atoms or electrons from antioxidant materials.[34] A color change in the ethanolic solution of DPPH is noticed once it reacts with an antioxidant, which is due to the scavenging action of antioxidant on DPPH by releasing or donating hydrogen to form the stable yellow-colored DPPH.[174] Reactive oxygen species (ROS) have the capability of stimulating both thermal and auto-oxidation of lipids, that have been implicated in ageing and initiation of membrane injury in living organisms.[62] Hydroxyl (OH·) radicals have small half-life and are one of the most reactive and lethal free radicals with immense oxidative power, which combines swiftly with almost all molecules in its direct vicinity. It can cause several biological disorders such as carcinogenesis, cataracts, inflammation, aging, atherosclerosis, mutation, and cell death.[143] It can readily react with superoxide radicals, leading to damage to vascular system with consequences of development of multiple sclerosis and juvenile diabetes.[151] Hydrogen peroxide being a weak oxidizing agent can cause inactivation of some enzymes by directly oxidizing the essential thiol groups of the enzymes. It is capable of crossing cell membrane swiftly, and once it enters the cell, it can most likely react with Cu^{2+} and Fe^{2+} to form potent hydroxyl radicals.[114]

Nitric oxide (NO) is involved in many biological functions, such as neurotransmission, smooth muscle relaxation, blood pressure regulation, antitumor, and antimicrobial activities. However, it contributes to oxidative damage, because of the ability to react with superoxide that leads to the formation of peroxynitrite anion, which may result to DNA fragmentation and instigate lipid peroxidation.[126,152] Therefore, NO generation must be tightly regulated to curtail the harmful effects. Moreover, oxidation of ABTS by potassium persulfate can lead to generation of ABTS radicals which is an exceptional means to estimate antioxidant activity of chain breaking and hydrogen-donating antioxidants that scavenge lipid peroxyl radicals.[97] Also, ferric ion reducing antioxidant power (FRAP) is another important method that can be used to evaluate the antioxidant competence of foods, nutritional supplements and beverages containing polyphenols. The FRAP assay is a simple and fast technique to estimate antioxidant activities.[121]

The blood coagulation system is a very intricate system that seeks to maintain fluidity of the blood on one hand (prevention of clot formation) and the prevention of excessive bleeding on the other hand (activation of clot formation). Malfunction in the process of fibrinolysis in the body can lead to devastatingly undesirable outcomes including deep vein thrombosis, stroke, myocardial infarction, and embolism of the lung.[107,108] Thrombus (blood clot) formation in the system can lead to huge harm by obstruction of cardiovascular system which may eventually result in death.[43] Uncontrolled formation of blood clot can result from infections,[98] that are often connected to the occurrence of autoimmune reactions, cardiovascular disorders, injuries, development of cancer, and allergic responses.[29,135] However, this anomaly can be remediated through the use of thrombolytic agents to aptly break up the blood clots.[124]

Clinically, conventional methods like the use of heparin have several challenges, including drug instability, extreme hemorrhage; little life of action as well as excessive expenditure of managing the treatment.[89] Also, streptokinase and urokinase are in use to preclude clot formation, and for lysis of thrombi. Nonetheless, there are reports of adverse effects with the use of these agents as they also trigger excessive bleeding linked with reocclusion and reinfarction, and this has necessitated the research into deployment of nanomaterials as both anticoagulation and thrombolytic agents.[107] The issues and complications related with the use of conventional anticoagulants can be resolved and avoided through the applications of nanotechnology. Nanoparticles can be contrived to specifically work together with the blood coagulation system to avoid various blood disorders.[55] The biomedical importance of AgNPs, AuNPs, and Ag–AuNPs as anticoagulants/thrombolytics has been recently reported by many researchers.[29,80,89,124]

1.4 BIOMEDICAL APPLICATIONS OF SILVER NANOPARTICLES (AgNPs): A CASE STUDY

1.4.1 ANTIMICROBIAL ACTIVITIES OF AgNPs

Singh et al.[166] developed suitable scheme to synthesize AgNPs through green route with the use of extract of new roots obtained from *Panax ginseng*; a renowned plant of medicinal importance with the major active constituents being ginsenosides. Synthesized AgNPs had spherical morphology and ranged from 10 to 30 nm in size and also monodispersed in nature. The synthesized AgNPs absorbed maximally at 412 nm in UV–Vis spectroscopy. Also, the selected-area electron diffraction (SAED) analysis showed the presence of crystalline silver. The EDX analysis of the AgNPs revealed strong signals of silver occurring at 3 keV that is typical of AgNPs as previously reported for EDX spectra of AgNPs.[45] In the XRD, the pure nature of the biosynthesized AgNPs was established. Moreover, the AgNPs suppressed the growth of pathogenic microorganisms: *Bacillus anthracis*, *Bacillus cereus*, *Vibrio parahaemolyticus*, *Staphylococcus aureus*, and *Escherichia coli*. Similarly, the leaf of the plant was investigated to synthesize AgNPs.[167] Brown colloidal solution of AgNPs that was formed absorbed maximally at 420 nm and the monodispersed spherical-shaped AgNPs were 5–15 nm in dimension. At 3 μg/mL, profound antimicrobial activities were displayed by the nanoparticles using *E. coli, V. parahaemolyticus, Salmonella enterica, S. aureus, B. cereus,* and *B. anthracis*. Moreover, at 4 μg/mL, the AgNPs completely prevented *P. aeruginosa* and *S. aureus* from producing biofilm.

Also, Prasannaraj et al.[136] used different parts of 10 medicinal plants, such as leaves of *Andrographis paniculata, Alstonia scholaris, Centella asiatica, Aegle marmelos, Moringa oleifera, Eclipta prostrata;* stem barks of *Terminalia arjuna, Thespesia populnea;* as well as root bark of *Semecarpus anacardium*, and *Plumbago zeylanica* to synthesize AgNPs. About 20 mL of each extract was reacted with 180 mL of 1 mM aqueous $AgNO_3$ and left in the dark overnight at ambient temperature. A white to reddish brown color for diverse forms of AgNPs such as sphere, fiber, rod, and cuboid were obtained after the biosynthesis. The AgNPs were averagely 34–98 nm in size. Also, EDX studies indicated the presence of Ag within 2.8 and 3.2 keV, which is found associated with the absorption characteristics of silver nanocrystallites. Furthermore, the AgNPs were found to be crystalline in nature as confirmed through XRD investigation. Moreover, the antimicrobial potentials of synthesized AgNPs were tested against six microbial pathogens: *S. epidermidis, S. aureus, K. pneumoniae, E. coli, P. vulgaris,* and *P. aeruginosa* in which case varying degrees of activities were displayed against the test isolates. Also, these strains were evaluated to examine the consequence of AgNPs on their growth by using the broth dilution technique. In this case, maximum inhibition of growth of 86.9% was observed in *P. aeruginosa* treated with TaAgNPs, while the lowest was 24.2% observed in *E. coli* treated with ApAgNPs. Furthermore, two biofouling strains of *S. epidermidis* and *P. aeruginosa* were exposed to the AgNPs with biofilm inhibition activities of 69.4–79.6% obtained for 50 μg/mL of CaAgNPs and SaAgNPs.

In another work, Qayyum and Khan[138] described the biosynthesis of AgNPs using ten different plants extracts (NP1-*Caryota urens*, NP2-*Pongamia glabra*, NP3-*Hamelia patens*, NP4-*Thevetia peruviana*, NP5-*Calendula officinalis*, NP6-*Tectona grandis*, NP7- *Ficus petiolaris*, NP8- *Ficus busking*, NP9-*Juniper communis*, and NP10-*Bauhinia purpurea*) in reactions that involved mixing of 1 mL of the plant extracts with 20 mL of 1 mM $AgNO_3$ kept overnight at 150 rpm incubation at 37°C. The color change indicating the reduction of silver was clear to light yellow to dark brown and UV–Vis spectra revealed absorption peaks of 415–452 nm for the ten AgNPs. The nanoparticles were monodispersed with little or no clumping and the shapes were observed to be spherical with presence of quasi-spherical, triangular, and pentagonal shapes through TEM imaging. The crystallinity of the AgNPs was established by XRD investigations. Capping and stabilization of biosynthesized AgNPs were predicted to be due to the occurrence of proteins molecules and phenolic groups, which were indicated by amide vibrations, stretching of the amines, and O–H stretching obtained in the IR spectra. Image J-software was used to estimate the size range and average size of the AgNPs. The size distribution graphs showed that majority of AgNPs were in the range of 1–70 nm. The antimicrobial potential of the AgNPs was tested against *E. coli, K. pneumoniae, E. cloacae, S. mutans, S. aureus*, and *C. albicans*. The range of minimum inhibitory concentration (MIC) was 16–64 μg/mL for *E. coli, K. pneumoniae,* and *E. cloacae*, 16–128 μg/mL for *S. mutans* and *S. aureus* and 32–256 μg/mL for *C. albicans*.

Lateef et al.[78] described the synthesis of AgNPs using crude keratinase that was produced extracellularly by a novel isolate of *Bacillus safensis* LAU 13 (GenBank accession No. KJ461434) that degraded keratins.[77,95] The biosynthesized AgNPs were described as spherical-shaped and 5–30 nm in size. The AgNPs absorbed maximally at 409 nm. The FTIR result indicated that proteins were the responsible biomolecules that capped and stabilized the biosynthesized AgNPs with peaks occurring at 3410, 2930, 1664, 1618, 1389, and 600 cm^{-1}. From XRD data, it was established that the AgNPs were of face-centered cubic phase, crystalline in nature, and averagely sized 8.3 nm. The antibacterial ability of the AgNPs (150 μg/mL) was tested against five clinical strains of *E. coli* whose growths were inhibited by 8.6–12.5 mm indicating the efficacy of the particles against test isolates.

Lateef and Adeeyo[96] have also synthesized AgNPs using crude laccase produced by a UV-induced mutant (UV10) strain of *L. edodes* that led to the development of yellowish brown colloidal solution through the synthesis of the AgNPs. The AgNPs exhibited SPR at 430. The results from the FTIR spectrum specified that proteins accounted for AgNPs synthesis, and SEM images established the biofabrication of AgNPs in the form of walnut that is of 50–100 nm in size. The laccase mediated AgNPs showed effective antibacterial potentials against six clinical isolates, two each of *E. coli*, *P. aeruginosa*, and *K. pneumonia* with zones of inhibition recorded as 15 and 20 mm for *E. coli* strains, 12 and 14 mm for *P. aeruginosa* strains, and 11 and 12 mm for *K. pneumoniae* strains. Total antimicrobial activity of 60% was achieved by AgNPs.

Raja et al.[141] demonstrated biosynthesis of AgNPs through the use of aqueous pod extract of *Peltophorum pterocarpum.* The AgNPs were uniformly spherical, 70–85 nm in diameter as revealed by SEM imaging and absorbed maximally with a peak at 407 nm. FTIR spectrum of the extract revealed peaks indicating the presence of N-H bond of amino acids, C–H stretching, as well as –OH stretching of phenolic compounds and alcohols. In EDS spectrum, the occurrence of silver was shown through the strong silver peak that occurred at 2.983 keV. Furthermore, XRD analysis established that the AgNPs were crystalline. The antibacterial activity of AgNPs was demonstrated against *E. coli* using the well-diffusion method. Also, AgNPs synthesized with the culture supernatant of *Pseudomonas aeruginosa* described by Jeyaraj et al.[58] led to the development to brown colloidal solution, which was an indication that AgNPs has been synthesized. A strong and broad absorption peak between 430 and 450 nm was exhibited by AgNPs. The biosynthesized AgNPs were revealed by SEM to be uniformly spherical and very small, while the DLS measurement showed that the particles ranged in size between 10 and 200 nm with the mean size being 80 nm. The biological activity of the AgNPs was demonstrated through antibacterial assay, whereby zones of inhibition of 8–10 mm were obtained against *E. coli* and *S. aureus*. Abdel-Aziz et al.[2] also synthesized AgNPs using extract of *Chenopodium murale*. The brownish colored AgNPs absorbed maximally at 440 nm. From the TEM images, a size range of 30–50 nm was obtained for the biosynthesized AgNPs. Moreover, biological investigations of the particles established its potency against *S. aureus* when compared with 5 mM $AgNO_3$ used in the synthesis and plant extract alone.

Furthermore, Lateef et al.[85] described the green biosynthesis of AgNPs in an experiment that involved exploitation of extracellular extract of *Bacillus safensis* LAU 13. The biosynthesized nanoparticles developed into dark brown colloidal solution indicating the synthesis of AgNPs that exhibited maximum absorbance at 419 nm, which is within the range of absorbance of 391–460 nm that have been reported for AgNPs.[64,81,85,96,137,178,184] In the

FTIR spectrum, there were strong peaks occurring at 3308, 2359, 1636, and 422 cm^{-1} that demonstrated the participation of proteinous molecules to cap and stabilize the biosynthesized AgNPs. According to TEM images, the particles whose sizes fell within 5 and 95 nm were of spherical morphology. Furthermore, silver occurred as major metal in EDX spectrum, while the selected area electron diffraction (SAED) demonstrated existence of crystalline particles that have face-centered cubic phase. The AgNPs was also investigated for antimicrobial properties and it was revealed that the AgNPs was capable of inhibiting growth of multidrug-resistant isolates of *S. aureus, E. coli, Klebsiella granulomatis,* and *P. aeruginosa* to the extent of 11–19 mm by using 10–100 μg/mL of AgNPs. In synergistic studies, the AgNPs improved the activities of some antibiotics such as cefixime, ofloxacin, and augmentin to the tune of 7.4–142.9%. Also, the AgNPs inhibited the growth of deterioration microorganisms in paints; by totally inhibiting the growth and activities of *Aspergillus fumigatus, A. flavus, S. aureus,* and *P. aeruginosa*. The AgNPs was investigated for anti-candida activity and results showed clearly that the growth of *C. albicans* was inhibited by the AgNPs (11–15 mm) with MIC recorded as 40 μg/mL.[90]

Nayak et al.[122] has also documented the green synthesis of AgNPs by utilizing the extracts that were obtained from the barks of *Azadirachta indica* and *Ficus benghalensis*. For the synthesis, $AgNO_3$ (1 mM) and the bark extract were reacted together in ratio 1:9 and held at varying temperatures until the development of dark reddish brown color, which indicated the synthesis of the AgNPs. While the biosynthesized *F. benghalensis* AgNPs maximally absorbed at 426 nm, that of *A. indica* had maximal absorption at 420 nm. Field emission scanning electron microscopy (FE-SEM) revealed that typical size of both nanoparticles was approximately 60 nm. Both AgNPs were clearly shown to be spherical in shape with rough surfaces by the FE-SEM images. FTIR peaks were found at 3590, 3340, 2310, 1693, 1519, and 615 cm^{-1} for *F. benghalensis* AgNPs, and 3617, 3332, 2319, 1663, 1523, and 635 cm^{-1} for AgNPs that were synthesized through the use of extract of the bark of *A. indica*. These peaks were ascribed to existence of functional groups such as O–H, N–H, C≡N, C=C, and C–H that indicated the participation of alcohols, phenols, primary and secondary amines, nitriles, aromatic compounds, and alkynes in the processes of bioreduction and biofabrication of the AgNPs. The XRD analysis was performed to characterize the synthesized nanoparticles with regard to crystallinity. Bragg's peaks (angle 2θ) corresponding to (111), (200), (220), (311), and (222) were obtained, which confirmed the existence of face centered cubic (fcc) crystalline elemental silver. The AgNPs synthesized showed potential antimicrobial activities against *V. cholera, P. aeruginosa, Bacillus subtilis*, and *E. coli*. The MIC values of biosynthesized AgNPs against these bacteria varied between 12.5 and 25 μg/mL, while MBC was obtained at 100 μg/mL.

Furthermore, RamKumar et al.[146] illustrated biosynthesis of AgNPs in an investigation that involved the exploit of aqueous extract of seaweed, *Enteromorpha compressa*. The AgNPs were found to be of spherical orientation having clear lattice fringes with size ranging between 4 and 24 nm as shown by high-resolution transmission electron microscope (HRTEM) images; and was reported to absorb maximally at 421 nm. The FTIR spectra revealed the occurrence of diverse functional groups such as C=O and O–H indicating carboxylic acid and hydroxyl groups respectively, as well as C–OH vibrations corresponding to the presence of strong primary alcohol groups. Also, the XRD analysis of AgNPs showed characteristic Ag peaks at 38.1°, 44.2°, 64.4°, and 77.1° that were indexed as (111), (200), (220), and (311) in the Bragg's planes. The biosynthesized AgNPs were investigated for both antibacterial and antifungal potentials. The zones of inhibition against the bacterial isolates were obtained in the range of 10.5–12.0 mm while that of the fungal strains was between 9.2 and 10.2 mm. In the antibacterial assay, the most inhibited bacterium was *E. coli*, and then *K. pneumoniae, P. aeruginosa, S. aureus*, and *S. paratyphi*, while in antifungal assay, *A. niger* was the most inhibited, and subsequently trailed by *A. flavus*, *A. ochraceus*, *Fusarium moniliforme*, and *A. terreus*.

Singh et al.[165] has documented the biogenic synthesis of AgNPs using the root extract of the Korean red ginseng, which belongs to the *Araliaceae* family. The reaction mixture was completely turned to brown within 1 h indicating synthesis of AgNPs, which was maximally absorbed in 410 nm in the UV–Vis spectrum. Images obtained from TEM revealed the photosynthesis of spherical-shaped and monodispersed AgNPs that were about 10–30 nm in size. Also, the elemental characterization of the particles using EDX led to the emergence of characteristic peak at 3 keV, which is definitive of nanosized metallic silver.[38] Elemental mapping analysis indicated that the silver was the predominant element, confirming the purity of the nanoparticles. Evaluation of the biological activity of AgNPs showed that it was active against *C. albicans, S. aureus*, *B. cereus,* and *V. parahaemolyticus* by inhibiting their growth to the extent of 16–27 mm. In addition, the AgNPs exhibited degradation of biofilms in *S. aureus* and *Pseudomonas aeruginosa*.

Bhakya et al.[23] investigated the suitability of extract of root of *Helicteres isora* to biofabricate AgNPs. The spherical-shaped AgNPs maximally absorbed at 450 nm through bioreduction process that changed color of the reactants from yellow to reddish brown. The IR spectrum of synthesized AgNPs showed strong peaks at 3434.59, 2927.41, 2842.56, 1630.51, 1385.60, and 1024.01 cm^{-1}. Five distinct peaks at 38.12°, 44.38°, 64.45°, and 77.41° indicating the (111), (200), (220), and (311) fingerprint of silver were obtained from the XRD analysis. The monodispersed spherical-shaped AgNPs were in the dimension of 16–95 nm as read from the TEM images. Moreover, the AgNPs displayed good antibacterial activity against *Salmonella typhi*, *Vibrio cholerae*, *P. aeruginosa*, *E. coli*, *B. subtilis*, and *Micrococcus luteus*.

Basu et al.[19] also worked on the synthesis of AgNPs using the aqueous and methanolic extracts of seeds of *Dolichos biflorus* Linn. The extracts served as sources of bio-reducing agents in the process. There was transformation in the color of reactants from transparent to grey when water extract was used, whereas, it turned to brown in the case of use of the methanolic extract of *D. biflorus*. The AgNPs absorbed maximally within 420–430 nm in the UV–Vis spectra. The nanoparticles were predominantly spherical in form also showed agglomeration as depicted in scanning electron microscopic analysis. The EDX spectra revealed the main peak as for Ag with the weak presence of elements such as O, C, Cl, K, Al, and Na. The XRD analysis indicated figures corresponding to Bragg reflections that are in conformity with the fcc structure of Ag, while SAED pattern established the polycrystallinity of the AgNPs. Also, HRTEM showed 0.223 nm as the line spacing for the biosynthesized AgNPs. It was shown that growth of *E. coli*, *B. subtilis*, *S. aureus*, and *P. aeruginosa* were effectively inhibited through the application of the synthesized AgNPs.

Lateef et al.[86] described the green synthesis of AgNPs through the exploitation of cobweb extract of spider as a novel biomaterial. The synthesis of AgNPs was noticed by dark brown color change noticed in the reaction solution. The cobweb-AgNPs absorbed maximally at 436 nm in the UV–Vis spectra. In the FTIR spectrum, there were presence of prominent peaks at 3298, 2359, 2089, and 1635 cm^{-1}, confirming that protein molecules were involved in the processes of bioreduction, capping of the nanoparticles, and subsequent stabilization of the biosynthesized cobweb-AgNPs. The nearly spherical-shaped particles had dimensions of 3–50 nm as revealed by TEM images. In the EDX spectral, silver was present as the most conspicuous element, while the SAED blueprint confirmed the characteristic crystalline property of cobweb-AgNPs. In antimicrobial investigations using the cobweb-AgNPs at concentration of 100 μg/mL, inhibition of growth of bacterial isolates including multidrug-resistant *K. granulomatis, S. aureus, E. coli*, and *P. aeruginosa* varied between 10 and 17 mm. Furthermore, it was established that cobweb-AgNPs improved performance of some antibiotics such as augmentin, cefixime, and ofloxacin in the AgNPs-antibiotic synergy investigations. In addition, cobweb-AgNPs were included in emulsion paint as antimicrobial additive. At the working concentration of 5 μg/mL, there was absolute suppression of growth of microorganisms such as *P. aeruginosa, E. coli, A. fumigatus*, and *A. niger*.

Similarly, Lateef et al.[80] in another investigation using the metabolite of arthropods in nanobiotechnology reported the use of nest extract of paper wasp (*Polistes* spp.) to synthesize AgNPs. Biofabrication of AgNPs was marked by a change in color to brownish color, which became stable after 10 min of reaction of $AgNO_3$ with the extract. The crystalline polydispersed AgNPs absorbed maximally at 428 nm in the UV–Vis spectrum and data from FTIR analysis showed outstanding peaks at 3426 and 1641 cm^{-1} indicating active participation of phenolics and protein molecules in biofabrication of AgNPs. TEM analysis showed well dispersed anisotropic composition of rod, triangle, sphere, hexagon, and rhombus particles whose dimensions are in the range of 12.5–95.55 nm. Through EDX spectra, the prominence of silver in the colloidal solution was established. The AgNPs were tested for antimicrobial properties against *K. granulomatis, P. aeruginosa, A. fumigatus, A. flavus*, and *A. niger*. The AgNPs showed potent antibacterial activities against multidrug strains of *K. granulomatis* and *P. aeruginosa* with zones of inhibition ranging from 12 to 35 mm. It also exhibited excellent antifungal activities by completely suppressing the growth of *A. flavus* and *A. niger*, while 75.61% growth inhibition was produced against *A. fumigatus* at 100 μg/mL.

Azeez et al.[15] has documented biosynthesis of AgNPs under ambient conditions through the use of cocoa bean extract (CBE). The synthesized CBE-AgNPs were fairly polydispersed and spherical with sizes ranging from 8.96 to 54.22 nm. In UV–Vis region, the CBE-AgNPs showed absorption peak at 438.5 nm. From the FTIR spectrum, conspicuous peaks occurring at 3275.13 and 1635.54 cm^{-1} indicated the participation of phenolics and protein molecules in the biosynthesis of AgNPs. The EDX analysis showed the prominent peak that indicated the major incidence of silver in colloidal solution; and the typical crystalline face-centered cubic phase of the AgNPs was confirmed by SAED pattern. The CBE-AgNPs displayed substantial actions toward multi-drug defiant *S. aureus, K. pneumonia*, and *E. coli* through inhibition by 10–14 mm, in addition to boosting the actions

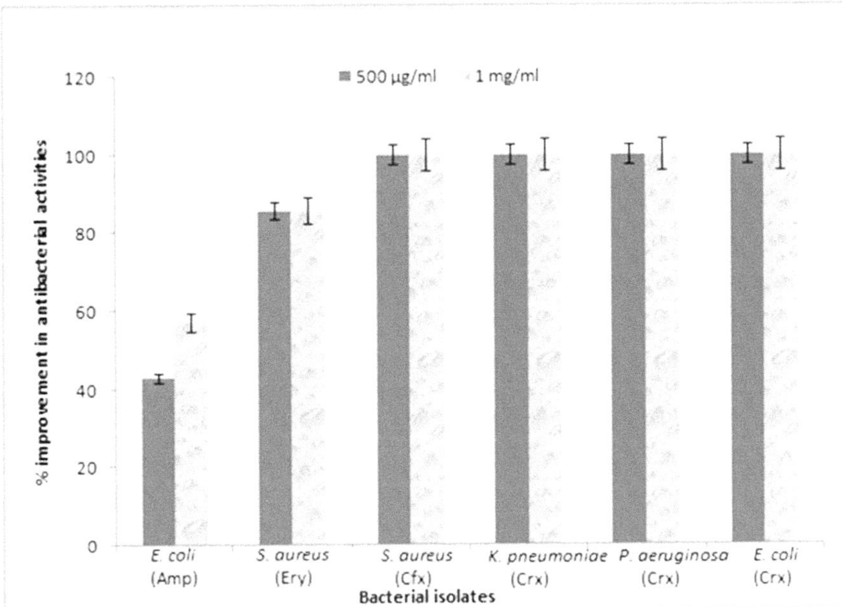

FIGURE 1.1 The synergistic antibacterial activities of cocoa bean extract-AgNPs with some antibiotics.

of ampicillin, erythromycin, cefixime, and cefuroxime by 42.9–100% in synergistic studies (Fig. 1.1). Also, the AgNPs produced pronounced antimicrobial activities against *S. aureus, E. coli, K. pneumoniae, S. pyogenes, P. aeruginosa, A. flavus, A. fumigatus*, and *A. niger* when incorporated into emulsion paint as antimicrobial additive at 5 µg/mL concentration.

Lateef et al.[82] in another study detailed the synthesis of AgNPs under ambient condition by means of an agro-waste: cocoa pod husk extract (CPHE). In the reaction mix, the transformation of color leading to the development brown colloids was the primary indication that the CPHE-AgNPs has been synthesized. The nanoparticles were observed to exhibit absorption peak at 428.5 nm in the UV–Vis region.

FTIR spectra of the CPHE-AgNPs showed strong peaks at 3294.42 and 1635.64 cm^{-1}, indicating that proteins and phenolics actively participated in the biogenic route which led to the production of the CPHE-AgNPs. The synthesized nanoparticles of 4–32 nm were adequately dispersed within the organic milieu, and they were predominantly spherical-shaped having a face-centered cubic phase. The crystalline structure was as shown by SAED and

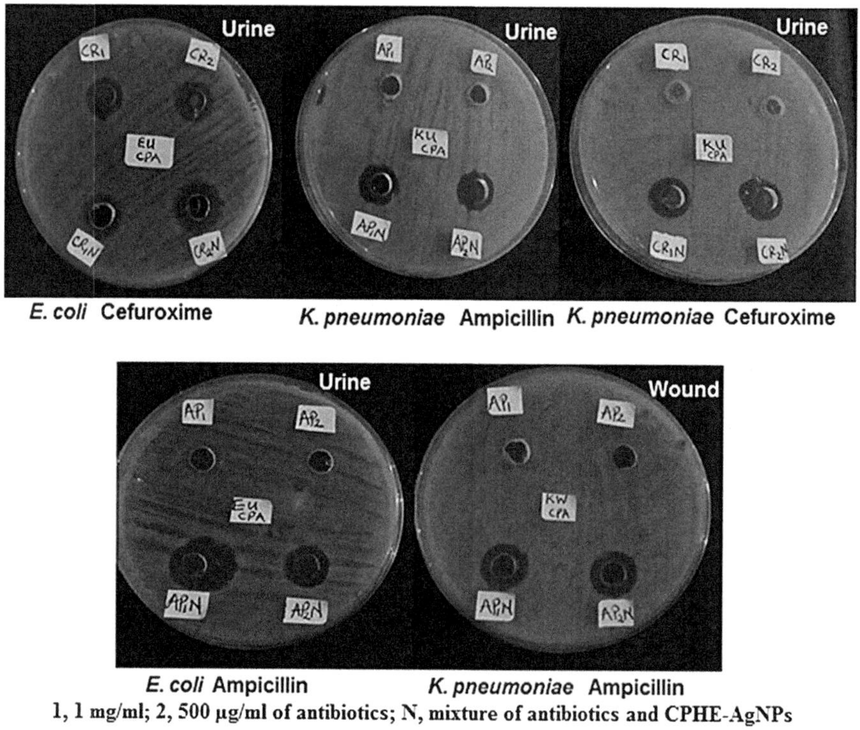

FIGURE 1.2 The synergistic activities of AgNPs synthesized using cocoa pod extract (CPHE) when combined with cefuroxime and ampicillin on some drug-resistant clinical isolates of bacteria.

EDX analysis indicated the presence of silver as the prominent element. The CPHE-AgNPs, when evaluated for antimicrobial actions, showed commendable performance against bacterial strains of *K. pneumoniae* and *E. coli* which were inhibited by 10–14 mm. In synergistic studies with cefuroxime and ampicillin, CPHE-AgNPs contributed to 42.9–100% improvement in the antibacterial activities of the antibiotics (Fig. 1.2). Moreover, the integration of CPHE-AgNPs into emulsion paint resulted in the efficient prevention of microbial growth of *P. aeruginosa, K. pneumoniae, S. pyogenes, E. coli, S. aureus, A. fumigatus*, *A. niger*, and *A. flavus* (Fig. 1.3).

Also, pod extract of *Cola nitida* was used to biofabricate AgNPs as described by Lateef et al.[83] A dark brown colloidal solution of AgNPs with maximum absorbance at 431.5 nm in the UV–Vis spectra was produced. The FTIR spectrum illustrated strong peaks at 3336.85, 2073.48, and 1639.49 cm^{-1}

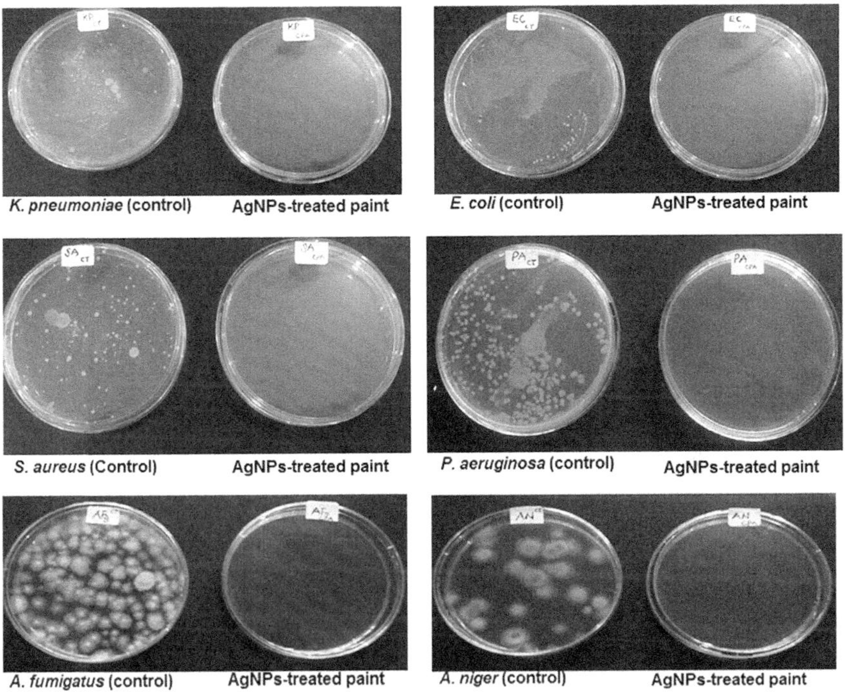

FIGURE 1.3 **(See color insert.)** Antimicrobial activities of biosynthesized CHPE-AgNPs in emulsion paint.

indicating functional groups relating to proteins as the molecules accountable for the biofabrication of the AgNPs and its protection against agglomeration. From TEM analysis, the AgNPs were spherical and polydispersed in nature having dimensions ranging from 12 to 80 nm. EDX analysis displayed silver as the prominent metal present; and SAED pattern of the AgNPs is in conformity with the face-centered cubic phase and crystalline attributes of AgNPs. Biosynthesized AgNPs showed good antimicrobial potentials. At the working concentrations of 50–150 μg/mL, the AgNPs successfully inhibited the growth of *P. aeruginosa, K. granulomatis*, and *E. coli* with zones of inhibition in the range of 12–30 mm. When incorporated in white emulsion paint, the AgNPs exhibited a total elimination of *P. aeruginosa* and *E. coli*, in addition to that of *A. niger*, *A. fumigatus*, and *A. flavus*.

Lateef et al.[81] in a similar study reported bio-inspired synthesis of AgNPs under ambient conditions with the use of extracts obtained from the seed shell (SS) and seed (S) of *Cola nitida*. Brown and yellowish orange colors were observed in the reaction mixture containing seed and seed shell extract, respectively after synthesis. Absorption peaks at 457.5 and 454.5 nm were obtained in the UV–Vis spectra for the S-AgNPs and SS-AgNPs, respectively. In the FTIR spectra, there were major peaks at 3292.49, 2086.98, and 1631.78 cm^{-1} obtained for S-AgNPs, and 3302.13, 2086.05, and 1633.71 cm^{-1} for SS-AgNPs. These observations pointed out that the presence of protein molecules in the extracts drove the subsequent biofabrication of AgNPs. Images from TEM revealed that both AgNPs were of spherical morphology and have dimensions of 5–50 nm. The EDX analysis confirmed the existence of silver as a prominent element in the colloidal solution, while the face-centered cubic phase and crystalline nature of AgNPs were confirmed by SAED pattern. In an antibacterial assay, both AgNPs (50–150 μg/mL) gave a good account of their efficacies by inhibiting the growth of *K. granulomatis, P. aeruginosa*, and *E. coli* by 10–32 mm. In comparison, SS-AgNPs demonstrated superior antibacterial potential with MIC of 50 μg/mL against all the three investigated isolates, while the MICs of 50–120 μg/mL were obtained for S-AgNPs.

The silver nanoparticles (PL-AgNPs) using aqueous *Piper longum* fruit extract (PLFE) reported by Reddy et al.[151] were studied for antibacterial activities using the standard agar diffusion method. It was found to exhibit antibacterial sensitivity against the pathogenic bacterial strains of *S. aureus, B. cereus, B. subtilis*, and *Pseudomonas* spp. At concentrations of 10 and 20 μg/mL, PL-AgNPs displayed zones of inhibition from 1.0 to 2.0 cm for the bacterial isolates. Singh et al.[169] also described the extracellular synthesis of AgNPs by culture supernatant of *Sporosarcina koreensis* DC4. The transformation in color of the reactant solution from yellow to deep brown within 48 h was observed for the synthesis. The synthesized AgNPs was absorbed maximally at 424 nm in the UV–Vis spectrum. The FE-TEM analysis revealed that the nature of the AgNPs was spherical in morphology with dimensions ranging from 30 to 50 nm. According to DLS, polydispersity index (PDI) was obtained to be 0.240 for the AgNPs. The EDX spectrum showed highest peak at 3 keV for AgNPs; and the XRD pattern of nanoparticles exhibited strong peaks in the entire spectrum of 2θ value within the range of 20–80°, and this outline is in similarity to the Braggs's fingerprint of silver. The AgNPs displayed excellent antimicrobial activities against *E. coli, Vibrio parahaemolyticus, B. anthracis, Salmonella enterica, S. aureus*, and *B. cereus*. The results clearly demonstrated that the AgNPs showed MIC values of 3–6.9 μg/mL against the tested pathogenic bacteria. Moreover, AgNPs enhancement of antibacterial activities of some commercially available antibiotics (such as rifampicin, vancomycin, penicillin G, oleandomycin, lincomycin, and novobiocin) revealed that 3 μg/mL of AgNPs satisfactorily enhanced the antimicrobial effectiveness of the tested antibiotics. Additionally, the AgNPs showed biofilm inhibition properties against *S. aureus, E. coil*, and *P. aeruginosa* depicting adequate action at 6 μg/mL.

Dhand et al.[31] illustrated phytosynthesis of AgNPs by employing extract of dried roasted seed of *Coffea arabica*. The biosynthesis was marked with color change, leading to development of dark brown solution that absorbed maximally at 459 nm. The XRD pattern revealed that the AgNPs were vastly crystalline and exhibited a cubic, face centered lattice having distinctive (111), (200), (220), and (311) orientations. The TEM images showed that the AgNPs were spherical and ellipsoidal shaped structures and the existence of elemental fingerprint of silver was confirmed by the EDX analysis. The FTIR spectra revealed a descending shift of absorption bands of 800–1500 cm^{-1} as responsible for bioformation of the AgNPs. Digital light scattering analysis of particles revealed their distribution around 20–30 nm. In the antibacterial assay, AgNPs exercised actions against *E. coli* and *S. aureus* with MIC of 0.2675 mg/L for both bacterial strains. The zones of inhibition were 2.7 and 3.1 cm for *S. aureus* and *E. coli*, respectively at 0.535 mg/L.

Gomathi et al.[47] also described the green method of AgNPs synthesis using *Datura stramonium* leaf extract. A color change to dark brown was observed during the synthesis and a maximal absorption at 444 nm was recorded in the UV–Vis spectrum indicating formation of isotropic spherical AgNPs. In the FTIR spectrum, prominent peaks occurring at 1023, 1110, 1380, 1460, 1517, 1647, and 1738 cm^{-1} were obtained. The spherical shape of the particles was confirmed through the TEM images, which also showed size range of 15–20 nm. SAED pattern revealed the crystallinity of silver in the solution, while EDAX report asserted the presence of Ag crystal through occurrence of a peak at 3 eV. The phytosynthesized AgNPs presented anti-growth action toward cells of *S. aureus* and *E. coli*. The antibacterial activity of AgNPs was increased gradually as the stock solution increased from 50 to 200 μL.

Furthermore, the spherical SEEr-AgNPs reported by Dhayalan et al.[32] exhibited excellent antibacterial activities against strains of *E. coli* and *S. aureus* with zones of inhibition in the range of 20–28 mm and 22–27 mm for both strains, respectively. The color of the solution was observed to change from colorless to yellowish-brown during the synthesis and characteristic peak between 400 and 440 nm was obtained for the AgNPs in the UV–Vis spectrum. The FTIR spectra showed peaks at 3645, 3543, 3479, 3427, and 1417 cm^{-1} confirming functional groups of phytochemicals as involved in the reduction and stabilization of AgNPs. TEM images revealed that the

AgNPs are spherical and polydispersed in nature and had size in the range of 5–35 nm. XRD pattern for the SEEr-AgNPs indicated various peaks at (220), (311), (400), (331), (422), (511), (440), (531), (620), and (533).

Banerjee et al.[17] studied the synthesis of AgNPs through the use of leaf extracts of *Ocimum tenuiflorum* (black tulsi), *Azadirachta indica* (neem), and *Musa balbisiana* (banana). In the investigation, leaf extracts of the three plants were reacted with aqueous 1 mM silver nitrate solution individually, leading to series of transition in change of color of the reaction mix from initial that culminated into colloidal brownish solution which signaled the bioformation of AgNPs. The AgNPs showed absorption spectra of the particles peaked at 425–475 nm which are in consistency with the established SPR of AgNPs. The broadening nature of the absorption peaks laid credence to the formation of large polydispersed nanoparticles as a result of the slow rates of bioreduction. The FTIR spectra of all the three AgNPs showed major bands at 1025, 1074, 1320, 1381, 1610, and 2263 cm^{-1}, which are typified by –C–OC–, ether linkages, –C–O–, germinal methyls, –C=C– groups and alkyne bonds, respectively; and they correspond to stretching vibrational bands that are characteristic of compounds such as flavonoids and terpenoids.[54] The SEM and TEM analyses confirmed that the tulsi-, neem-, banana-, and mediated AgNPs were cuboidal, triangular, and spherical shapes, respectively and have sizes up to 200 nm. The EDS report showed a strong signal for Ag with some minor peaks that are indicative of carbon and oxygen that possibly arose from the compounds that are associated with the plant extracts and subsequently borne on the AgNPs. The biosynthesized AgNPs also exhibited excellent antibacterial activities against *Bacillus* spp. and *E. coli* strains. For banana extract-mediated AgNPs, zones of inhibition for *Bacillus* spp. and *E. coli* were 16 and 14 mm, respectively; for neem extract-mediated AgNPs, zones of inhibition were 16 and 12 mm for *Bacillus* spp. and *E. coli*, respectively; while for tulsi extract-mediated AgNPs, zone of inhibition of 14 mm was recorded.

Lateef et al.[79] investigated the synthesis of AgNPs mediated by leaf and seed extracts of *Synsepalum dulcificum* (miracle fruit plant). The formation of brown colloidal silver nanoparticles was noticed during the exposure of the solution to sunlight for photoactivated synthesis. The polydispersed leaf and seed AgNPs were absorbed maximally at 440 and 438.5 nm, respectively. The FTIR spectra laid credibility to participation of phenolic compounds and proteins in the phytosynthesis of the AgNPs with prominent peaks at 3408, 2357, 2089, and 1639 cm^{-1} for leaf extract-mediated AgNPs; and 3404, 2368, 2081, and 1641 cm^{-1} for the seed-AgNPs. Morphological studies of the AgNPs through TEM indicated the presence of fairly spherical-shape and crystalline particles having sizes of 5–22 and 4–26 nm for leaf extract-mediated AgNPs and seed extract-mediated AgNPs, respectively. Little agglomeration was noticed in seed extract-mediated AgNPs, while the leaf extract-mediated AgNPs was reportedly well dispersed. The EDX spectrum revealed the prominence of Ag in both AgNPs. The two investigated AgNPs exhibited antimicrobial activities when tested against some pathogenic bacteria as well as fungal isolates. The AgNPs inhibited the growth of *P. aeruginosa* and *K. granulomatis* with zone of inhibition of 11–24 mm at MIC of 60 μg/mL. Also, the AgNPs showed excellent potencies against the fungal isolates by inducing 100% suppression of the growth of *A. flavus* and *A. niger*. However, 75.60% and 73.17% growth inhibition was produced against *A. fumigatus* by leaf extract-mediated AgNPs and seed extract-mediated AgNPs, respectively.

Ajitha et al.[11] also described the phytosynthesis of AgNPs using the aqueous leaf extract of *Sesbania grandiflora*. The solution containing the AgNPs turned yellow during the synthesis and the AgNPs was reported to maximally absorb at 416 nm in the UV–Vis spectrum. The FTIR spectroscopy acknowledged that carbonyl groups, as well as amide I bands of proteins, were involved in the nanoparticles formation. The morphology of the AgNPs as described by the TEM was spherical and well dispersed, with mean particle diameter of 16 nm. The XRD study revealed the fcc configuration of AgNPs with good crystalline nature; while the EDX established the incidence of metallic Ag at ~3 keV, and SAED pattern showed bright circular fringes indexed to different planes of fcc structure of silver, which agree with the result of the XRD analysis. The synthesized AgNPs displayed antimicrobial potentials against four bacterial strains (*Staphylococcus* spp., *Pseudomonas* spp., *Bacillus* spp., and *E. coli*), as well as four fungi (*Penicillium* spp., *A. flavus*, and *A. niger*). The zones of inhibition recorded against the bacterial strains ranged from 3 to 8 mm, while that of the fungal strains were 3–7 mm.

AgNPs were synthesized through the exposure of cells of *Microbacterium marinilacus* and *Stenotrophomonas maltophilia* to the solution of Ag^+ ions by Mukherjee.[112] The bioformation of AgNPs was discernible through the appearance of dark brown color and this was further confirmed by obtaining an absorbance peak at 400 nm. AgNPs produced peaks at elevated wavenumbers of 3652, 3737, 3750, and 3842 cm^{-1} in the FTIR spectrum, which are distinctive of –OH stretch of alcohol. SEM images displayed the AgNPs as tiny cubes with particle sizes ranging from 10 to 50 nm. The XRD analysis showed strong peaks values corresponding to planes of Ag. These peaks are substantiated with the standard powder diffraction card of Ag_2O sample with JCPDS cards No. 76-1393. The synthesized AgNPs was tested against *E. coli* and *B. cereus* with the use of agar well diffusion method. The AgNPs produced higher growth inhibition against *E. coli* as opposed to results obtained with *B. cereus.*

Also, the spherical biosynthesized AgNPs using *Plumeria alba* (Frangipani) flower extract (FFE) by Mata et al.[105] was also investigated for antibacterial properties. Different concentrations at 100–400 µg/mL of the AgNPs were used against bacterial strains of *S. aureus, E. coli, B. subtilis*, and *B. cereus*. The AgNPs produced significant inhibitory activities against *E. coli* (10–18 mm), *B. subtilis* (8–16 mm), *S. aureus* (5–13 mm), and *B. cereus* (3–11 mm). The formation of the AgNPs was evident due to transformation in color from pale yellow to yellowish brown of the solution and the AgNPs showed maximum absorption peak at 445 nm. The FTIR spectrum showed bands that indicated the active phytochemicals like carboxylic, sulfhydryls, and amino groups present in plant extracts as responsible for reducing, capping, and subsequent bio-inspired production of AgNPs. The XRD pattern indicated Bragg's reflections with 2θ values of 38.2°, 44.3°, and 64.6° that were ascribed to (111), (200), and (220) facets of Ag^0. Morphological studies from TEM images revealed that the AgNPs were of spherical morphology having mean size of 36.19 nm and the EDX spectrum established the existence of silver in the AgNPs.

Also, Shanmugam et al.[159] described the synthesis of AgNPs through the use of the extract obtained from the leaves of *Aristolochia indica* that performed the roles of bioreduction as well as capping agent. The phytosynthesis of AgNPs was signaled through change in color from red to brown in a reaction that took about 24 h. The FTIR spectra of the synthesized AgNPs depicted peaks at 1549.86, 1438.77, 1384.04, 1321.35, 1238.66, 1073.73, 896.44, 816.99, and 781.38 cm^{-1}. SEM and TEM images revealed that the AgNPs were spherical geometry as well as polydispersed in distribution. The XRD pattern established that the synthesized AgNPs has fcc structure and the EDS spectrum affirmed the existence of elemental Ag. The particles were found to display excellent growth inhibitory activities against five bacterial strains (*S. epidermidis*, *E. coli, S. aureus, B. subtilis,* and *E. faecalis*) and five fungi (*Trichophyton simii, A. flavus*, *T. mentagrophytes, C. albicans*, and *T. rubrum*). At concentration of 500 µg/disc, the AgNPs was found potent with zones of inhibition in the range of 7–11 and 9–13 mm for the bacterial and fungal strains, respectively.

Phull et al.[134] described the synthesis of BC-AgNPs from crude extract of *Bergenia ciliata*. Color change of the reacting solution to brownish indicated the synthesis of the BC-AgNPs, which had utmost absorption at 425 nm. The FTIR spectrum revealed absorption bands at 3016, 2990.4, 2862.3, 1699.5, 1542, 1456, 1304.8, 1054.9, and 1033.1 cm^{-1} which indicated the possible involvement of phyto-constituents to synthesize AgNPs. Morphologically, BC-AgNPs were reported to be spherical-shaped having average particle size of 35 nm. The BC-AgNPs exhibited good antimicrobial activities against bacterial strains (*S. aureus, E. aerogenes*, and *B. bronchiseptia*) and fungal strains (*A. niger, A. fumigatus*, and *F. solani*). The zones of inhibition against the bacterial strains ranged from 7.5 to 11 mm and 7.5 to 9 mm against the fungal strains.

Elemike et al.[39] have also shown that the leaf extract of *Talinum triangulare* (TT) can facilitate the production of AgNPs. Transformation in color from light yellow to reddish brown marked the synthesis of the TT-AgNPs which showed maximum absorbance between 417 and 430 nm in the UV–Vis spectrum. The FTIR results projected that proteins and enzymes probably aided bioformation of AgNPs through bioreduction of the silver ions. TEM analysis clearly showed that the TT-AgNPs are spherically shaped with mean dimension of 13.86 nm. XRD showed prominent peaks at 32.32°, 37.50°, 54.11°, and 67.72° corresponding to (111), (200), (220), and (222) planes of fcc of AgNPs, and which are in conformity with JCPDS file No. 004-0783. EDAX analysis of TT-AgNPs confirmed the percentage of elemental silver as 43.14%, which was highest in the composition. The TT-AgNPs also displayed excellent antimicrobial activities which are significantly higher than that obtained with the leaf extract and $AgNO_3$ against *C. albicans, S. typhi, S. aureus, B. subtilis,* and *E. coli*. The zones of inhibition obtained against the microorganisms ranged from 13 to 23 mm at 100 μg/mL.

Jinu et al.[60] likewise employed the extract of leaves of AgNPs *Solanum nigrum* for the phytosynthesis of AgNPs that resulted in the development of yellowish brown color signaling the biofabrication of AgNPs in the solution. Absorbance peak at 428 nm was observed for the synthesized AgNPs while the FTIR spectrum showed prominence for bands that are indicative of amide I and II. These bands represented the presence of proteins/enzymes and these are concluded to be accountable for bioreduction of silver ions to AgNPs and further stabilized the particles.[8] The cuboidal-shaped AgNPs had average particle size of 20 nm as shown by SEM. XRD results of bioengineered AgNPs established crystalline nature of the particles, while the EDX spectrum depicted existence of conspicous signal for silver ions in AgNPs. The synthesized AgNPs showed excellent antibacterial activities when tested against six pathogenic bacterial strains; *P. vulgaris, K. pneumoniae, S. epidermidis, P. aeruginosa, S. aureus*, and *E. coli* using 50 μL of AgNPs. These bacteria were inhibited to varying degrees by 9.6–21.5 mm. Also, the phytosynthesized AgNPs displayed antibiofilm activities against *S. epidermidis* and *P. aeruginosa* by 34.8% to 61.5% and 41.2% to 68.6%, respectively.

Also, Ravichandran et al.[148] described the synthesis of *Artocarpus altilis* silver nanoparticles (BAgNPs) by utilizing aqueous extract obtained from the leaves of *Artocarpus altilis,* in which case the color changed to brown to signal the formation of the BAgNPs, which had maximum absorbance peak at 432 nm. Results obtained from the FTIR spectrum of BAgNPs were depictive of C–N and C–C bonds that suggest the presence of polyphenols in the leaf extract of *A. altilis.* Thus, polyphenols were implicated in the synthesis of BAgNPs and also responsible to have capped and stabilized the particles. The BAgNPs were shown to be spherical and polydispersed; and particle sizes of 34 and 38 nm were obtained with the SEM and TEM analysis, respectively. The XRD pattern obtained for the BAgNPs had five prominent peaks observed at $2\theta = 38.24°$, 44.39°, 64.57°, 77.55°, and 81.57° corresponding to (111), (200), (220), (311), and (222) Bragg's reflections of fcc arrangement of metallic Ag. The AgNPs were investigated for antimicrobial activities against three bacteria; *E. coli, P. aeruginosa*, and *S. aureus*; and a fungus; *A. vesicolor.* Results showed that *P. aeruginosa* and *E. coli* were susceptible to BAgNPs compared with *S. aureus,* while *A. vesicolor* displayed a minimal level of susceptibility toward BAgNPs.

Moreover, eco-friendly synthesis of Siberian ginseng AgNPs (Sg-AgNPs) using pharmacologically active stem of Siberian ginseng (*Eleutherococcus senticosus*) was reported by Abbai et al.[1] In this investigation, it was noticed that dark brown color developed from light yellow solution which signaled synthesis of AgNPs. Sg-AgNPs absorbed maximally at 440 nm; and FTIR examination signified proteins and aromatic hydrocarbons as major players in bioformation and stability of Sg-AgNPs. FE-TEM indicated that the AgNPs were mainly spherical-shaped having average diameter of 126 nm as shown by DLS. It also has polydispersity index of 0.25, while the EDX spectrum of the AgNPs showed peak at ~3 keV. The XRD pattern of AgNPs revealed peaks at $2\theta = 38.62°$, 44.72°, 65.02°, 77.84°, and 81.90° that match (111), (200), (220), (311), and (222) planes of fcc-shaped crystalline AgNPs. The antibacterial effectiveness of the Sg-AgNPs was assessed against *B. anthracis* (NCTC 10340), *S. aureus* (ATCC 6538), *E. coli* (BL21), and *V. parahaemolyticus* (ATCC 33844). Using concentration of 10–30 μg/mL of the Sg-AgNPs, these bacteria were inhibited by 10.7–13.8 mm.

Niraimathi et al.[123] described phytosynthesis of AgNPs by utilizing aqueous extract of *Alternanthera sessilis*, which was investigated for antimicrobial activities against pathogenic strains of *E. coli* and *S. aureus*. The bacteria were found to be sensitive to the AgNPs. Moteriya and Chanda[110] also synthesized AgNPs that was mediated by the flower extract of *Caesalpinia pulcherrima*. The development of a dark brown color from the initial pink color confirmed the phytosynthesis of AgNPs. Absorbance peak at 410 nm was obtained for the dark brown solution; and the FTIR spectrum showed peaks at 2916.37, 2848.86, 1741.72, 1602.85, 1357.89, 1205.51, 1033.85, and 669.30 cm^{-1}. The morphology of the AgNPs studied by TEM analysis was reported to be spherical in shape and uniformly distributed without agglomeration that are averagely 12 nm in size. The XRD spectrum showed different peaks that are in consonance with fcc structure of AgNPs. The synthesized AgNPs displayed good antimicrobial activities against several microorganisms. *S. aureus* and *C. glabrata* had the MIC of 2.5 mg/mL, followed by *B. cereus, E. coli, S. typhimurium, C. albicans*, and *Cryptococcus neoformans*, which exhibited slightly higher MIC value of 5 mg/mL; and *B. subtilis* and *C. rubrum*, which exhibited the highest MIC value (10 mg/mL). The strains of *P. aeruginosa* and *K. pneumoniae* displayed resistance to the actions of AgNPs.

From the foregoing, it is evidently clear that green and eco-friendly synthesized AgNPs using different biological components such as plant, algal, fungal, and bacterial extracts have demonstrated noteworthy activities against number of bacterial and fungal pathogens. Evidences also exist for the increased investigations of other novel bioresources that include enzymes, pigments, and arthropods' metabolites to synthesize AgNPs. The combative activities displayed by some AgNPs against multi-drug resistant isolates indicate the potentials that these nanoparticles hold in the war against drug resistance. Suffice to say that these particles can serve as veritable tools against MDR strains in both clinical setting and personal care products. The antimicrobial activities of AgNPs are executed through a number of phenomena, which include: the generation reactive oxygen species,[28] denaturation of DNA, proteins and enzymes by combining with sulfur and phosphorus,[3] destruction of cell walls and membranes as well as promotion of intracellular ion efflux[63] to lead to death of the cell. With these mechanisms of action, nanoparticles have positioned themselves to exercise various actions toward drug-resistant microorganisms for diverse applications.[186]

1.4.2 LARVICIDAL ACTIVITIES OF SILVER NANOPARTICLES

Govindarajan et al.[50] demonstrated green biosynthesis of AgNPs by employing aqueous extract of leaves of *Carissa spinarum* which produced brown-yellowish colloid that indicates the bioformation of the AgNPs, which absorbed maximally at 447 nm. Morphologically, the AgNPs was cubic- and spherical-shaped with dimensions between 40 and 100 nm according to SEM images. The mean size observed according to TEM results was 40 nm. The FTIR spectrum showed bands corresponding to the stretching of –OH, the amide I and amide II band which may signify the presence of enzymes, proteins, or polysaccharides in the extract and these may be responsible for the phytosynthesis of AgNPs. The XRD pattern exhibited four peaks at 38.22°, 44.37°, 64.54°, and 77.47° that represented the (111), (200), (220), and (311) reflections of fcc structure of metallic Ag. Larvicidal activities of AgNPs were investigated with the biosynthesized AgNPs depicting excellent toxicity against larva of *A. subpictus*, *A. albopictus*, and *C. tritaeniorhynchus* with LC_{50} values of 8.37, 9.01, and 10.04 μg/mL, respectively. However, the particles could be described as non-toxic in non-target organisms that include *Gambusia affinis*, *Anisops bouvieri*, and *Diplonychus indicus* where very high values of LC_{50} in the range of 424.09–647.45 μg/mL were obtained.

Also, Rajakumar and Rahuman[142] described the synthesis of AgNPs utilizing aqueous extract from *Eclipta prostrata* that led to generation of a brown colored AgNPs solution, which exhibited surface plasmon resonance at 420 nm. Absorption bands were located at 1079, 1383, 1627, and 1729 cm^{-1} that correspond to the occurrence of carbonyl groups, fatty acids, flavonoids, and proteins in the extract. The TEM analysis revealed the AgNPs to bespherical with some elongated particles with sizes ranging between 35 and 60 nm and mean size of 45 nm. The XRD profile showed a degree of Bragg reflections indexed to the (111), (200), (220), and (311) facets of AgNPs. In the investigations of larvicidal activities of the AgNPs, larvae of organisms employed were studied for exposure to different concentrations of both extract and the synthesized AgNPs for a period of 24 h. The LC_{50} obtained for the aqueous extract and the AgNPs were 27.49 and 4.56 mg/L, and 27.85 and 5.14 mg/L against *C. quinquefasciatus* and *A. subpictus*, respectively. Also, the LC_{90} obtained for the aqueous extract and the AgNPs were 70.38 and 13.14 mg/L, and 71.45 and 25.68 mg/L against *C. quinquefasciatus* and *A. subpictus*, respectively.

In another study, Anbazhagan et al.[13] described the green-synthesized AgNPs using stem extract of *Musa paradisiaca* that led to the formation of brown–yellow solution which showed absorbance peak at 410 nm. Major peaks at 464.74, 675.61, 797.07, 1059.42, 1402.58, 1639.69, 2115.61, and 3445.75 cm^{-1} in the FTIR spectrum and the results from TEM revealed that the AgNPs were mostly spherical-shaped and polydispersed with dimensions of 5–35 nm. Also, EDX profile has two peaks shown which were found between 2.8 and 4 keV that indicate the presence of silver. The larvicidal investigations involved exposure of 25 samples of *A. stephensi* larvae (I–IV instars) and pupae to AgNPs for 24 h by placing them in 250-mL capacity beaker with the use of dechlorinated water. The AgNPs were toxic to the various developmental forms of *A. stephensi*, with LC_{50} of 3.642, 5.497, 8.561, 13.477, and 17.898 ppm obtained for first instar, second instar, third instar, fourth instar and pupae, respectively.

In addition, Roopan et al.[154] biosynthesized AgNPs by utilizing the extract of mesocarp of *Cocos nucifera*, which led to the formation of dark brown colored AgNPs after 4 h that exhibited surface plasmon resonance at 433 nm. The TEM results showed that the AgNPs were nearly spherical-shaped with mean size of 23 nm; and EDS revealed presence of elemental silver metal signal. Also, the XRD spectrum showed characteristic Bragg peaks facets of fcc AgNPs, confirming the crystallinity of the nanoparticles. Effective larvicidal activities of the AgNPs were investigated in the larvae of *A. stephensi* and *C. quinquefasciatus*. The comparative LC_{50} and LC_{90} of AgNPs against instars of *A. stephensi* were 87.24 ± 4.75 mg/L, and 230.90 ± 17.10 mg/L, respectively; while in *C. quinquefasciatus*, the values were 49.89 ± 2.42 mg/L and 84.85 ± 6.50 mg/L for LC_{50} and LC_{90}, respectively.

Biosynthesized AgNPs produced using the leaf extract of *Euphorbia hirta*[137] were evaluated for larvicidal in *A. stephensi*, and the uppermost larval death occurred in the I, II, III, and IV instar larvae and pupae with values of LC_{50} being 10.14, 16.82, 21.51, 27.89, and 34.52 ppm, respectively. The corresponding LC_{90} against these developmental stages were also 31.98, 50.38, 60.09, 69.94, 79.76 ppm, respectively. Also, AgNPs phytosynthesized using the leaf extract of *Rhizophora mucronata* displayed activities toward larvae of *A. aegypti* and *C. quinquefasciatus*. LC_{50} values of the synthesized AgNPs were 0.585 and 2.615 mg/L; while LC_{90} were obtained as 0.891 and 6.291 mg/L, respectively.[46] It was also reported by Govindarajan and Benelli[49] that *Hymenodictyon orixense*-mediated synthesis of AgNPs exhibited 449 nm surface plasmon resonance and they were confirmed to be spherical, and polydispersed with sizes ranging from 25 to 35 nm by SEM, AFM, and TEM analysis, respectively. The FTIR analysis showed bands that correspond to stretching of –OH, –NH stretch vibrations in the amide, C–C stretching of aromatic amines and C–O stretching of alcohols, enzymes, proteins, or polysaccharides that were present in the extract. Also, the XRD spectrum showed four diffraction peaks that depicted fcc structure of metallic silver. EDX profile equally affirmed the existence of silver in the sample. In larvicidal studies, *H. orixense*-synthesized AgNPs showed very high level of toxicity against larvae of *A. subpictus*, *A. albopictus*, and *C. tritaeniorhynchus* with LC_{50} and LC_{90} values being 17.10 and 33.16; 18.74 and 35.25; and 20.08 and 36.94 μg/mL. Conspicuously, the AgNPs were found to be non-toxic in non-target mosquito predator, *Diplonychus indicus* with extremely high LC_{50} of 833 μg/mL.

AgNPs synthesized with aqueous leaf extract of *Zornia diphylla* as bioreducing and capping agent was reported by Govindarajan et al.[51] The AgNPs exhibited absorbance peak at 418.5 nm; and transmittance peaks 3327.63, 2125.87, 1637.89, 644.35, 597.41, and 554.63 cm^{-1} were obtained with the FTIR spectrum that corresponds to CN stretch, C=N, $-NH_2$, –NH, –H stretch, and –OH groups, respectively. The peaks are characteristically found to be associated with terpenoids and flavonoids.[54] SEM analysis revealed the AgNPs as spherical in shape and with high degree of aggregation possessing mean size of 30–60 nm, while TEM analysis revealed triangle, truncated triangles, decahedral, and spherical morphologies that were averagely 37 nm. Bragg reflections that correspond to (111), (200), (220), (311), and (222) were encountered in XRD spectrum, while EDX profile indicated a prominent signal within silver domain that confirmed existence of AgNPs. The acute toxicity of leaf extract of *Z. diphylla* and phytosynthesized AgNPs were assessed using larvae of *A. subpictus*, the Japanese encephalitis vector *Culex tritaeniorhynchus*, and dengue vector *Aedes albopictus*. In contrast to the aqueous leaf extract, the biosynthesized AgNPs displayed more potent toxicity in *A. subpictus*, *A. albopictus*, and *C. Tritaeniorhynchus* with corresponding LC_{50} of 12.53, 13.42, and 14.61 μg/mL.

Likewise, AgNPs synthesized using peel extracts of *Arachis hypogaea* that showed increased brown coloration with increased incubation time as reported by Velu et al.[180] The AgNPs showed greatest absorbance at 450 nm; and FTIR analysis indicated absorption bands at 3836, 3749, 3419, 2921, 2851, 2286, 1632, 1384, 1220, 1175, 1087, 868, 706, 662, 640, 623, and 558 cm^{-1}, which strongly points to the participation of polyphenols, amino groups, carboxyl groups, and amino acid residues in AgNPs synthesis. The AgNPs were spherical-shaped and have size distribution ranging from 20 to 50 nm. Discrete signals and elevated atomic strength were obtained for silver in the EDX spectrum, while XRD profile displayed strong peaks that correspond to (110), (111), (121), and (200) Bragg's reflection of fcc structure of AgNPs. Also, SAED pattern showed that the AgNPs were crystallized. The AgNPs were investigated for larvicidal potency using the fourth instar larvae of *A. aegypti*, and *A. stephensi*, which showed high susceptibility to the AgNPs. Mortality of 100% was recorded after 24 h treatment with 15 mg/L concentration of the AgNPs. The LC_{50} and LC_{90} of synthesized AgNPs against fourth instar larvae of *A. aegypti* were 1.85 and 10.36 mg/L, respectively, while values of 3.13 and 11.15 mg/L, were obtained against *A. stephensi*.

Also, seaweed (*Codium tomentosum*) was exploited to synthesize AgNPs as reported by Murugan et al.[116] Yellowish-brown color was developed from the synthesized Seaweed-mediated silver nanoparticles, which showed absorbance peak at 420 nm. Results obtained from the FTIR analysis revealed peaks at 3496, 2970, 1498, 1479, and 593 cm^{-1} that correspond to geminal methyls, aromatic rings, and ether linkages that may point toward the existence of flavones and terpenoids in the extract and being responsible for the stabilization of the AgNPs. SEM micrographs showed AgNPs as predominantly irregular in shape that were of 20–40 nm in dimension. EDX spectrum displayed very sharp and strong peak at 3 keV distinctive of metallic Ag nanocrystallites, while XRD profile showed Bragg's reflections of (111), (200), (220), (311), and (222). The AgNPs was tested for larvicidal activities using *A. stephensi* and the consequences indicated that AgNPs possessed significantly higher larvicidal potential compared with the *C. tomentosum* extract. The extract of *C. tomentosum* produced LC_{50} of 255.1 to 487.1 ppm against *A. stephensi* larva and pupa, respectively, while those of the AgNPs were 18.1 to 40.7 ppm, respectively.

In a related study, AgNP biosynthesized using the seed kernel extract of *A. indica* as bioreducing and stabilizing agents was documented by Murugan et al.[115] The phytosynthesized AgNPs absorbed maximally at 450 nm, while peaks of 449.00, 597.99, 660.46, 1016.71, 2137.89, and 3361.62 cm^{-1} obtained in FTIR spectrum pointed toward the occurrence of diverse groups that include alkane, alkene, carboxylic acid, and amine in the extract. Compounds that are rich in these groups have earlier been implicated as bioreductants to biosynthesize AgNPs.[22] The SEM analysis showed the AgNPs as spherical in shape, having diameter of 35–60 nm; while EDX spectrum deep-rooted the existence of elemental Ag alongside oxygen, which is related to the extracellular natural moieties adsorbed on the surface of AgNPs. The XRD profile showed that AgNPs formed were crystalline. In the larvicidal property studies, instar I–IV larvae and pupa of *A. stephensi* were subjected to treatments with the seed kernel extract of *A. indica* and its AgNPs. The LC_{50}

obtained for the instar I, II, III, IV larvae and pupa were 232.8, 260.6, 290.3, 323.4, and 348.4 ppm, respectively for the *A. indica* seed kernel extract, while that of the AgNPs were 3.9, 4.9, 5.6, 6.5, and 8.2 ppm, respectively.

Additionally, fern-mediated AgNPs synthesized using the extract of *Pteridium aquilinum*, acted as a bioreductant as well as a capping agent was reported by Panneerselvam et al.[127] Change in color from yellow to brownish solution as a function of catalytic reduction of Ag^+ to Ag^0 was observed for the AgNPs, which absorbed maximally at 420 nm. The results from the FTIR analysis hinted that phenols, flavonoids, carbohydrates, alkaloids, proteins, tannins, glycosides, and saponins possibly accounted for bioreduction and stability of the fern-AgNPs. The AgNPs was generally spherical-shaped, polydispersed in distribution and having sizes considered to be 35–65 nm according to SEM analysis. The EDX spectrum confirmed the existence of pure silver and weak signals of other elements; and XRD revealed diffraction patterns that matched the lattices of crystalline AgNPs. The synthesized AgNPs displayed potent larvicidal activities against both larvae and pupae of *Anopheles stephensi*. The LC_{50} obtained for the instar I, II, III, IV larvae and pupa were 7.48, 10.68, 13.77, 18.45, and 31.41, ppm respectively for the *P. aquilinum*-synthesized AgNPs. In addition, the deployment of the fern extract and the AgNPs ($10 \times LC_{50}$) resulted in total dealth of larvae after 72 h of exposure. It was concluded that both fern extract and the AgNPs have the capacity to reduce life expectancy and productiveness in the adults of *A. stephensi*.

Similarly, the green biosynthesized AgNPs that involved the use of culture filtrate of *Bacillus safensis* has been reported by Lateef et al.[85] The larvicidal potential of AgNPs was investigated through the exposure of ten larvae of *Anopheles* mosquito to 10 mL of different quantities of AgNPs. The results of larvicidal activity showed that 100% mortality was achieved at the end of 12 h for all concentrations, and the LC_{50} was 42.19 μg/mL. Also, the spherical CBE-AgNPs synthesized by Azeez et al.[15] displayed larvicidal properties. It was reported to show high-quality potency against larvae of *A. gambiae* projecting LC_{50} of 44.37 μg/mL. The CPHE-AgNPs synthesized with the aid of extract of cocoa pod also displayed effective larvicidal actions against the larvae of *Anopheles* mosquito.[82] Using AgNPs of 10–100 μg/mL, the percentage larvicidal activity obtained was 70%–100% within 2 h of exposure. The LC_{50} recorded was 43.52 μg/mL for the biosynthesized AgNPs.

The abundance of literature on the potentials of green AgNPs as larvicidal agents in controlling insects of medical and veterinary significance is an indication that these particles may be deployed as potent nanotools to combat vectors of several diseases in the near future.

1.4.3 ANTIOXIDANT ACTIVITIES OF AgNPs

Prasannaraj and Venkatachalam[136] reported the synthesis of medicinal plant-mediated AgNPs namely: *Andrographis paniculata* AgNPs (ApAgNPs), *Moringa oleifera* AgNPs (MoAgNPs), *Centella asiatica* AgNPs (CaAgNPs), *Aegle marmelos* AgNPs (AmAgNPs), *Plumbago zeylanica* AgNPs (PzAgNPs), *Eclipta prostrata* AgNPs (EpAgNPs), *Terminalia arjuna* AgNPs (TaAgNPs), *Alstonia scholaris* AgNPs (AsAgNPs), *Thespesia populnea* AgNPs (TpAgNPs), and *Semecarpus anacardium* AgNPs (SaAgNPs). The antioxidant activities of biosynthesized AgNPs was assayed through the release of reactive oxygen species (ROS) using XTT (sodium 2,3,-bis(2-methoxy-4-nitro-5-sulfophenyl)-5-[(phenylamino)-carbonyl]-2H-tetrazolium inner salt in *E. coli* model. It is a known fact that the generation of ROS is a familiar means of action through which antibiotics can induce killing of cells among bacteria.[36] In the experiment, there was 300% increase in the generation of ROS in *E. coli* that were treated with SaAgNPs and EpAgNPs, whereas similar treatments with SaAgNPs and AmAgNPs led to about 400% increase in the amount ROS generated in cells of *P. aeruginosa* cells. Furthermore, the exposure of cells of *K. pneumoniae* to ApAgNPs and SaAgNPs also produced 400% increase in the amount of ROS generated, while in cells of *P. vulgaris,* exposure to ApAgNPs and EpAgNPs increased the amount of ROS generated by about 300%.

Moreover, higher antioxidant activity compared to leaf extract alone or silver nitrate solution was exhibited by the AgNPs synthesized using extract of *Chenopodium murale* as reported by Abdel-Aziz et al.[2] For DPPH scavenging potential, the percentage scavenging recorded was about 59.43% and 65.43% at 20 mg/L for the plant-extract and the AgNPs, respectively indicating that plant-AgNPs possessed a higher scavenging activity. Besides, the results of β-carotene oxidation revealed that about 51.13% and 53.38% oxidation at 20 mg/L for the plant-extract and the AgNPs, respectively were recorded. Thus, the AgNPs demonstrated higher antioxidant activity than the extract alone. Likewise, AgNPs were biofabricated using the cell filtrate of *Bacillus safensis* by Lateef et al.[85] to investigate antioxidant activities. In the DPPH radical scavenging study, activities of 40.56–89.40% at concentrations of 20–100 μg/mL, which were evidently better than values obtained for standards (quercetin and *β*-carotene) were recorded. The IC_{50} was 15.99 μg/mL. Also, the ferric ion reducing power of AgNPs was calculated as 1.84–2.42 at 20–100 μg/mL.

Bhakya et al.[23] in a study reported the use of root extract of *Helicteres isora* for the synthesis of 16–95 nm sized spherical AgNPs that were maximally absorbed at 450 nm in the UV–Vis spectra. In the DPPH scavenging studies, the AgNPs exhibited 90% scavenging activity. For the hydrogen peroxide scavenging potential, 85.35% and 93.31% activities were reported for ascorbic acid and AgNPs, respectively at 100 μg/mL. Also, the activity of AgNPs was concentration dependent in nitric oxide scavenging assay with the maximum activity of 80.46% scavenging recorded at concentration of 100 μg/mL. The antioxidant ability of AgNPs may be ascribed to the biomolecules that are borne on them, that had their origin in the root extract. Also, CPHE-AgNPs, as described by Lateef et al.[82], was shown to possess excellent antioxidant activities in both DPPH and ferric ion reducing assays. By evaluating AgNPs concentration of 20–100 μg/mL, DPPH scavenging activities of 32.62–84.50% were obtained. The IC_{50} of CPHE-AgNPs was 49.70 μg/mL, while that of quercetin and β-carotene were 430 and 710 μg/mL, respectively. Also, ferric ion reducing activities of 14.44–83.94% were obtained with AgNPs at the investigated concentrations as compared to β-carotene and quercetin that gave values of 11.53–65.38%, and 12.05–100% at higher concentrations of 0.2–1.0 mg/mL.

The spherical-shaped AgNPs fabricated by pod extract of *Cola nitida*[83] were investigated for potential antioxidant activities. The AgNPs exhibited a strong antioxidant activity with an IC_{50} of 43.98 μg/mL against DPPH, and a ferric ion reduction of 13.62–49.96% at concentrations of 20–100 μg/mL. Reddy et al.[151] elucidated the biofabrication of AgNPs using aqueous *Piper longum* fruit extract (PLFE). The synthesized PL-AgNPs displayed good antioxidant activities in various assays. PL-AgNPs showed higher reducing activity compared with the PLFE (fruit extract) which was found to dose-dependent in the reduction of ferric ion. However, in the DPPH assay, the average percentage inhibition of synthesized PL-AgNPs was 67%. Also, PL-AgNPs quenched superoxide radical in dose-response manner whereby the average inhibition was about 60% higher than activity displayed by the extract alone. In the nitric oxide quenching activity, the average inhibition of PL-AgNPs was found to be 70%. Furthermore, PL-AgNPs showed comparable effectiveness with the extract alone to quench radicals generated from H_2O_2 with approximately 96% rate of inhibition.

Furthermore, Dhayalan et al.[32] biofabricated AgNPs by employing extract of the seed of *Embelia ribes* (SEEr). The synthesized SEEr-AgNPs demonstrated commendable antioxidant activities as revealed by DPPH and the phosphomolybdenum assays. The IC_{50} of 100 μg/mL was obtained in the DPPH radical scavenging activity, while 60 μg/mL was obtained as the IC_{50} in the phosphomolybdenum assay, which involved the reduction of Mo (VI) to Mo (V). Mata et al.[105] have also produced AgNPs through the utilization of *Plumeria alba* (Frangipani) flower extract (FFE). The AgNPs was analyzed for antioxidant activities using the DPPH, the reducing power, superoxide, H_2O_2 scavenging, and nitric oxide scavenging assays. The AgNPs exhibited IC_{50} at 100 μg/mL in the DPPH assay, while high-quality activities at concentrations of 25–100 μg/mL were recorded in the reducing power assay. Also, percentage scavenging of 91.8, 77.19, and 78.01% were obtained at dose of 60 μg/mL of the AgNPs, extract alone and rutin (standard), respectively in the hydrogen peroxide radical scavenging test, whereas 51% inhibition was reported at concentration of 60 μg/mL for the AgNPs in the nitric oxide scavenging assay. However, FFE (extract) was found to have higher activity than the AgNPs in the superoxide radical quenching assay and this may be ascribed to the high amount of phytoconstituents that were present in the extract.

AgNPs have been biosynthesized by employing extract of leaves of *Aristolochia indica* as described by Shanmugam et al.,[159] with excellent capability in scavenging DPPH and ABTS. Highest activity of 81.19% was obtained at concentration of 100 μg/mL in DPPH assay, while the scavenging activity was highest at 64.01% at 100 μg/mL concentrations of the AgNPs in the ABTS scavenging assay. Phull et al.[134] synthesized BC-AgNPs, whose morphology and properties were shown to exhibit good antioxidant activities. In the DPPH free radical scavenging test, the BC-AgNPs displayed maximum activity of 59.31% as against 51.29% by the BC extract alone. Also, in the total antioxidant test expressed as ascorbic acid equivalent (AAE), the BC-AgNPs displayed activity of 60.48 AAE against 38.8 AAE by the BC extract alone. In another study, the spherical TT-AgNPs synthesized through the use of the leaf extract of *Talinum triangulare* (TT) by Elemike et al.[39] showed scavenging of 30% –88% at concentrations between 25 and 100 μg/mL in the DPPH free radical scavenging test.

In addition, Kanipandian et al.[62] described the phytoreduction process to synthesize AgNPs by exploring the extract of *Cleistanthus collinus*. The synthesis was marked by a color change from white to yellowish brown, and absorption peak at 420 nm was obtained in the UV–Vis spectroscopic studies. The bands in the FTIR spectrum indicated the role of protein molecules by acting as ligand to increase the stability of AgNPs. Morphological studies of the AgNPs by TEM revealed the bioformation of spherical-shaped AgNPs that were 20 to 40 nm in dimension. Dominant peak at 3 KeV that is typical of Ag was obtained from the phytosynthesized AgNPs as revealed by EDX; and SAED blueprint of AgNPs established poly-crystallinity of the particles. The AgNPs showed good antioxidant activities based on DPPH, OH^-, ferric reducing power and the hydrogen peroxide scavenging assays. In DPPH assay, the radical was scanged in increasing manner as the concentration of AgNPs increased. At concentrations of 50–1000 μg/mL, AgNPs scavenged the generated radicals by 20%–69%. The hydroxyl radical scavenging action of the AgNPs was 79% at a concentration of 1000 μg/mL. Also, for AgNPs concentrations of 50–1000 μg/mL, ferric ions were reduced by 41.83–84.64%, while inhibition was found to be 85.05% at concentration of 1000 μg/mL in the hydrogen peroxide scavenging assay.

Nakkala et al.[119] described the phytosynthesis of AgNPs using leaf extract of *Costus pictus*. The CPAgNPs absorbed maximally at 430 nm in UV–Vis spectrum; and FTIR bands indicated active participation of amine and carbonyl groups in the formation of CPAgNPs. The morphology of the CPAgNPs as determined by SEM showed that they were spherical in shape and polydispersed in nature. The DLS results implied that CPAgNPs were averagely 46.7 nm in size and the EDX spectrum of the CPAgNPs showed strong signals of elemental silver. The results on synthesized CPAgNPs for potent antioxidant activities indicated that CPAgNPs showed better DPPH free radical scavenging activity compared to CPLE (plant extract). It also exhibited higher ferric ion reducing ability for the plant extracts and the same result was recorded for the hydrogen peroxide scavenging activity. CPAgNPs displayed higher superoxide scavenging activity at 90% compared to CPLE at various concentrations, and also 49.36% nitric oxide quenching activity.

In a related investigation, Nakkala et al.[118] elucidated biological production of AgNPs from leaf extract of *Gymnema sylvestre*. Color change from colorless to brownish yellow was an indication of the formation of the GYAgNPs which was maximally absorbed at 435 nm. The FTIR spectral of GYAgNPs produced shifting in several peaks, that occurred at 3535, 2090, 1645, 1379, and 1035 cm^{-1}; thus predicting the presence of active principles that have abundance of carbonyl groups in GYLE extract and implicated in catalytic reduction process that culminated in bioformation of GYAgNPs. Spherical-shaped GYAgNPs were produced as shown by TEM, and DLS determined average size of phytosynthesized GYAgNPs as 33 nm. Also, a strong silver signal was produced in the EDX spectrum. Furthermore, the results on antioxidant potentials of GYAgNPs indicated that GYAgNPs showed an average quenching activity of 82% which is higher than GYLE in the nitric oxide scavenging activity. Also, maximum percentage inhibition of 47% obtained by GYAgNPs at 60 μg/mL in the hydrogen peroxide scavenging activity was similar to that obtained by GYLE. GYAgNPs exhibited fair reduction of ferric ion compared to quercetin (standard used), but GYLE had lower performance. GYAgNPs showed substantial radical quenching activity for the superoxide scavenging activity, while GYAgNPs demonstrated superior scavenging of DPPH free radical when compared with the performance of GYLE at all concentrations that were employed.

Also, Jinu et al.[60] documented phytosynthesis of AgNPs by employing aqueous leaf extracts of *Solanum nigrum,* which was assessed for potent antioxidant activities by investigating its potential for ROS generation. The buildup of ROS was found to increase by 3-folds in AgNPs-exposed cells of *S. epidermidis*, *K. pneumoniae*, and *P. aeruginosa* compared to strains of *S. aureus*. Also, generation of ROS at AgNPs concentration of 50 mg/L was reportedly elevated in comparison to 25 mg/L concentration. Ravichandran et al.[148] have shown the practicability of synthesizing silver nanoparticles (BAgNPs) through the aid of aqueous extract of leaves of *Artocarpus altilis*. The synthesized BAgNPs displayed good ability at scavenging DPPH. Results showed that highest inhibition of 79.79% was obtained at 100 μg/mL, while IC_{50} value was 51.17 μg/mL. Moreover, Abbai et al.[1] described the eco-friendly phytosynthesis of Sg-AgNPs using the pharmacologically active stem of Siberian ginseng (*Eleutherococcus senticosus*). The Sg-AgNPs was evaluated for antioxidant activities in DPPH assay, with IC_{50} value of 100 μg/mL obtained. Also, the AgNPs displayed significantly higher antioxidant activity than its corresponding salt.

AgNPs synthesized with lipid that was extracted from a microalgae, *Acutodesmus dimorphus* that was grown using wastewater of dairy processing was described by Chokshi et al. [25] The colorless solution of the reaction mixture changed to dark brown to signify the synthesis of the AgNPs with absorbance peak at 420 nm. The FTIR spectrum showed that amide linkages in association with other functional groups perhaps took part in interacting with AgNPs to ensure that the particles were stabilized. TEM micrograph established the development of spherical-shaped as well as polydispersed AgNPs that were 2–20 nm in dimension; while EDX displayed noticeable peak at 3 KeV confirming the presence of elemental Ag in the particles. The synthesized AgNPs in investigations for antioxidant activities produced an IC_{50} value of 14.41 and 6.91 μg/mL for ABTS and DPPH free radical scavenging assays, respectively. Niraimathi et al.[123] described synthesis of AgNPs using aqueous extract of *Alternanthera sessilis* Linn. The change in color from yellowish to reddish brown color indicated the synthesis of AgNPs with maximum absorbance at 435.04 nm. The FTIR results are suggestive of the involvement of ascorbic acid and protein that were in the plant extract to facilitate the bioreduction process and capping of the particles respectively. The particles formed were in the form of aggregates as obtained from SEM images with diameter of 20–30 nm. Thermo gravimetric differential scanning colorimeter (TG-DSC) results showed that unrecompensed residue of Ag was 65.1%. Also, phase transition occurred at a temperature of 951°C that was very near the melting point of Ag. The AgNPs displayed good antioxidant activities in DPPH assay showing maximum activity of 62% at 500 μg/mL, while IC_{50} of 300.6 μg/mL was obtained.

Azeez et al.[14] biosynthesized AgNPs with the aid of extract of pod of *Cola nitida*. A dark brown change in color was obtained in the synthesis of the AgNPs that asorbed maximally at 431.5 nm. Results of FTIR spectrum elaborated bands that attest to the fact that proteins in the extract facilitated the reduction reaction of transforming Ag^+ to Ag^0, along with provision of stability of the particles. The nearly spherical-shaped AgNPs as revealed by TEM analysis had particle sizes of 12–80 nm, while its crystallinity was also established by SAED. The synthesized AgNPs were reported to impact improvements on the DPPH antioxidant activities of *A. caudatus* planted with 25, 50, 75, and 100 ppm of the AgNPs by 43.3%, 38.7%, 26.7%, and 6.48%, respectively. Also, it was found out that *A. caudatus* that was planted with 50 ppm of the synthesized AgNPs had the highest potency with smallest IC_{50} of 0.67 mg/mL. The study concluded that AgNPs might be a good material to potentiate improved production of phytomedicinal chemicals in plants.

Additionally, Vizuete et al.[181] has documented the eco-friendly phytosynthesis of AgNPs by employing fruit extracts of Shora (*Capparis petiolaris*), which displayed a change in color from a colorless to a brownish color. The AgNPs were maximally absorbed at 423 nm. The TEM analysis displayed the AgNPs as less polydispersed, spherical-shaped with varying dimension of 10–30 nm; and the SAED pattern clearly revealed that synthesized AgNPs are spherical and moderately crystalline in nature. The XRD spectrum showed four distinct diffraction peaks which originated from the (111), (200), (220), and (311) Bragg reflections of crystalline AgNPs. TheAgNPs showed modest antioxidant activity against DPPH with 38.98% activity compared to 25.94% by the extract at quantity of 0.5 mL. Moteriya and Chanda[110] synthesized

AgNPs mediated by flower extract of *Caesalpinia pulcherrima*. The AgNPs was shown to possess excellent antioxidant activities as evidenced by various tests conducted. The IC_{50} was 70, 38.5, and 55 μg/mL in the DPPH assay, superoxide ion scavenging activity, and ABTS cation radical assay, respectively. The reduction of ferric was obtained at 8.8 mg^{-1}.

Moreover, Parveen et al.[129] described a microwave assisted phytosynthesis of AgNPs by exploiting extract of leaves of *Fraxinus excelsior*. Color transformation from yellow to brown signaled synthesis of the AgNPs which absorbed maximally at 425 nm. The FTIR spectra showed strong absorption bands at 3438, 2923, 1637, 1376, and 1078 cm^{-1} for both extract of *F. excelsior* and phytofabricated AgNPs. The TEM analysis revealed primarily spherical-shaped and polydispersed particles that were of 25–40 nm diameter. Also, EDX profile displayed a prominent signal at 3 keV that is characteristic of Ag with few low signals obtained for O, Cl, and C that have their origin in biomolecules of the *F. excelsior* extract that have capped the AgNPs. The synthesized AgNPs displayed excellent antioxidant activity against DPPH free radicals with IC_{50} of 5.71 μg/mL, which is significantly higher than 8.4 μg/mL for the leaf extract.

More recently, Lateef et al.[88] examined the biosynthesis of AgNPs by exploiting different types of bioresources that included alkaline extract of spider cobweb (CB), as well as aqueous extracts of pod (KP), seed (KS), and seed shell (KSS) of *C. nitida*. Development of color observed in each of the silver nanoparticles biosynthesized was dark brownish in CB and KP-AgNPs, brownish in KS-AgNPs and yellowish orange in KSS-AgNPs. The absorbance peaks obtained from the UV–Vis spectra were at 431.5, 436, 454.5, and 457.5 nm for KP-, CB-, KSS-, and KS-AgNPs, respectively. The FTIR spectra indicated that the nanoparticles were produced and stabilized by proteins in the extracts. This was evidenced by the occurrence of well-known bands at 3292–3336 and 1631–1639 cm^{-1} that are indicative of the presence of N–H, C=C and C=O of protein molecules. The nanoparticles were observed to be nearly spherical-shaped, and polydispersed in distribution. Prominence of silver was confirmed by peak produced in the various spectra of EDX. SAED patterns indicated that the AgNPs were crystalline particles and the variations in sizes were obtained along these lines: 8–50 nm (KS-AgNPs), 12–80 nm (KP-AgNPs), 3–50 nm (CB-AgNPs), and 5–40 nm (KSS-AgNPs). All the biosynthesized NPs displayed excellent antioxidant activities with respect to the hydrogen peroxide scavenging activity obtained in the range of 77–99.8%, with the immediate clearance of the cloudy phosphate buffered H_2O_2 solution.

It can be concluded from the discussion in this section that AgNPs have extended their frontiers of application as scavenger of free radicals, which has a lot of relevance in nanomedicine, environmental cleansing, food production, and agriculture.

1.4.4 ANTICOAGULANT AND THROMBOLYTIC ACTIVITIES OF SILVER NANOPARTICLES

The spherical phytosynthesized AgNPs facilitated by the aqueous extract of pods of *Peltophorum pterocarpum* by Raja et al.[141] was investigated for anticoagulant properties. The anticoagulant potential of the AgNPs was examined by reacting 0.5% v/v of the biosynthesized AgNPs with adult blood collected in a vial. The anticoagulant activity was confirmed by the absence of clot formation in the reaction mixture.

Also, the AgNPs synthesized with the culture supernatant of *P. aeruginosa* by Jeyaraj et al.[58] were tested for potential anticoagulation properties. Blood plasma was obtained from hale and hearty volunteers (aged 20–40 years). The test sample contained blood with AgNPs at 0.5% (v/v) respectively in a vial. It was reported that blood clots were not formed in the vial that contained AgNPs. Also, a stable anti-coagulant effect of the AgNPs was reported. Moreover, the spherical CBE-AgNPs synthesized by Azeez et al.,[15] were investigated for ability to prevent the coagulation of human blood. CBE-AgNPs at concentration of 170 μg/mL was mixed in equal proportion with blood unreservedly bequeathed by a healthy volunteer at room temperature (30 ± 2°C). In this in vitro investigation, the CBE-AgNPs successfully protected the blood from coagulation, and also preserved the morphological feature of the red blood cells as observed in the freshly collected blood (Fig. 1.4). Furthermore, AgNPs synthesized by fresh leaves of *Panax ginseng* reported by Singh et al.[167] were also investigated for anticoagulant activities. The results demonstrated that the development of blood clots was subdued effectively in the presence of the AgNPs, thus confirming its anticoagulant potency.

In another report, the biosynthesized miracle fruit plant (*Synsepalum dulcificum*) leaf extract-mediated AgNPs and seed extract-mediated AgNPs described by Lateef et al.[79] were investigated for anticoagulant and thrombolytic activities. Both leaf extract-mediated and seed extract-mediated AgNPs displayed excellent blood anticoagulant activities, where clotting of blood samples was effectively prevented. The presence of well-dispersed red

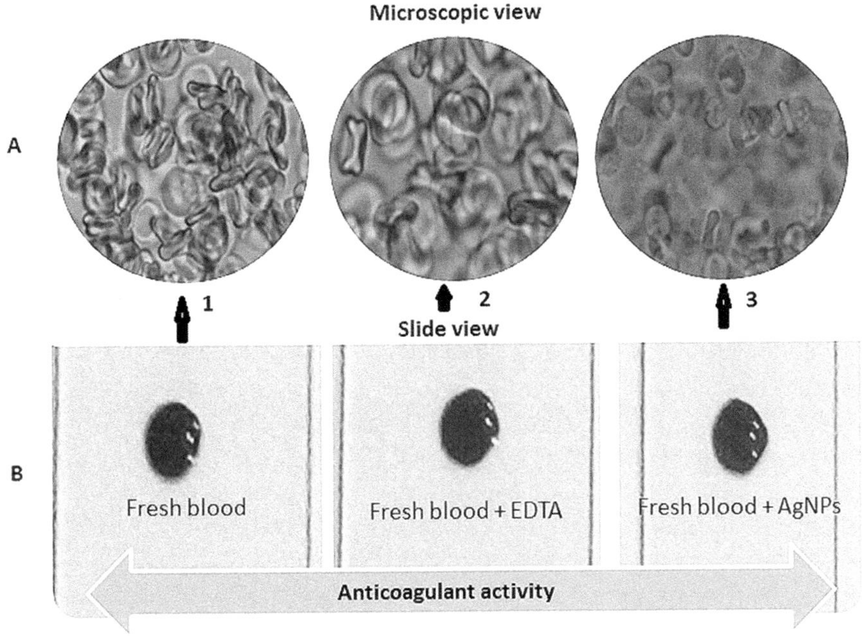

FIGURE 1.4 (See color insert.) Anticoagulant activity of AgNPs biosynthesized using the extract of cocoa beans.

blood cells was revealed by microscopic examinations, and the micrographs obtained were comparable with those obtained using the conventional EDTA blood anticoagulant. Moreover, both AgNPs caused dissolution of pre-formed blood clots within 2 min, providing clear blood fluid spread out on microscopic slides. When observed using optical microscope, red blood cells in AgNPs-treated blood clots were seen as well dispersed as opposed to the clumps seen in the negative controls. Also, AgNPs biosynthesized using nest extract of paper wasp (*Polistes* spp.) by Lateef et al.[80] were investigated for anticoagulation and thrombolysis. The AgNPs did not permit clotting of blood in comparison with what was obtained in the positive control with EDTA; and complete dissolution of pre-formed blood clots within 5 min of addition of the AgNPs was achieved. The light microscopic image obtained for both the anticoagulant and thrombolytic tests confirmed that the disc-shaped morphology of the red blood cells was preserved.

Kalishwaralal et al.[61] demonstrated the bacteriogenic synthesis of AgNPs in biomass of *Brevibacterium casei*. Aqueous solution of $AgNO_3$ was treated with *B. casei* biomass with the appearance of yellowish brown color after 24 h reaction. The AgNPs were synthesized and maximally absorbed at 420 nm. The FTIR spectroscopy gave indication for the occurrence of proteins as probable bioorganic agents accountable for reducing silver ion and capping of the particles, which increased the stability of the synthesized AgNPs. The TEM analysis revealed that AgNPs were relatively uniform in dimension and spherical-shaped with sizes ranging from 10 to 50 nm. The XRD pattern specified the occurrence of powerful peaks of nanoparticles (111), (200), (220), and (311) that are interpreted as fcc crystalline Ag to confirm the crystallinity of the particles. The ability of the AgNPs to inhibit coagulation of blood plasma was examined. The AgNPs showed excellent anticoagulant activity that was established by inhibiting the development of blood clots when blood was mixed with AgNPs. The stability of AgNPs anticoagulated blood was established in an experiment that involved extended contact of blood plasma with the particles for up to 24 h without showing any noteworthy decline in activity.

Similarly, the CB-, KP-, KS-, and KSS-AgNPs produced by Lateef et al.[88] as earlier reported were further tested for anticoagulant and thrombolytic activities. All the synthesized AgNPs displayed excellent anticoagulant activities and microscopic examination showed that the characteristic disc-shaped morphology of the red blood cells was essentially conserved in all experiments that were treated with AgNPs. Also, all the AgNPs were able to lyse already formed blood clots and the percentage thrombolysis obtained were; 55.76%, 60.46%, 72.73%, and 89.83% for CB-AgNPs, KP-AgNPs, KS-AgNPs, and KSS-AgNPs, respectively (Fig. 1.5). In another report, Lateef et al.[90] detailed the biosynthesis of AgNPs using extracellular extract of *Bacillus safensis*, which heralded the development of dark brown colored AgNPs with absorption peak at 419 nm. The synthesized AgNPs displayed excellent anticoagulation and thrombolysis of blood. The AgNPs barred the generation of blood clot as an anticoagulation mediator and this compared satisfactorily with EDTA that was used as positive control in the investigation, while the microscopic view also displayed red blood cells that were well dispersed. There existed no noticeable coagulation of blood samples that were exposed to AgNPs when held up to 24 h, in contrary to the negative control where total blood coagulation occurred within 10 min. Furthermore, the AgNPs caused dissolution of the pre-formed blood clots almost immediately; indicating incredibly high thrombolytic activity and microscopic images of the dissolved clot obtained showed unambiguous diffusion of the blood clot by AgNPs.

Paul et al.[131] described the biosynthesis of AgNPs by exploring powder of leaves of *Diplazium esculentum* (retz.) sw. which led to the development of deep grey solution indicating formation of colloidal AgNPs. Surface plasmon resonance at 439 nm was exhibited by the synthesized AgNPs, and FTIR analysis gave obvious sign of the occurrence of proteins and other organic molecules suggested to have been produced extracellularly by *D. esculentum*. TEM images demonstrated the biofabrication of anisotrophic AgNPs that were mostly spherical, triangular and oval-shaped in dimension of 10–45 nm. Also, the XRD spectrum evidenced the development of fcc structure of Ag whose crystallites possessd mean diameter of 9.71 nm. The synthesized AgNPs demonstrated anticoagulation activity as substantiated by prevention of development of blood clots in the reaction mix that contained blood alongside AgNPs. However, blood clots were formed in distinctive manner in the blood sample that was not exposed to AgNPs. The AgNPs performed excellently as anticoagulant to the extent that there was no observable reduced anticoagulative activity following 24 h of exposure of blood to the AgNPs.

Harish et al.[52] have demonstrated simplistic synthesis of vastly stable AgNPs by exploiting biopolymer of xylan for the purpose of both reduction and stabilization. The Ag^+ was reduced to AgNPs and subsequently stabilized by xylan, leading to the appearance of brown color that absorbed maximally at 405 nm.

TEM results revealed that the AgNPs were plydispersed and spherical in morphology having dimensions that varied from 20 to 45 nm. The Zeta potential approximated for the particles was −17.5 mV, which indicated that the nanoparticles were well distributed in the colloidal solution as a result of electrostatic repulsion occurring between the adjoining particles. The WB-xylan AgNPs was investigated for fibrinolytic activities by fibrin plate assay and it was found that the particles induced dispersion of fibrin clots. The diameter of zone of clearance of the fibrin clot was found to be linearly dependent on the concentration of the AgNPs. Furthermore, the introduction of the particles to pre-formed blood clot led to its dispersal within 5 min of reaction, which was similar to the result of sodium citrate that was used as positive control in the experiment. Also, optical microscopic images undoubtedly showed that blood clots were dissolved by the particles into fine discrete cells, laying credence to thrombolysis.

Lateef et al.[70,90] have proposed mechanism of the thrombolytic activities of AgNPs (Fig. 1.6). It was inferred that the nanoparticles could directly act on fibrin (Mechanism 2), thereby breaking it as substantiated in the plate

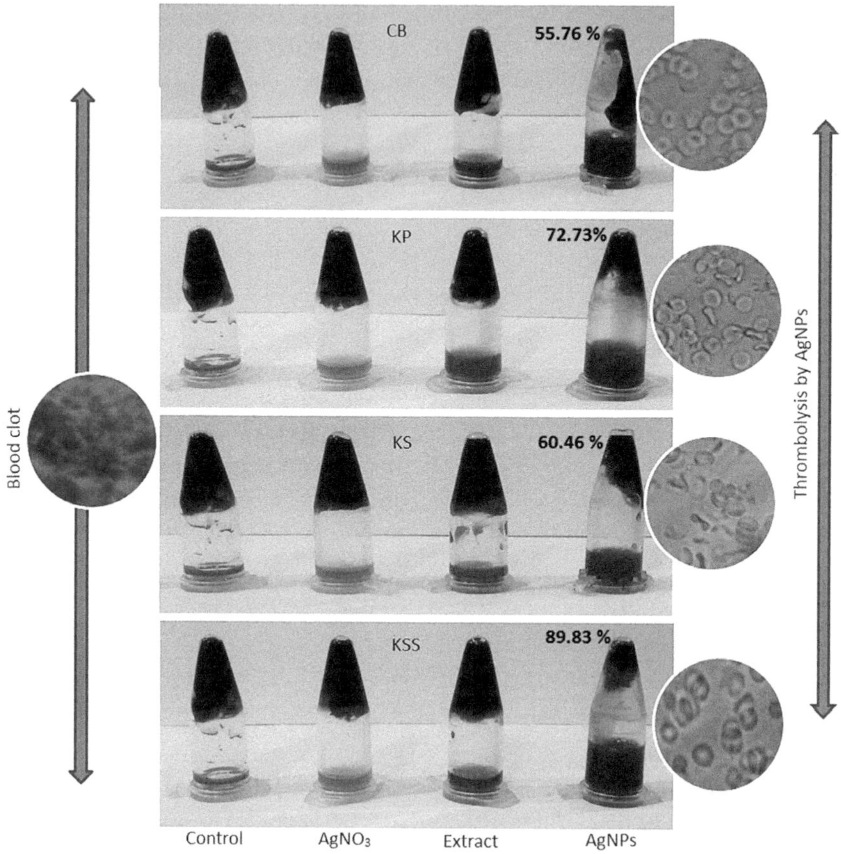

FIGURE 1.5 (See color insert.) Thrombolytic activities of CB-, KP-, KS-, and KSS-AgNPs.
Source: Reprinted with permission from Ref [88].

assessment that was demonstrated by Harish et al.,[52] or the particles may serve as plasminogen activator, causing it to liberate plasmin that inturn sever the blood clot (Mechanism 1). Furthermore, the inhibitors may be acted upon by AgNPs thereby preventing the activation of plasminogen and plasmin. It was suggested that the pronounced thrombolytic activities of AgNPs may be due to result of simultaneous exhibition of the two mechanisms.

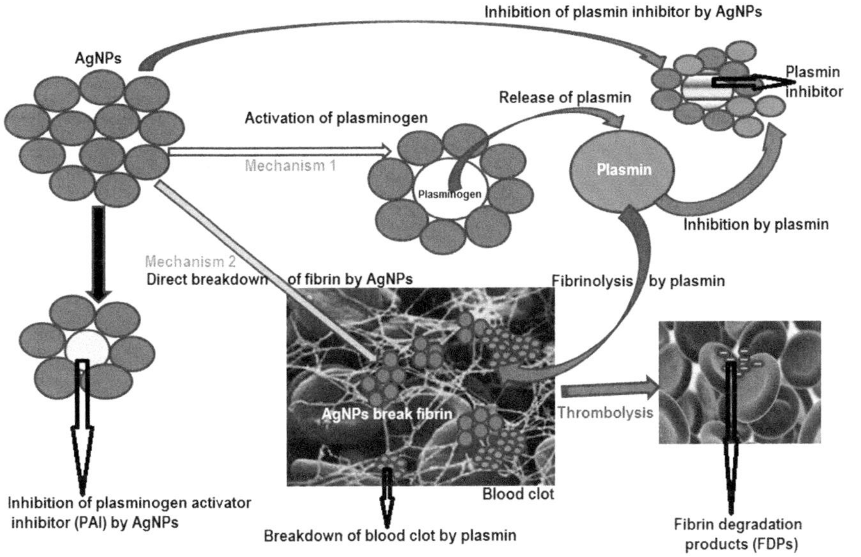

FIGURE 1.6 **(See color insert.)** The possible mechanisms of thrombolytic activity of AgNPs.

1.5 BIOMEDICAL APPLICATIONS OF GOLD NANOPARTICLES: A CASE STUDY

1.5.1 ANTIMICROBIAL ACTIVITIES OF GOLD NANOPARTICLES

Ojo et al.[124] reported green biosynthesis of AuNPs by exploiting the extracellular fraction of *Bacillus safensis*. The synthesized AuNPs showed a characteristic purple color with maximum absorbance at 561 nm. FTIR spectroscopic analysis of the AuNPs revealed peaks at 3318, 2378, 2114, 1998, 1636, 1287, 446, and 421 cm^{-1} showing the responsibility of proteins as chemical principles that capped and stabilized biosynthesized nanoparticles. The TEM images revealed the anisotropic nature of the uniform spherical AuNPs biosynthesized. The particles were polydispersed in distribution, with the sizes ranging from 10 to 45 nm. Moreover, the EDX examination of AuNPs colloids showed Au as most prevalent metal; while SAED displayed biosynthesized nanoparticles as crystals that are ring-shaped. Furthermore, at concentration of 200 μg/mL, the AuNPs demonstrated good growth inhibitions of 66.67% and 75.32% in assessment that involved *A. fumigatus* and *A. niger,* respectively.

Also, Naveena and Prakash[121] described the biosynthesis of AuNPs utilizing the aqueous extract of red marine algae (*Gracilaria corticata*). The change in color of yellow to pink–red in the reaction mixture amounted to the synthesis of AuNPs. The UV–Vis spectra revealed that the AuNPs were absorbed at 540 nm and the nanoparticles were found to be well distributed without aggregation and with sizes ranging from 45 to 57 nm as observed in the TEM images. The synthesized AuNPs and ciprofloxacin-conjugated AuNPs were evaluated for antibacterial activities against some pathogenic bacterial strains of *Enterococcus faecalis, S. aureus, Enterobacter aerogenes*, and *E. coli* by agar well diffusion technique. Minimal inhibitory zones were obtained from the AuNPs, but these were improved in the AuNPs conjugated with ciprofloxacin which were slightly greater than zones obtained with the use of antibiotics alone. The highest antimicrobial action was obtained in the antibiotic conjugated AuNPs against *E. coli* (24 mm), followed by *E. aerogenes* (21 mm), while modest action was produced against *S. aureus* (19 mm); and the least action was exerted toward *E. faecalis* (14 mm).

Basu et al.[19] documented the synthesis of AuNPs by utilizing aqueous and methanolic extracts of seeds of *Dolichos biflorus* as bioreducing agents. The biofabrication of AuNPs in the two investigations was heralded by transformation in color, leading to development of ruby red from the initial yellowish solution. The AuNPs showed prominent absorption band at 530 nm. The nanoparticles were predominantly spherical in shape with some irregular form with agglomeration as observed in SEM analysis. The EDX spectra revealed main peak for Au with some minor detections for O, C, K, Al, Cl, and Na. The XRD blueprint depicted information corresponding to Bragg reflections that established fcc structure of Au; and SAED results also laid credence to the polycrystalline nature of the particles. Also, high resolution TEM showed 0.24 nm as the line spacing for the biosynthesized AuNPs. The synthesized AuNPs exhibited poor antibacterial potential against human pathogenic bacteria such as *E. coli, P. aeruginosa, S. aureus*, and *B. subtilis*.

In another study, AuNPs were synthesized using the extracts of *Inonotus obliquus* (Chaga mushroom) at room temperature by Lee et al.[97] There was manifestation of dark brown color from the initial light brown within 30 min, indicating the synthesis of AuNPs. The AuNPs showed absorption peak at 532 nm within the typical range for AuNPs in the UV–Vis spectrum. The FTIR spectrum showed prominent bands at 3434, 1636, and 695 cm^{-1} indicating O–H stretching vibrations of alcohols and/or phenols, N–H bend of the amines and feeble intensity for the aromatic C–H bending, respectively. The TEM analysis showed that the AuNPs had spherical, rod, and triangle shapes in the range of 11.0–37.7 nm with a mean diameter of 22.8 nm. The crystalline makeup of the mycosynthesized AuNPs was confirmed by SAED pattern showing bright circular rings. The EDX spectrum revealed a prominent Au signal along with weak signals from C, Ca, Al, O, Cl, Si, K, Mg, and Na. The AuNPs were also tested for antibacterial activities. The best growth inhibition was observed in cells *S. aureus* (16 mm), trailed by *E. coli* (14 mm) and *B. subtilis* (12 mm).

The AuNPs synthesized in an environmentally benign process using the aqueous extract of *Elettaria cardamomum* seeds by Rajan et al.[143] was studied for antibacterial activities using the standard agar diffusion method. *Elettaria cardamomum* is a perennial herbaceous plant known as the "queen of spices" belonging to the ginger family *Zingiberaceae*.[157] The biosynthesized AuNPs were maximally absorbed at 527 nm. The FTIR spectra of AuNPs indicated prominent bands at 1635, 1536, 1402, and 1030 cm^{-1}. The AuNPs were reportedly spherical particles and are crystalline in nature as confirmed from XRD analysis. The XRD pattern showed typical diffraction peaks at 38.47°, 44.53°, 64.80°, 77.86°, and 82.65° that is equivalent to 1 (111), (200), (220), (311), and (222) planes, respectively of fcc structure of Au. It was found to exhibit antibacterial activity toward the pathogenic strains of *S. aureus, E. coli,* and *P. aeruginosa*.

Moreover, the gold bionanoparticles synthesized by *Nocardiopsis* sp. MBRC-48 reported by Manivasagan et al.[102] were studied for antimicrobial activities using the well diffusion method. The color of the solution containing the cell-free supernatant of *Nocardiopsis* sp. MBRC-48 and 1×10^{-3} M $HAuCl_4.3H_2O$ changed to pinkish indicating the synthesis with absorption peak at 530 nm. The FTIR spectrum of the AuNPs showed bands at 3431, 2937, 1643, 1462, and 1031 cm^{-1} indicating that the AuNPs were surrounded by some proteins, enzymes, and metabolites. The TEM analysis proved the AuNPs to be spherical-shaped with some degrees of polydispersity but without significant agglomeration. The average sizes of the AuNPs were 11.57 nm by DLS analysis. XRD investigation elaborated strong signals that match (111), (200), (220), and (311) of Bragg's planes on the basis of fcc structure of AuNPs. The EDX spectroscopy proved the presence of elemental gold peak as a major signal. It exhibited significant antimicrobial potency against *P. aeruginosa* (ATCC 27853), *B. subtilis* (ATCC 6633), *E. coli* (ATCC 10536), *S. aureus* (ATCC 6538), *C. albicans* (ATCC 10231), *A. fumigatus* (ATCC 1022), *A. niger* (ATCC 1015), and *A. brasiliensis* (ATCC

16404). Highest antimicrobial action was exerted toward *S. aureus* and *C. albicans*, but the lowest action was found in *A. brasiliensis* at both 50 and 100 μL.

Muthukumar et al.[117] described the syntheses of AuNPs by utilizing the extracts of leaves of *Carica papaya* (CP) and *Catharanthus roseus* (CR) as well as the blend of the two extracts (CPCRM). The change in color of the reactant solution from yellow to dark ruby red color confirmed the synthesis. The existence of extensive optical absorption peak encountered between 500 and 600 nm revealed that the surface plasmon resonance and intensity increased with increasing concentration of extracts. The FTIR confirmed the occurrence of functional groups of plant phytochemicals in the phytosynthesized AuNPs. The form of CP-, CR-, and CPCRM-mediated AuNPs was spherical in addition to presence of other morphologies that include triangle and hexagonal structures based on SEM analysis. Generally, the particles were not agglomerated. HR-TEM revealed that the CP-, CR-, and CPCRM-AuNPs were generally spherical and triangular with sizes ranging from 2 to 20 nm, 3.5 to 9 nm, and 6 to 18 nm, respectively. The AuNPs samples were indexed with the characteristic fcc configuration of Ag that exhibited (111), (200), (220), and (311) planes as revealed by the XRD analysis. The synthesized AuNPs displayed excellent antibacterial activities against pathogenic bacterial strains of *S. aureus, E. coli, B. subtilis*, and *P. vulgaris.* The highest onslaught was detected in *E. coli* (20 mm), then *P. vulgaris* (18 mm); and generally CPCRM-AuNPs showed greater antibacterial activities against all the bacterial strains than CP-AuNPs and CR-AuNPs. The minimum inhibitory concentration against the pathogenic bacterial strains ranged from 62.5 to 250 μg/mL for CP and CR-AuNPs, and 15.625 to 125 μg/mL for CPCRM-AuNPs.

Swain et al.[176] in a study demonstrated the use of extracts obtained from roots and leaves of *Vetiveria zizanioides* and *Cannabis sativa* to synthesize AuNPs. Transformation in color from whitish solution to ruby red was noticed for both AuNPs synthesis. Both AuNPs showed absorption peak at 538 nm within the typical range for AuNPs in the UV–Vis spectrum. Both VZ-AuNPs and CS-AuNPs were observed to be spherical-shaped and aggregated with mean dimension of 40 nm by the SEM analysis. The particle size distributions were 2.99 and 2.11 nm for VZ-AuNPs and CS-AuNPs, respectively by the DLS analysis. The zeta potential for both AuNPs synthesized was −57.48 mV showing that they were stable. The in vitro antifungal assessment of the phytosynthesized AuNPs was carried out by agar disc diffusion technique against *Penicillium* spp., *Aspergillus* spp., *A. flavus, A.*

fumigatus, Fusarium spp., and *Mucor* spp. The zones of inhibition ranged from 3.8 to 4.8 mm. The zone of inhibition was highest against *A. flavus* and the least against *Penicillium* spp. and *A. fumigatus*.

Jayaseelan et al.[57] also reported the phytosynthesis of AuNPs using aqueous extract of seeds of *Abelmoschus esculentus*. An absorption peak observed at 536 nm revealed the surface plasmon resonance of biofabricated AuNPs. The FTIR spectrum visibly showed that the components of seeds of *A. esculentus* that were rich in –OH facilitated the phytosynthesis and also stabilized the AuNPs that were produced. The XRD analysis confirmed the bioformation of crystalline AuNPs by showing peaks at 38°, 44°, 64°, and 77° that match the (111), (200), (220), and (311) of cubic structure of Au. Images from FE-SEM indicated the presence of spherical-shaped nanoparticles that are distributed within 45–75 nm. The AuNPs exhibited excellent antifungal activities against *Puccinia graminis tritici, A. flavus, C. albicans*, and *A. niger* by inhibiting the fungal growth by 18, 17, 16, and 15 mm against *C. albicans*, *P. graminis*, *A. flavus*, and *A. niger*, respectively. Dhayalan et al.[32] elucidated the synthesis of AuNPs by employing extract of the seeds of *Embelia ribes* (SEEr). The reaction mix was transformed in color to produce wine red colloid during the synthesis and characteristic peak between 500 and 550 nm was obtained for the AuNPs. The FTIR spectra showed peaks at 3462, 1649, 1566, and 1425 cm^{-1} confirming functional groups of phytochemicals that took part in processes leading to formation of reduced and stabilized AuNPs. TEM images revealed that the AuNPs are spherical and polydispersed in nature and were distributed over 10–30 nm. XRD pattern for the SEEr-AuNPs showed peaks at (111), (200), (220), and (311). The spherical SEEr-AuNPs exhibited excellent antibacterial activities against strains of *E. coli* and *S. aureus* displaying zones of inhibition that varied from 28 to 34 mm and 22 to 27 mm, respectively.

Patra and Baek[130] described the synthesis of AuNPs through the exploitation of the rind extract of *Citrullus lanatus* (watermelon). The colorless solution containing the AuNPs turned dark brown within 1 h of synthesis and the AuNPs absorbed maximally at 560 nm. The FTIR spectrum of the synthesized AuNPs showed distinct absorption bands at 3548, 3464, 2367, 1654, 1106, and 526 cm^{-1}. Through SEM images, the AuNPs were observed to be spherical-shaped with size distribution of 20–140 nm. The EDX spectrum showed that elemental composition of AuNPs was largely of gold that was tainted with some traces of oxygen, chlorine, and potassium. The XRD analysis of the AuNPs enunciated peaks at 38.14°, 44.26°, 64.54°, and 77.51° that are in agreement with (111), (200), (220), and (311) fcc structure of Au. The synthesized AuNPs exhibited potential antibacterial activities against five food-borne pathogens; namely *E. coli* ATCC 43890, *B. cereus* ATCC 13061, *S. typhimurium* ATCC 43174, *S. aureus* ATCC 49444, and *Listeria monocytogenes* ATCC 19115) with zones of inhibition that varied from 9.23 to 11.58 mm. Further, the AuNPs showed effective antibacterial activities through synergy in combined with rifampicin and kanamycin.

Basavegowda et al.[18] synthesized AuNPs at ambient temperature by employing aqueous extract of fruits of *Hovenia dulcis*. The color of the solution containing the AuNPs turned reddish brown after 30 min of synthesis, and the AuNPs showed absorption peak at 536 nm. The FTIR spectrum of the synthesized AuNPs revealed absorption bands at 3372, 2924, 1620, 1399, and 1050 cm^{-1}. The disappearance of C=O group and shift of a C=C stretching to 1620 cm^{-1} by comparison with FTIR spectrum of crude extracts revealed the developments in bioformation of capped and stabilized AuNPs. The morphology of the AuNPs as revealed by TEM micrographs displayed polydispersed spherical and hexagonal-shaped AuNPs whose particles were distributed from 15 to 20 nm. The EDX spectrum showed the appearance of an intense absorption peak at 3 keV representing the composition of gold phase. The face-centered cubic structure of AuNPs was established by XRD peaks at 38°, 44°, 64°, and 77°, which correspond to the (111), (200), (220), and (311) planes with clear circular ringed spots in the SAED. The AuNPs were investigated for antibacterial activities against pathogenic *E. coli* and *S. aureus*; and it displayed good activities that produced inhibition of 18–19 mm at 100 μg/mL.

In another study, Ehmann et al.[37] reported AuNPs synthesized by reacting 6-O chitosan sulfate solution (10 mg/mL) with $HAuCl_4{\cdot}3H_2O$, under microwave irradiation at 100°C for 8 min. The AuNPs exhibited absorbance peak at 530 nm and TEM analysis revealed the AuNPs as spherical particles with average geometric diameter of approximately 15 nm. The EDX spectrum showed that the nanoparticles were extremely negatively charged (−35 mV) that established the existence of negatively charged sulfated chitosan on the Au core. The S-ChiAuNPs was investigated for antibacterial potentials toward *E. coli* MG 1655 (R1-16). The particles exhibited tremendous antibacterial activity, as revealed in SEM images whereby the membranes of bacterial cells that were treated with S-ChiAuNPs were fragmented and the lysed cells were clearly seen.

Although AuNPs generally have lower antimicrobial activities when compared with AgNPs, it is apparent from the reports that were discussed above that they also exhibit some levels of activities against bacteria and fungi. Such activities may extend the relevance of applications of AuNPs as theranostic agent, for simultaneous imaging and antimicrobial therapy in nanomedicine.

1.5.2 LARVICIDAL ACTIVITIES OF GOLD NANOPARTICLES

Subramaniam et al.[173] demonstrated biosynthesis of AuNPs through the use of extract obtained from flowers of *Couroupita guianensis* to biofabricate AuNPs and also stabilized them. The change in color from colorless to pale yellow and finally dark brown during synthesis indicated formation of AuNPs, which absorbed maximally at 560 nm. The FTIR spectrum showed conspicuous bands at 422.41, 3421.72, 2362.80, 1641.42, 1514.12, and 1456.26 cm^{-1}, which are indicative of C=C stretch nitro groups of aromatics, amine N–H stretching, nitrile C≡N stretching, N–H bending in amines I, N–O asymmetric stretching in nitro compounds, and C–H bending in alkanes, respectively. In TEM, it was shown that spherical and oval-shaped AuNPs were formed, with sizes in the range of 29.2–43.8 nm. The XRD analysis depicted strong peaks analogous to (111), (200), and (220) Bragg's reflection on the basis of fcc structure of AuNPs; and the EDX spectrum revealed strong presence of Au with unique absorption peak that occurred between 2 and 3 keV, indicating the existence of gold nanocrystallites. The AuNPs demonstrated toxicity toward larvae, pupae and adult of *Anopheles stephensi* larvae, pupae, and adults with LC_{50} estimated as 17.36 ppm for first instar larvae, 19.79 ppm for second instar larvae, 21.69 ppm for third instar larvae, 24.57 ppm for fourth instar larvae, 28.78 ppm for pupae, and 11.23 ppm for the adult mosquito. Within 72 h, single treatment with the AuNPs (10 × LC_{50}) led to 100% larval mortality under field application. At the moment, there appears to be paucity of reports on the use of AuNPs for larvicidal activities.

1.5.3 ANTIOXIDANT ACTIVITIES OF GOLD NANOPARTICLES

The biosynthesized AuNPs obtained through the use of aqueous extract of red marine algae *Gracilaria corticata* by Naveena and Prakash[121] was tested for antioxidant potentials using the DPPH and the ferric ion reducing antioxidant power (FRAP) assays. In DPPH scavenging test, the highest activity of 80.6% at 200 μg/mL was recorded; while in the FRAP, highest activity of 237.3 mm/

mL was obtained at 200 μg/mL. Also, the AuNPs synthesized by Lee et al.[97] using the extracts of *Inonotus obliquus* (Chaga mushroom) at room temperature were investigated for possible antioxidant activity using the ABTS radical scavenging assay. The ABTS scavenging result reportedly improved with rising concentrations of AuNPs. The maximum and minimum ABTS radical scavenging activities were obtained at 1 mM and 0.125 mM, respectively.

Rajan et al.[143] reported the environmentally friendly rapid synthesis of AuNPs using the aqueous seeds extract of *Elettaria cardamomum*. The antioxidant activities of AuNPs were studied with the DPPH, nitric oxide and OH radicals scavenging assays. In the DPPH radical scavenging assay, percentage activities ranging from 19.87% to 62.18% were obtained at concentrations between 1.25 and 20 μL. The AuNPs exhibited activity of about 64.44% at concentration of 200 μL in the nitric oxide radical scavenging assay. Also, 67.5% inhibition, which indicated strong radical scavenging activity of synthesized AuNPs for the hydroxyl radical scavenging activity was reported. Manivasagan et al.[102] reported the biosynthesis of AuNPs by *Nocardiopsis* sp. MBRC-48, which demonstrated high-quality antioxidant activity using DPPH assay. The average inhibition of DPPH by biosynthesized AuNPs was 69% compared to the standard (ascorbic acid). Also in the reducing power assay, the results obtained were constantly superior to those attained during DPPH scavenging. The total antioxidant capacity was analyzed using the method of Ravikumar et al.[149] and the AuNPs were found to possess higher level of antioxidant activity than ascorbic acid.

Dhayalan et al.[32] elucidated the synthesis of AuNPs using extract obtained from the seeds of *Embelia ribes* (SEEr). The synthesized SEEr-AuNPs exhibited excellent antioxidant activities as revealed by through DPPH free radical scavenging and the phosphomolybdenum assays. The IC_{50} of 20 μg/mL was obtained in the DPPH radical scavenging activity, while 40 μg/mL was obtained as the IC_{50} in the phosphomolybdenum assay, in which case Mo (VI) was reduced to Mo (V). Also, the synthesized AuNPs that were facilitated by exploring rind extract of *Citrullus lanatus* as described by Patra and Baek[130] was investigated for antioxidant activities. The AuNPs displayed strong reducing power in the reducing power assay. The percentage scavenging activity obtained in the DPPH, nitric oxide and ABTS scavenging assays were 24.69%, 25.62%, and 29.42%, respectively. Additionally, Basavegowda et al.[18] have documented synthesis of AuNPs at room temperature through the exploitation of aqueous fruit extract of *Hovenia dulcis*. The AuNPs were investigated for antioxidant properties using the DPPH, H_2O_2, nitric oxide (NO), and ferric reducing antioxidant power (FRAP) methods.

In the DPPH free radical scavenging assay, there were increased activities with rise in concentration; with the peak performance of 59.17% obtained at concentration of 500 μg/mL. Also, the hydrogen peroxide scavenging activity of the AuNPs increased with rising concentration, and the highest activity obtained was 48.60% at AuNPs concentration of 100 μg/mL. The AuNPs showed high response of 88.75% at concentration of 500 μg/mL for nitric oxide assay, while it displayed a relatively low reducing abilities at concentrations of 100–500 μg/mL when compared with a standard sample (ascorbic acid) in the FRAP assay.

Sathishkumar et al.[156] has documented the biosynthesis of gold nanoparticles (CGAuNPs) within a little period of time by utilizing the aqueous extract of fruit of *Couroupita guianensis*. The CGAuNPs were observed to display absorbance peak at 530 nm. FTIR confirmed that water-soluble phenolic compounds played prominent roles in the reducing and stabilizing the CGAuNPs. Also, TEM images showed the morphology of the particles as uniformly sized anisotropic CGAuNPs, which were reported to be triangular, spherical, and hexagonal having the average size of about 25 nm. The anisotropic nature of CGAuNPs was further established through the XRD studies that showed face-centered cubic crystalline particles; while DLS and EDAX results established that the synthesized CGAuNPs were contaminants free, stable, negatively charged and without aggregation. The CGAuNPs displayed excellent antioxidant activities with the IC_{50} estimated to be 37 μg/mL for the DPPH free radical scavenging assay, and 36 μg/mL for the hydroxyl radical scavenging effect. Also, the superoxide scavenging activity of CGAuNPs was reported to increase with rising concentrations with maximum inhibition rate of 89.8% obtained.

Similarly, Abbai et al.[1] described the eco-friendly phytosynthesis of Siberian ginseng gold nanoparticles (Sg-AuNPs) using the pharmacologically active stem of Siberian ginseng (*Eleutherococcus senticosus*), which is an oriental herbal adaptogen. Change in color from light yellow to dark purple was noticed during the synthesis. Sg-AuNPs showed maximum absorbance at 575 nm, and FTIR analysis indicated that phenolic compounds were responsible for the biosynthesis and stability of Sg-AuNPs. FE-TEM results indicated that the AuNPs were predominantly spherical in shape with a Z-average hydrodynamic diameter of 189 nm revealed by DLS. The polydispersity index of Sg-AuNPs was 0.10, while the EDX spectrum of the AuNPs showed an optical absorption band peak at ~2.3 keV. The XRD pattern of the AuNPs displayed diffraction peaks at 38.62°, 44.72°, 65.02°, 77.84°, and 81.90°, which correspond to (111), (200), (220), (311), and (222) planes of fcc crystalline AuNPs. The Sg-AuNPs was studied for antioxidant activities using the DPPH free radical scavenging method and IC_{50} of 250 μg/mL was established. Also the AuNPs displayed significantly higher antioxidant activity than its corresponding salt.

Abel et al.[4] reported the biosynthesis of AuNPs by straightforward addition and stirring of *C. tora* leaf powder and $HAuCl_4$ solution that subsequently yielded a dispersion of AuNPs. Color transformation of the solution that manifested in the appearance of light purple colloid from initial gold color indicated the synthesis of AuNPs. The biosynthesized AuNPs showed absorption peak at 538 nm and the FTIR spectrum revealed conspicuous peaks around 3464.15, 2926.01, 2395.59, 2011.76, 1622.13, 993.34, 844.82, and 561.29 cm^{-1}. The TEM images revealed that the AuNPs were nearly spherical-shaped with size of about 57 nm. The antioxidant potential of the AuNPs was assessed using the catalase activity assay and the nitric oxide scavenging assay. H_2O_2 is very hazardous to the cells and tissues, but catalase (which is an ever-present antioxidant enzyme) is capable of degrading hydrogen peroxide to consequently produce water and oxygen. Thus, the increase in catalase activity corresponds to increase in antioxidant activity. About 60% increase in catalase activity was obtained in treatment with 100 μg/mL concentration of the AuNPs. Also, 70% nitric oxide inhibition was recorded with the AuNPs in the concentration range of 25–75 μg/mL.

Nakkala et al.[119] documented the synthesis of AuNPs utilizing extract of leaves of *Costus pictus*. The CPAuNPs absorbed maximally at 530 nm in UV–Vis spectrum and FTIR bands indicated the participation of amine and carbonyl groups in the formation of CPAuNPs. The morphology of the CPAuNPs as determined by SEM showed that they were spherical-shaped and polydispersed in nature. The DLS particle size analyzer revealed that the CPAuNPs had a mean diameter of 37.2 nm and EDX spectrum showed strong signals of elemental gold. The synthesized CPAuNPs was investigated for potent antioxidant activities and results indicated that CPAuNPs exhibited more ferric ion reducing ability than CPLE (plant extract). The same trend was recorded for the hydrogen peroxide scavenging activity, though its DPPH free radical scavenging activity was lower in comparison with plant extract alone. Also, CPAuNPs displayed higher superoxide scavenging activity of 90% compared to CPLE at various concentrations and 62% nitric oxide quenching activity compared to 50% activity that was shown by CPLE.

In another report, Nakkala et al.[118] elucidated the biological synthesis of AuNPs from *Gymnema sylvestre* leaf extract. Color change from colorless to red was an indication of the formation of the GYAuNPs which was maximally absorbed at 536 nm. The FTIR spectral analysis of GYAuNPs demonstrated shifts in bands which occurred at 3543, 2671, 2090, 1715, 1644, 1416, and 1025 cm^{-1}; and this revealed that compounds present in GYLE (extract) that were rich in carbonyl groups could be implicated in the bioreduction and biofabrication of GYAuNPs. The TEM images specified the formation of spherical-shaped GYAuNPs, while the DLS particle size analyzer revealed normal size of synthesized GYAuNPs as 26 nm. Also, a strong gold signal was produced in the EDX spectrum. Furthermore, the results on antioxidant potential of GYAuNPs indicated that GYAuNPs showed an average quenching activity of 58% which is higher than GYLE in the nitric oxide scavenging activity. Also, maximum percentage inhibition of 47% obtained by GYAuNPs at 60 μg/mL in the hydrogen peroxide scavenging activity was similar to that obtained by GYLE. GYAuNPs demonstrated fair ferric ion reducing in comparison with quercetin (standard used); while GYLE displayed lower activity. GYAuNPs showed substantial radical quenching activity for the superoxide scavenging activity, while GYAuNPs exhibited better DPPH free radical scavenging activity than GYLE at different concentrations employed.

Markus et al.[104] reported synthesis of AuNPs using novel probiotic *Lactobacillus kimchicus* DCY51^{T}. The change in color from yellowish-white to deep purple indicated the formation of AuNPs, which was further confirmed by the display of characteristic peak at 540 nm. The FTIR spectrum showed bands at 3422.71, 2935, 1233.92, 1048.49 cm^{-1}, which are ascribed to the broadening of the vibrations of 1° amines (N–H), alkane (C–H), amine (C–N), and alcohol (C–O) groups, respectively. The incidence of prominent peaks of 1° amines and carbonyl stretch in the amide linkages is an indication that free amino groups owing to amino acid residues and surface-bound proteins were responsible for the formation of protein capping layers on the nanoparticles to prevent their agglomeration.[161] FE-TEM analysis showed that the AuNPs were entirely monodispersed, spherical-shaped with varying sizes of 5–30 nm, while the EDX spectrum displayed the maximum absorbance peak at 2.3 keV that matched the distinguishing peak of elemental gold. The XRD spectrum revealed four characteristic peaks of AuNPs that are consistent with (111), (200), (220), and (311) lattice plane of Bragg's reflection. The synthesized AuNPs displayed antioxidant activities when tested in the DPPH free radical scavenging assay producing IC_{50} value of 233.75 μg/mL.

Additionally, Tahir et al.[177] described the synthesis of AuNPs using leaf extract of *Nerium oleander* which showed color change from yellow to black within 2 h of reaction of gold salt precursor and the leaf extract. The synthesized AuNPs showed maximum absorbance at 560 nm and the

FTIR analysis displayed absorption bands at 3433, 2923, 1626, 1385, 1054, and 535 cm^{-1} responsible for O–H stretching, aldehydic C–H stretching, C=C aromatic, C–N stretching vibrations of straight chain amines and alkyl halides, respectively. The TEM analysis revealed nearly spherical-shaped AuNPs which are highly dispersed and had small sizes ranging from 2 to 10 nm. The XRD blueprint established the crystalline structure of AuNPs and the EDX displayed strong signals of gold atoms. Moreover, the biosynthesized AuNPs displayed effective free radical inhibition activity against DPPH radicals.

Also, Balasubramani et al.[16] established the synthesis of AuNPs with powdered leaf extract (decoction) of *Antigonon leptopus*. The AuNPs produced a purple color within 4 min of synthesis and maximum absorption at 557 nm was observed in the UV–Vis spectrum. The FTIR spectrum revealed the presence of diverse functional groups depicted by infrared band located at 3372, 2924, 2208, 1723, and 1601 cm^{-1} which correspond to amide (N–H) group, C–H stretching vibrations of alkanes, C=C stretching of alkynes, carboxylic group and N–H bend of amine (I) band. HR-TEM analysis showed that the nanoparticles were nearly spherical, with a small number of triangular shapes, and with sizes that ranged from 13 to 28 nm. EDX spectrum showed Au as the highest simple composition along with occurrence of some traces of Cu; and SAED pattern established the crystalline nature of synthesized AuNPs through the formation of bright circular rings. The XRD showed peaks at 38.04°, 44.04°, 64.52°, and 77.68° that are analogous to (111), (200), (220), and (311) Bragg's reflections of fcc structure of elemental gold. The AuNPs displaced significantly superior DPPH free radical scavenging activity than the plant extract to the tune of 87.33% for the AuNPs and 69.33% for the plant extract at working concentration of 100 μg/mL.

1.5.4 ANTICOAGULANT AND THROMBOLYTIC ACTIVITIES OF GOLD NANOPARTICLES

The high compatibility of gold has enabled its utilization in nanomedicine. It has been demonstrated that AuNPs can act as blood anticoagulant using the green route of synthesis.[124,167] In these cases, the blood was prevented from coagulation in a similar way, which AgNPs also act as anticoagulant. Some studies have documented the importance of AuNPs as drug carriers and as theranostic agents to manage blood coagulation disorders.[53,68,163] In addition, recent investigations have shown tremendous anticoagulant activities of

AuNPs that were biosynthesized using the cell-free extracts of non-pathogenic strains of *Enterococcus* species. All these investigations lay credence to the potential application of AuNPs as anticoagulant agents to herald a new dawn in nanomedicine, whereby the use of nanoagents can alleviate the negative side effects that accompany the utilization of conventional drugs.

Singh et al.[167] described the synthesis of AuNPs by fresh leaves of *Panax ginseng*. The color of the solution containing the AuNPs developed to dark purple indicating the synthesis. The synthesized AuNPs absorbed maximally at 578 nm in the UV–Vis spectra. FE-TEM analysis illustrated spherical-shaped AuNPs that were monodispersed in distribution with sizes that varied from 10 to 20 nm. The EDX and elemental mapping results revealed maximum circulation of elemental gold in the nanoproduct, which confirmed its purity. The XRD results established the crystalline nature of AuNPs. The results for anticoagulant activities of AuNPs demonstrated that the development of blood clots was inhibited effectively in the presence of the AuNPs, thus confirming its anticoagulant potency. Also, the green biosynthesis of spherical AuNPs by employing extracellular fraction of *Bacillus safensis* by Ojo et al.[124] was shown to possess anticoagulation and thrombolytic activities. It was reported that the AuNPs disallowed coagulation of blood whose stability was ensured over an extended period. Further, the nanoparticles caused lysis of blood clot within 5 min of reaction, and the nature of the nanoparticles-anticoagulated blood; and dissolved blood clot was observed by optical microscope which showed good comparison with fresh blood sample.

Moreover, Kalishwaralal et al.[61] demonstrated synthesis of AuNPs using *Brevibacterium casei* biomass. Aqueous solution of $HAuCl_4$ was treated with *B. casei* biomass with the appearance of dark purple color after 24 h reaction. The synthesized AuNPs were maximally absorbed at 540 nm. The FTIR gave indication for the existence of proteins as probable biomolecules that accounted for reduction and capping process which increased the stability of synthesized AuNPs. The TEM analysis showed that the AuNPs were reasonably uniform in diameter and spherical-shaped with sizes ranging from 10 to 50 nm. The XRD pattern indicated strong peaks that are consistent with (111), (200), (220), and (311) Bragg's reflection, which are interpreted as crystalline fcc gold and this confirmed the crystalline nature of the AuNPs. The AuNPs showed excellent anticoagulant activity that was established by inhibiting the development of blood clots in the experiment where blood was held along with the AuNPs. The stability of AuNPs-anticoagulated blood was established by extended contact of the blood plasma with the particles for approximately 24 h, without showing any noteworthy reduction in anticoagulation.

Kim et al.[69] also studied the green synthesis of AuNPs with aqueous extract of earthworm. The development of a wine red color during synthesis indicated the formation of the EW-AuNPs with surface plasmon resonance at 533 nm. The FTIR spectrum based on the shifts in band for both the aqueous extract and the synthesized EW-AuNPs suggested that the proteins/peptides in the extract most probably accounted for the reduction of Au^{3+} to produce the EW-AuNPs. FE-SEM analysis revealed that the EW-AuNPs were cubic and block-shaped, while the TEM analysis indicated the particles as mainly spherical-shaped possessing mean size of 6.13 nm. The XRD pattern showed peaks at 38.3°, 44.7°, 64.7°, and 77.4°, which are typical of the (111), (200), (220), and (311) planes of crystalline Au. The particles improved anticoagulant activity of heparin in activated partial thromboplastin time (aPTT) assay. The clotting times of the deionized water (negative control) and heparin (positive control) were 44.1 and 50.8 s, respectively. No considerable anticoagulant activities were prominent in the extract (47.2 s), the EW-AuNPs (44.8 s), or in combination of heparin with extract (50.9 s). However, in experiment whereby heparin was combined with EW-AuNPs, there was prolonged clotting time (60.4 s), which indicated performance improvement of 118.9% and 134.8% compared with clotting times at the same concentrations for heparin and EW-AuNPs alone, respectively.

Jian et al.[59] have documented synthesis of fibrinogen-modified AuNPs (Fib-AuNPs). AuNPs was first synthesized by reacting 1% trisodium citrate (0.5 mL) with 0.01% $HAuCl_4$ (50 mL) and heated under reflux which led to a development of pink coloration. TEM analysis confirmed the AuNPs to be monodispersed that possessed mean diameter of 32 nm. The preparation of Fib-AuNPs involved reaction between 32 nm AuNPs and fibrinogen in Tris-HCl buffer (20 mM, pH 7.4) for 30 min. Thereafter, a solution of bovine serum albumin was added and subsequently maintained at ambient temperature for 10 min. Fibrinogen molecules easily conjugated with the AuNPs through electrostatic and hydrophobic interactions. The Fib-AuNPs so produced were studied for detection of plasmin and plasminogen in blood serum samples at nanomolar range. It was concluded that detection of plasmin, urokinase, and plasminogen by Fib-AuNPs showed that the method has immense prospect for application as diagnostic procedure to unravel diseases that are related to abnormalities in fibrinolysis such as severe and chronic stages of cerebral thrombosis and embolism.

In another investigation, Ehmann et al. [37] synthesized AuNPs by employing 6-O chitosan sulfate solution. Small volumes of S-ChiAuNPs were thereafter mixed with the blood plasma that was gotten from patients afflicted with different forms of sickness to determine the prothrombin time in each case. The diseases of the three patients that were used enhanced their possibility to develop thrombosis and blood clotting. Consequently, any positive outcome toward increasing the clotting time in these patients may give hope to deployment of the AuNPs system in anticoagulant dosage therapy. On the introduction of a little amount (0.25 μL, c = 0.1 mg/mL) of S-ChiAuNPs to appropriate blood plasma, the values of aPTT practically doubled in all blood samples. At higher working volume using 0.50 μL of 0.1 mg/mL of the S-ChiAuNPs, there was a further increament in aPTT. A comparable extension in the clotting progression also manifested in the procedure involving PT. In all the samples, the PT was appreciably increased. The increase in both aPTT and PT in the blood samples of all the patients showed that the particles exercised effects on both extrinsic and intrinsic blood coagulation pathways in a non-specific activity form.

Paul et al.[132] also described one-pot green synthesis of AuNPs with the use of dried biomass of *Momordica cochinchinensis*. It led to development of deep red solution which was absorbed maximally at 552 nm. Bands which correspond to amide I of polypeptides and symmetric stretching of carboxylate groups in the amino acid residues of the protein molecules were clearly shown in the FTIR spectrum and this clearly indicated the presence of proteins and additional organic compounds that were extracellularly produced by *Momordica cochinchinensis*. The TEM results showed that the AuNPs were mostly spherical, oval and triangular with sizes of 10–80 nm, while the XRD revealed peaks which correspond to fcc structure of elemental gold. The EDX pattern confirmed the formation of AuNPs. The anticoagulation properties of the AuNPs were demonstrated by inhibiting the formation of blood clots which was evident in test tube that contained blood in the presence of AuNPs, while blood clots were formed in the control experiment. The inhibition of clot formation was extended for about 24 h, which suggested that the AuNPs displayed long-term outcome to prevent clotting. Furthermore, Kim et al.[70] reported the green synthesis of heparin-reduced AuNPs. Two types of AuNPs that were synthesized, such as: Hep-AuNPs-I and Hep-AuNPs-II; and these were investigated for anticoagulation properties. Hep-AuNPs-I displayed relatively mild anticoagulant activity, with a PT prolonged by approximately 13% compared to the positive control (heparin). But no blood anticoagulant activity was detected when the TT (Thrombin time) and aPTT were estimated. Meanwhile, Hep-AuNPs-II displayed strong anticoagulant activity, with a prolonged PT (26.7%), TT (21.8%), and aPTT (23.2%) compared to the positive control.

All these investigations confirm potential applications of AuNPs as anticoagulant agents to herald a new dawn in nanomedicine, whereby the use of nanoagents can alleviate the negative side effects that accompany the utilization of conventional drugs. Although the use of AuNPs as thrombolytic agents is at infancy, yet there are only few reports on these applications.[124] The ability of AuNPs to dissolve blood clot is envisaged to follow the same mechanism shown by AgNPs. Recent investigations in the laboratory of authors of this chapter have further shown that more biosynthesized AuNPs can efficiently dissolve thrombi, without any distortion in the morphology of red blood cells.

1.6 BIOMEDICAL APPLICATIONS OF SILVER–GOLD ALLOY NANOPARTICLES: A CASE STUDY

1.6.1 ANTIMICROBIAL ACTIVITIES OF SILVER–GOLD ALLOY NANOPARTICLES

Ojo et al.[124] reported the green biosynthesis of silver–gold alloy nanoparticles (Ag–AuNPs) with extracellular fraction of *Bacillus safensis*. The bacteriogenic formation Ag–AuNPs showed a characteristic purple color that maximally absorbed at 545 nm. FTIR spectroscopic analysis of the Ag–AuNPs revealed peaks at 3310, 2345, 2203, 2033, 1636, 1273, 502, 453, 424 cm^{-1} indicating the contribution of proteins toward the bioformation of capped and stabilized Ag–AuNPs. The TEM images revealed the anisotropic nature of the uniform spherical Ag–AuNPs biosynthesized with some irregular aggregation to produce some rod-shaped particles. Further, the Ag–AuNPs were polydispersed in distribution, with sizes ranging from 13 to 80 nm. Moreover, the EDX analysis of Ag–AuNPs indicated that silver and gold were the most abundant element in the colloid solution, while SAED confirmed the biosynthesized nanoparticles as crystal structures that were ring-shaped. Furthermore, Ag–AuNPs demonstrated more excellent antifungal property than AuNPs biosynthesized from the same cell-free extract. Growth inhibitions of about 83.33% and 90.78% against *A. niger* and *A. fumigatus,* respectively were achieved at working concentration of 200 μg/mL of Ag–AuNPs.

Lateef et al.[89] described the phytosynthesis of Ag–AuNPs by employing aqueous fractions of various parts of *Cola nitida*, such as leaves (LF), seeds (SE), seed shell (SS), and pod (KP). For the biosynthesis, about 1 mL of the supernatant of each extract was reacted with 10 mL of a mixture of 1 mM of both $AgNO_3$ and $HAuCl_4$ (4:1). Within 5 min of reaction, there was emergence of dark brown color that developed from the initial light orange as a result of biosynthesis of Ag–AuNPs that absorbed maximally between 497 and 531 nm. The FTIR spectra revealed prominent peaks at 3209, 2088, and 1641 cm^{-1} for LFAg–AuNPs, 3329, 2129, and 1635 cm^{-1} for SSAg–AuNPs; 3396, 2088, and 1647 cm^{-1} for SEAg–AuNPs; and 3302, 2098; and 1635 cm^{-1} for KPAg–AuNPs. The prevalent functional groups were reported to be O–H stretch of carboxylic acid, N–H of amines, C=O stretch of amides and C=C stretch of alkenes. According to TEM analysis, the LF-, SE and SSAg–AuNPs were nearly spherical in morphology having dimensions of 17–90 nm; while anisotropic structures of hexagon, rod, triangle, and sphere of about 12–91 nm were obtained in KPAg–AuNPs. SAED results showed distinct and poly-crystalline configurations of Au and Ag; and the EDX spectra revealed the prominence of Au and Ag. The four samples of Ag–AuNPs were investigated for potential antifungal activities using the methods of Khatami et al.[67] At concentration of 150 μg/mL, 100% (total) inhibition of growth of *A. flavus* was accomplished by all the phytosynthesized Ag–AuNPs. However, growth inhibitions varying from 76.83% to 100%, and 69.51% to 75.61% were obtained for *A. fumigatus* and *A. niger*, respectively.

1.6.2 LARVICIDAL ACTIVITIES OF SILVER–GOLD ALLOY NANOPARTICLES

Lateef et al.[89] reported the biosynthesis of LF-, SE-, SS, and KPAg–AuNPs as earlier described in section 1.6.1. The four samples of Ag–AuNPs were investigated for potential larvicidal activities by exposing ten larvae of *Anopheles gambiae* to different concentrations of the Ag–AuNPs (60–100 μg/mL) at ambient temperature. It was reported that 100% larval death occurred at 3, 24, 48, and 72 h for the SE-, LF-, KP-, and SSAg–AuNPs at least concentration of 60 μg/mL, respectively.

1.6.3 ANTIOXIDANT ACTIVITIES OF SILVER–GOLD ALLOY NANOPARTICLES

Kumari et al.[72] described the synthesis of Ag–AuNPs at ambient temperature using the fruit juice of pomegranate. The formation of Ag–AuNPs was marked by the light reddish color developed by reacting solution during synthesis. The Ag–AuNPs maximally absorbed at 479 nm in the UV–Vis spectrum. The FTIR spectrum bands suggested that the relative reduction in the strength of phenolic hydroxyl stretching band in the spectrum of Ag–AuNPs indicated preferential role of phenolic compounds as reductants in the biosynthesis by releasing electrons to form quinones. The emergence of band at 1380 cm^{-1} in the spectrum of alloy indexed to –C–O– stretching resulting from water-soluble compounds like terpenoids and flavonoids showed their presence in the fruit juice.[133] TEM images confirmed the formation of core-shell structured Ag–AuNPs with average size of 12 nm; and SAED pattern established the crystalline nature of the bimetallic nanoparticles. XRD analysis showed four characteristic peaks for the Ag–AgNPs, which revealed the crystallized fcc structure. The synthesized Ag–AgNPs was investigated for potential antioxidant activities using the nitric oxide and OH scavenging assay. In both assays, the scavenging activities were enhanced with increasing concentration of Ag–AuNPs. Concentrations of 25, 50, and 100 μL were recorded to exhibit percentage inhibition comparable with that of gallic acid, which was the standard in the nitric oxide radical scavenging assay. Also in the OH^- radical scavenging assay, percentage inhibition of the Ag–AuNPs increased from 36% to 76%, which was similar to that of gallic acid.

1.6.4 ANTICOAGULANT AND THROMBOLYTIC ACTIVITIES OF SILVER–GOLD ALLOY NANOPARTICLES

While Ag–AuNPs have been used for different purposes that include antimicrobial agents,[41,89,124,183] dye degradation,[89,124] photothermal,[41] larvicidal,[89] control of biofilms[101] and as sensors,[139] there are limited studies on evaluation of biosynthesized Ag–AuNPs for anticoagulation and thrombolysis.[89,124] As the developments in the use of Ag–AuNPs unfold, it may present a great opportunity in nanomedicine in the diagnosis and treatment of cardiovascular pathologies. The combination of antimicrobial efficacy of AgNPs and biocompatibility and bioimaging potentials of AuNPs can promote the use of Ag–AuNPs as a very good theranostic agent.

The green biosynthesized polydispersed anisotropic Ag–AuNPs using the extracellular fraction of *Bacillus safensis* by Ojo et al.[124] has shown to possess anticoagulation and thrombolytic activities. It was reported that the nanoparticles prevented coagulation of blood, which enjoyed stability for extended period of time. The particles also caused dissolution of blood clot within 5 min of action. The nature of the nanoparticles-anticoagulated blood and dissolved blood clot were observed by optical microscope, which showed good comparison with fresh blood sample.

Similarly, Lateef et al.[89] demonstrated the biosynthesis of Ag–AuNPs using the leaf, seed, seed shell and pod extracts of *Cola nitida*. The four samples of Ag–AuNPs were investigated for potential anticoagulant and thrombolytic activities, and all of them were reported to prevent coagulation of fresh human blood, which was similar to the result obtained with EDTA. The microscopic investigation also confirmed that the shapes of the red blood cells were similar to that of the fresh blood sample. This is suggestive of their potential in the management of blood coagulation disorders. Further, the four samples of Ag–AuNPs reportedly dissolved pre-formed human blood clots rapidly in 5 min, and microscopic investigations established that the blood clots were lysed.

1.7 PROSPECTIVE FUTURE AND RESEARCH OPPORTUNITIES

The synthesis of MeNPs through green approach has continued to court the interests of scientists from diverse backgrounds due to numerous benefits. The process is eco-friendly, characterized with the lack of use of harmful chemicals and techniques in the synthesis, which promotes their biocompatibility and low toxicity. It is also economical, rapid, cost-effective, and can be accomplished under benign conditions. The abundance of several biomolecules in biological entities that can concomitantly serve as both reduction and capping agents, has also fueled the growing trends in one-pot synthesis of MeNPs. Amongst the MeNPs, silver (Ag), gold (Au), and their bimetallic alloy (Ag–AuNPs) have been vividly studied owing to their novel optical, physical, chemical, photothermal, catalytic, and electrical attributes for multiple applications. Some of the important usefulness of these nanoparticles is in their use as antimicrobial, larvicidal, antioxidant, anticoagulant, and thrombolytic agents.

These biomedical applications are envisaged to combat myriads of diseases facing mankind; particularly the antimicrobial resistance phenomena, control of vector-borne diseases, mitigation of the deleterious activities of free radical species, and control/management of blood coagulation disorders. It is evidently clear that properties exhibited by nanoparticles aptly positioned them to be used as vital tools in the development of new generation of nanomedical. Therefore, this review chapter presents the contributions of green synthesized Ag, Au, and Ag–AuNPs for biomedical applications with due diligence to antimicrobial, larvicidal, antioxidant, anticoagulant, and thrombolytic activities. Until now, there is no review that summarizes the biomedical applications of Ag, Au, and Ag–AuNPs as a compendium. The review underscores the importance of these particles in the emerging disciplines of nano- and biomedicine.

1.8 SUMMARY

This chapter has reviewed the recent trends in green synthesis of Ag, Au, and Ag–AuNPs for antimicrobial, larvicidal, antioxidant, anticoagulant, and thrombolytic applications in nanotechnology. The involvement of diverse materials of biological origin in the green synthesis of the nanoparticles is rapidly increasing, thereby creating bountiful production of nanoparticles. This can pave way for the large-scale biofabrication of nanoparticles of unique attributes and applications.

Within the scope of this chapter, the green synthesized Ag, Au, and Ag–AuNPs have shown tremendous activities that are poised to be milestones in the development of nano- and biomedicine. While several biomedical applications have been established for Ag and AuNPs, literature survey has shown that similar investigations on Ag–AuNPs are limited.

ACKNOWLEDGMENT

Authors acknowledge the authority of LAUTECH, Ogbomoso for the provision of the facilities used in the research cited in this chapter.

KEYWORDS

- **biosynthesis**
- **green synthesis**
- **nanomedicine**
- **plasmon resonance**
- **zone of inhibition**

REFERENCES

1. Abbai, R.; Mathiyalagan, R.; Markus, J.; Kim, Y. J.; Wang, C.; Singh, P.; Ahn, S.; Farh, M. E. A.; Yang, D. C. Green Synthesis of Multifunctional Silver and Gold Nanoparticles from the Oriental Herbal Adaptogen: Siberian Ginseng. *Int. J. Nanomed.* **2016,** *11*, 3131.
2. Abdel-Aziz, M. S.; Shaheen, M. S.; El-Nekeety, A. A.; Abdel-Wahhab, M. A. Antioxidant and Antibacterial Activity of Silver Nanoparticles Biosynthesized using *Chenopodium murale* Leaf Extract. *J. Saudi Chem. Soc.* **2014,** *18*, 356–363.

3. Abdel-Hafez, S. I.; Nafady, N. A.; Abdel-Rahim, I. R.; Shaltout, A. M.; Daròs, J. A.; Mohamed, M. A. Assessment of Protein Silver Nanoparticles Toxicity Against Pathogenic Alternaria Solani. *3 Biotech.* **2016,** *6* (199), 1–12.
4. Abel, E. E.; Poonga, P. R. J.; Panicker, S. G. Characterization and *In vitro* Studies on Anticancer, Antioxidant Activity Against Colon Cancer Cell Line of Gold Nanoparticles Capped with *Cassia tora* SM Leaf Extract. *Appl. Nanosci.* **2016,** *6* (1), 121–129.
5. Adelere, I. A.; Lateef, A. A Novel Approach to the Green Synthesis of Metallic Nanoparticles: The Use of Agro-wastes, Enzymes and Pigments. *Nanotechnol. Rev.* **2016,** *5*, 567–587.
6. Adewoye, S. O.; Lateef, A. Assessment of the Microbiological Quality of *Clarias gariepinus* Exposed to an Industrial Effluent in Nigeria. *Environmentalist* **2004,** *24*, 249–254.
7. Ahamed, M.; AlSalhi, M. S.; Siddiqui, M. K. J. Silver Nanoparticles Applications and Human Health. *Clin. Chim. Acta* **2010,** *411* (23), 1841–1848.
8. Ahmad, A.; Mukherjee, P.; Senapati, S.; Mandal, D.; Khan, M. I.; Kumar, R.; Sastry, M. Extracellular Biosynthesis of Silver Nanoparticles Using the Fungus *Fusarium oxysporum*. *Coll. Surf. B Biointerfaces* **2003,** *28* (4), 313–318.
9. Ahmad, T.; Wani, I. A.; Manzoor, N.; Ahmed, J.; Asiri, A. M. Biosynthesis, Structural Characterization and Antimicrobial Activity of Gold and Silver Nanoparticles. *Coll. Surf. B Biointerfaces* **2013,** *107*, 227–234.
10. Ahmed, S.; Ahmad, M.; Swami, B. L.; Ikram, S. Review on Plants Extract Mediated Synthesis of Silver Nanoparticles for Antimicrobial Applications: A Green Expertise. *J. Adv. Res.* **2016,** *7* (1), 17–28.
11. Ajitha, B.; Reddy, Y. A. K.; Rajesh, K. M.; Reddy, P. S. *Sesbania grandiflora* Leaf Extract Assisted Green Synthesis of Silver Nanoparticles: Antimicrobial Activity. *Mater. Today Proc.* **2016,** *3* (6), 1977–1984.
12. Anand, K.; Gengan, R. M.; Phulukdaree, A.; Chuturgoon, A. Agroforestry Waste *Moringa oleifera* Petals Mediated Green Synthesis of Gold Nanoparticles and Their Anti-cancer and Catalytic Activity. *J. Ind. Eng. Chem.* **2015,** *21*, 1105–1111.
13. Anbazhagan, P.; Murugan, K.; Jaganathan, A.; Sujitha, V.; Samidoss, C. M.; Jayashanthani, S.; Amuthavalli, P.; Higuchi, A.; Kumar, S.; Wei, H.; Nicoletti, M. Mosquitocidal, Antimalarial and Antidiabetic Potential of *Musa paradisiaca*-synthesized Silver Nanoparticles: *In vivo* and *In vitro* Approaches. *J. Clus. Sci.* **2017,** *28* (1), 91–107.
14. Azeez, L.; Lateef, A.; Adebisi, S. A. Silver Nanoparticles (AgNPs) Biosynthesized Using Pod Extract of *Cola nitida* Enhances Antioxidant Activity and Phytochemical Composition of *Amaranthus caudatus* Linn. *Appl. Nanosci.* **2017,** *7* (1–2), 59–66.
15. Azeez, M. A. Lateef, A.; Asafa, T. B.; Yekeen, T. A.; Akinboro, A.; Oladipo, I. C.; Gueguim-Kana, E. B.; Beukes, L. S. Biomedical Applications of Cocoa Bean Extract-mediated Silver Nanoparticles as Antimicrobial, Larvicidal and Anticoagulant Agents. *J. Clust. Sci.* **2016,** *28* (1), 149–164.
16. Balasubramani, G.; Ramkumar, R.; Krishnaveni, N.; Pazhanimuthu, A.; Natarajan, T.; Sowmiya, R.; Perumal, P. Structural Characterization, Antioxidant and Anticancer Properties of Gold Nanoparticles Synthesized from Leaf Extract (Decoction) of *Antigonon leptopus* Hook& Arn. *J. Trace Elem. Med. Biol.* **2015,** *30*, 83–89.
17. Banerjee, P.; Satapathy, M.; Mukhopahayay, A.; Das, P. Leaf Extract Mediated Green Synthesis of Silver Nanoparticles from Widely Available Indian Plants: Synthesis,

Characterization, Antimicrobial Property and Toxicity aAnalysis. *Bioresour. Bioprocess.* **2014,** *1* (3), 2–10.
18. Basavegowda, N.; Idhayadhulla, A.; Lee, Y. R. Phyto-synthesis of Gold Nanoparticles Using Fruit Extract of *Hovenia dulcis* and Their Biological Activities. *Indus. Crops Prod.* **2014,** *52*, 745–751.
19. Basu, S.; Maji, P.; Ganguly, J. Biosynthesis, Characterisation and Antimicrobial Activity of Silver and Gold Nanoparticles by *Dolichos biflorus* Linn Seed Extract. *J. Exp. Nanosci.* **2016,** *11* (8), 660–668.
20. Benelli, G.; Mehlhorn, H. Declining Malaria, Rising of Dengue and Zika Virus: Insights for Mosquito Vector Control. *Parasitol. Res.* **2016,** *115* (5), 1747–1754.
21. Benelli, G. Green Synthesized Nanoparticles in the Fight Against Mosquito-borne Diseases and Cancer-a Brief Review. *Enz. Microb. Technol.* **2016,** *95*, 58–68.
22. Benelli, G. Plant-mediated Biosynthesis of Nanoparticles as an Emerging Tool Against Mosquitoes of Medical and Veterinary Importance: A Review. *Parasitol. Res.* **2016,** *115* (1), 23–34.
23. Bhakya, S.; Muthukrishnan, S.; Sukumaran, M.; Muthukumar, M. Biogenic Synthesis of Silver Nanoparticles and Their Antioxidant and Antibacterial Activity. *Appl. Nanosci.* **2016,** *6* (5), 755–766.
24. Bogireddy, N. K. R.; Anand, K. K. H.; Mandal, B. K. Gold Nanoparticles-synthesis by *Sterculiaacuminata* Extract and its Catalytic Efficiency in Alleviating Different Organic Dyes. *J. Mol. Liq.* **2015,** *211*, 868–875.
25. Chokshi, K.; Pancha, I.; Ghosh, T.; Paliwal, C.; Maurya, R.; Ghosh, A.; Mishra, S. Green Synthesis, Characterization and Antioxidant Potential of Silver Nanoparticles Biosynthesized from De-oiled Biomass of Thermotolerant Oleaginous Microalgae *Acutodesmus dimorphus*. *RSC Adv.* **2016,** *6* (76), 72269–72274.
26. Dar, M. A.; Ingle, A.; Rai, M. Enhanced Antimicrobial Activity of Silver Nanoparticles Synthesized by *Cryphonectria* sp. Evaluated Singly and in Combination with Antibiotics. *Nanomedicine* **2013,** *9* (1), 105–110.
27. Das, B.; Dash, S. K.; Mandal, D.; Ghosh, T.; Chattopadhyay, S.; Tripathy, S.; Das, S.; Dey, S. K.; Das, D.; Roy, S. Green Synthesized Silver Nanoparticles Destroys Multidrug Resistant Bacteria via Reactive Oxygen Species Mediated Membrane Damage. *Arab. J. Chem.* **2015,** E-article; http://dx.doi.org/10.1016/j.arabjc.2015.08.008.
28. Das, S.; Dhar, B. B. Green Synthesis of Noble Metal Nanoparticles Using Cysteine-modified Silk Fibroin: Catalysis and Antibacterial Activity. *RSC Adv.* **2014,** *4*, 46285–46292.
29. Davalos, D.; Akassoglou, K. Fibrinogen as a Key Regulator of Inflammation in Disease. *Sem. Immunopathol.* **2012,** *34* (1), 43–62.
30. Devi, P. S.; Banerjee, S.; Chowdhury, S. R.; Kumar, G. S. Eggshell Membrane: A Natural Biotemplate to Synthesize Fluorescent Gold Nanoparticles. *RSC Adv.* **2012,** *2*, 11578–11585.
31. Dhand, V.; Soumya, L.; Bharadwaj, S.; Chakra, S.; Bhatt, D.; Sreedhar, B. Green Synthesis of Silver Nanoparticles Using *Coffea arabica* Seed Extract and its Antibacterial Activity. *Mater. Sci. Eng. C* **2016,** *58*, 36–43.
32. Dhayalan, M.; Denison, M. I. J.; Anitha-Jegadeeshwari, L.; Krishnan, K.; Gandhi, N. *In vitro* Antioxidant, Antimicrobial, Cytotoxic Potential of Gold and Silver Nanoparticles Prepared Using *Embelia ribes*. *Nat. Prod. Res.* **2016,** *31* (4), 465–468.

33. Dinesh, D.; Murugan, K.; Madhiyazhagan, P.; Panneerselvam, C.; Kumar, P. M.; Nicoletti, M.; Jiang, W.; Benelli, G.; Chandramohan, B.; Suresh, U. Mosquitocidal and Antibacterial Activity of Green-synthesized Silver Nanoparticles from *Aloe vera* Extracts: Towards an Effective Tool Against the Malaria Vector *Anopheles stephensi*? *Parasitol. Res.* **2015,** *114* (4), 1519–1529.
34. Dolai, N.; Karmakar, I.; Kumar, R. S.; Kar, B.; Bala, A.; Haldar, P. K. Free Radical Scavenging Activity of *Castanopsis indica* in Mediating Hepatoprotective Activity of Carbon Tetrachloride Intoxicated Rats. *Asian Pacific J. Trop. Biomed.* **2012,** *2* (1), S243–S251.
35. Duy, N. N.; Du, D. X.; Van Phu, D.; Du, B. D.; Hien, N. Q. Synthesis of Gold Nanoparticles with Seed Enlargement Size by γ-irradiation and Investigation of Antioxidant Activity. *Coll. Surf. A Physicochem. Eng. Aspects* **2013,** *436*, 633–638.
36. Dwyer, D. J.; Camacho, D. M.; Kohanski, M. A.; Callura, J. M.; Collins, J. J. Antibiotic-induced Bacterial Cell Death Exhibits Physiological and Biochemical Hallmarks of Apoptosis. *Mol. Cell* **2012,** *46* (5), 561–572.
37. Ehmann, H. M.; Breitwieser, D.; Winter, S.; Gspan, C.; Koraimann, G.; Maver, U.; Sega, M.; Köstler, S.; Stana-Kleinschek, K.; Spirk, S.; Ribitsch, V. Gold Nanoparticles in the Engineering of Antibacterial and Anticoagulant Surfaces. *Carbohyd. Polym.* **2015,** *117*, 34–42.
38. Elavazhagan, T.; Arunachalam, K. D. *Memecylon edule* Leaf Extract Mediated Green Synthesis of Silver and Gold Nanoparticles. *Int. J. Nanomed.* **2011,** *6*, 1265–1278.
39. Elemike, E. E.; Onwudiwe, D. C.; Fayemi, O. E.; Ekennia, A. C.; Ebenso, E. E.; Tiedt, L. R. Biosynthesis, Electrochemical, Antimicrobial and Antioxidant Studies of Silver Nanoparticles Mediated by *Talinum triangulare* Aqueous Leaf Extract. *J. Clust. Sci.* **2017,** *28* (1), 309–330.
40. Eugenio, M.; Müller, N.; Frasés, S.; Almeida-Paes, R.; Lima, L. M. T.; Lemgruber, L.; Farina, M.; de Souza, W.; Sant'Anna, C. Yeast-derived Biosynthesis of Silver/Silver Chloride Nanoparticles and Their Antiproliferative Activity Against Bacteria. *RSC Adv.* **2016,** *6*, 9893–9904.
41. Fasciani, C.; Silvero, M. J.; Anghel, M. A.; Argüello, G. A.; Becerra, M. C.; Scaiano, J. C. Aspartame-stabilized Gold–Silver Bimetallic Biocompatible Nanostructures with Plasmonic Photothermal Properties, Antibacterial Activity, and Long-term Stability. *J. Am. Chem. Soc.* **2014,** *136*, 17394–17397.
42. Fazal, S.; Jayasree, A.; Sasidharan, S.; Koyakutty, M.; Nair, S. V.; Menon, D. Green Synthesis of Anisotropic Gold Nanoparticles for Photothermal Therapy of Cancer. *ACS Appl. Mater.* **2014,** *6*, 8080–8089.
43. Furie, B.; Furie, B. C. Mechanisms of Thrombus Formation. *New Eng. J. Med.* **2008,** *359* (9), 938–949.
44. Garg, S., Chandra, A.; Mazumder, A.; Mazumder, R. Green Synthesis of Silver Nanoparticles Using *Arnebia nobilis* Root Extract and Wound Healing Potential of its Hydrogel. *Asian J. Pharm.* **2014,** *8* (2), 95–101.
45. Geethalakshmi, R.; Sarada, D. V. L. Gold and Silver Nanoparticles from *Trianthema decandra*: Synthesis, Characterization, and Antimicrobial Properties. *Int. J. Nanomed.* **2012,** *7*, 5375–5384.
46. Gnanadesigan, M.; Anand, M.; Ravikumar, S.; Maruthupandy, M.; Vijayakumar, V.; Selvam, S.; Dhineshkumar, M.; Kumaraguru, A. K. Biosynthesis of Silver Nanoparticles

by Using Mangrove Plant Extract and Their Potential Mosquito Larvicidal Property. *Asian Pac. J. Trop. Med.* **2011,** *4* (10), 799–803.

47. Gomathi, M.; Rajkumar, P. V.; Prakasam, A.; Ravichandran, K. Green Synthesis of Silver Nanoparticles using *Datura stramonium* Leaf Extract and Assessment of Their Antibacterial Activity. *Resource-Eff. Technol.* **2017,** In Press, http://doi.org/10.1016/j.reffit.2016.12.005.
48. Govindarajan, M.; Benelli, G. Facile Biosynthesis of Silver Nanoparticles Using *Barleria cristata*: Mosquitocidal Potential and Biotoxicity on Three Non-target Aquatic Organisms. *Parasitol. Res.* **2016,** *115* (3), 925–935.
49. Govindarajan, M.; Benelli, G. One-pot Green Synthesis of Silver Nanocrystals Using *Hymenodictyon orixense*: A Cheap and Effective Tool Against Malaria, Chikungunya and Japanese Encephalitis Mosquito Vectors? *RSC Adv.* **2016,** *6* (64), 59021–59029.
50. Govindarajan, M.; Nicoletti, M.; Benelli, G. Bio-physical Characterization of Poly-dispersed Silver Nanocrystals Fabricated Using *Carissa spinarum*: A Potent Tool Against Mosquito Vectors. *J. Clust. Sci.* **2016,** *27* (2), 745–761.
51. Govindarajan, M.; Rajeswary, M.; Muthukumaran, U.; Hoti, S. L.; Khater, H. F.; Benelli, G. Single-step Biosynthesis and Characterization of Silver Nanoparticles Using *Zornia diphylla* Leaves: A Potent Eco-friendly Tool Against Malaria and Arbovirus Vectors. *J. Photochem. Photobiol. B Biol.* **2016,** *161*, 482–489.
52. Harish, B. S.; Uppuluri, K. B.; Anbazhagan, V. Synthesis of Fibrinolytic Active Silver Nanoparticle Using Wheat Bran Xylan as a Reducing and Stabilizing Agent. *Carbohydr. Polym.* **2015,** *132*, 104–110.
53. Hsu, C. L.; Chang, H. T.; Chen, C. T.; Wei, S. C.; Shiang, Y. C.; Huang, C. C. Highly Efficient Control of Thrombin Activity by Multivalent Nanoparticles. *Chemistry* **2011,** *17*, 10994–11000.
54. Huang, J.; Li, Q.; Sun, D.; Lu, Y.; Su, Y.; Yang, X.; Wang, H.; Wang, Y.; Shao, W.; He, N.; Hong, J. Biosynthesis of Silver and Gold Nanoparticles by Novel Sundried *Cinnamomum camphora* Leaf. *Nanotechnology* **2007,** *18* (10), 105104.
55. Ilinskaya, A. N.; Dobrovolskaia, M. A. Nanoparticles and the Blood Coagulation System. Part II: Safety Concerns. *Nanomedicine* **2013,** *8* (6), 969–981.
56. Jaganathan, A.; Murugan, K.; Panneerselvam, C.; Madhiyazhagan, P.; Dinesh, D.; Vadivalagan, C.; Chandramohan, B.; Suresh, U.; Rajaganesh, R.; Subramaniam, J.; Nicoletti, M. Earthworm-mediated Synthesis of Silver Nanoparticles: A Potent Tool Against Hepatocellular Carcinoma, *Plasmodium falciparum* Parasites and Malaria Mosquitoes. *Parasitol. Inter.* **2016,** *65* (3), 276–284.
57. Jayaseelan, C.; Ramkumar, R.; Rahuman, A. A.; Perumal, P. Green Synthesis of Gold Nanoparticles Using Seed Aqueous Extract of *Abelmoschus esculentus* and its Antifungal Activity. *Ind. Crops Prod.* **2013,** *45*, 423–429.
58. Jeyaraj, M.; Varadan, S.; Anthony, K. J. P.; Murugan, M.; Raja, A.; Gurunathan, S. Antimicrobial and Anticoagulation Activity of Silver Nanoparticles Synthesized from the Culture Supernatant of *Pseudomonas aeruginosa*. *J. Ind. Eng. Chem.* **2013,** *19* (4), 1299–1303.
59. Jian, J. W.; Chiu, W. C.; Chang, H. T.; Hsu, P. H.; Huang, C. C. Fibrinolysis and Thrombosis of Fibrinogen-modified Gold Nanoparticles for Detection of Fibrinolytic-related Proteins. *Anal. Chimica Acta* **2013,** *774*, 67–72.
60. Jinu, U.; Jayalakshmi, N.; Anbu, A. S.; Mahendran, D.; Sahi, S.; Venkatachalam, P. Biofabrication of Cubic Phase Silver Nanoparticles Loaded with Phytochemicals from

Solanum nigrum Leaf Extracts for Potential Antibacterial, Antibiofilm and Antioxidant Activities Against MDR Human Pathogens. *J. Clust. Sci.* **2017,** *28* (1), 489–505.

61. Kalishwaralal, K.; Deepak, V.; Pandian, S. R. K.; Kottaisamy, M.; BarathManiKanth, S.; Kartikeyan, B.; Gurunathan, S. Biosynthesis of Silver and Gold Nanoparticles Using *Brevibacterium casei*. *Coll. Surf. B Biointerfaces* **2010,** *77* (2), 257–262.
62. Kanipandian, N.; Kannan, S.; Ramesh, R.; Subramanian, P.; Thirumurugan, R. Characterization, Antioxidant and Cytotoxicity Evaluation of Green Synthesized Silver Nanoparticles Using *Cleistanthus collinus* Extract as Surface Modifier. *Mater. Res. Bull.* **2014,** *49*, 494–502.
63. Kanmani, P.; Lim, S. T. Synthesis and Structural Characterization of Silver Nanoparticles Using Bacterial Exopolysaccharide and its Antimicrobial Activity Against Food and Multidrug Resistant Pathogens. *Proc. Biochem.* **2013,** *48*, 1099–1106.
64. Kannan, R. R. R.; Arumugam, R.; Ramya, D.; Manivannan, K.; Anantharaman, P. Green Synthesis of Silver Nanoparticles Using Marine Macroalga *Chaetomorpha linum*. *Appl. Nanosci.* **2013,** *3* (3), 229–233.
65. Khalil, K. A.; Fouad, H.; Elsarnagawy, T.; Almajhdi, F.N. Preparation and Characterization of Electrospun PLGA/Silver Composite Nanofibers for Biomedical Applications. *Int. J. Electrochem. Sci.* **2013,** *8*, 3483–3493.
66. Kharlampieva, E.; Zimnitsky, D.; Gupta, M.; Bergman, K. N.; Kaplan, D. L.; Naik, R. R.; Tsukruk, V. V. Redox-active Ultrathin Template of Silk Fibroin: Effect of Secondary Structure on Gold Nanoparticle Reduction. *Chem. Mater.* **2009,** *21* (13), 2696–2704.
67. Khatami, M.; Pourseyedi, S.; Khatami, M.; Hamidi, H.; Zaeifi, M.; Soltani, L. Synthesis of Silver Nanoparticles Using Seed Exudates of *Sinapis arvensis* as a Novel Bioresource, and Evaluation of Their Antifungal Activity. *Bioresour. Bioprocess.* **2015,** *2* (19), 1–7.
68. Khlebtsov, N.; Bogatyrev, V.; Dykman, L.; Khlebtsov, B.; Staroverov, S.; Shirokov, A.; Matora, L.; Khanadeev, V.; Pylaev, T.; Tsyganova, N.; Terentyuk, G. Analytical and Theranostic Applications of Gold Nanoparticles and Multifunctional Nanocomposites. *Theranostics* **2013,** *3*, 167–180.
69. Kim, H. K.; Choi, M. J.; Cha, S. H.; Koo, Y. K.; Jun, S. H.; Cho, S.; Park, Y. Earthworm Extracts Utilized in the Green Synthesis of Gold Nanoparticles Capable of Reinforcing the Anticoagulant Activities of Heparin. *Nanoscale Res. Lett.* **2013,** *8* (1), 542.
70. Kim, H. S.; Jun, S. H.; Koo, Y. K.; Cho, S.; Park, Y. Green Synthesis and Nanotopography of Heparin-reduced Gold Nanoparticles with Enhanced Anticoagulant Activity. *J. Nanosci. Nanotechnol.* **2013,** *13* (3), 2068–2076.
71. Kravets, V.; Almemar, Z.; Jiang, K.; Culhane, K.; Machado, R.; Hagen, G.; Kotko, A.; Dmytruk, I.; Spendier, K.; Pinchuk, A. Imaging of Biological Cells Using Luminescent Silver Nanoparticles. *Nanoscale Res. Lett.* **2016,** *11*, 1–9.
72. Kumari, M. M.; Jacob, J.; Philip, D. Green Synthesis and Applications of Au–Ag Bimetallic Nanoparticles. *Spectrochim. Acta Part A Mol. Biomol. Spectrosc.* **2015,** *137*, 185–192.
73. Lara, H. H.; Ayala-Núñez, N. V.; Turrent, L. D. C. I.; Padilla, C. R. Bactericidal Effect of Silver Nanoparticles Against Multidrug-resistant Bacteria. *World J. Microbiol. Biotechnol.* **2010,** *26* (4), 615–621.
74. Lateef, A.; Ojo, M. O. Public Health Issues in the Processing of Cassava (*Manihot esculenta*) for the Production of '*lafun*' and the Application of Hazard Analysis Control Measures. *Qual. Assur. Saf. Crops Foods* **2016,** *8*, 165–177.

75. Lateef, A.; Yekeen, T. A. Microbial Attributes of a Pharmaceutical Effluent and its Genotoxicity on *Allium cepa*. *Int. J. Environ. Stud.* **2006,** *63* (5), 534–536.
76. Lateef, A. The Microbiology of a Pharmaceutical Effluent and its Public Health Implications.*World J. Microbiol. Biotechnol.* **2004,** *20*, 167–171.
77. Lateef, A.; Adelere, I. A.; Gueguim-Kana, E. B. *Bacillus safensis* LAU 13: A New Source of Keratinase and its Multi-functional Biocatalytic Applications. *Biotechnol. Biotechnol. Equip.* **2015,** *29*, 54–63.
78. Lateef, A.; Adelere, I. A.; Gueguim-Kana, E. B.; Asafa, T. B.; Beukes, L. S. Green Synthesis of Silver Nanoparticles Using Keratinase Obtained from a Strain of *Bacillus safensis* LAU 13. *Int. Nano Lett.* **2015,** *5*, 29–35.
79. Lateef, A.; Akande, M. A.; Azeez, M. A.; Ojo, S. A.; Folarin, B. I.; Gueguim-Kana, E. B.; Beukes, L. S. Phytosynthesis of Silver Nanoparticles (AgNPs) Using Miracle Fruit Plant (*Synsepalum dulcificum*) for Antimicrobial, Catalytic, Anticoagulant, and Thrombolytic Applications. *Nanotechnol. Rev.* **2016,** *5* (6), 507–520.
80. Lateef, A.; Akande, M. A.; Ojo, S. A.; Folarin, B. I.; Gueguim-Kana, E. B.; Beukes, L. S. Paper Wasp Nest-mediated Biosynthesis of Silver Nanoparticles for Antimicrobial, Catalytic, Anticoagulant, and Thrombolytic Applications. *3 Biotech.* **2016,** *6* (2), 140. DOI: 10.1007/s13205-016-0459-x
81. Lateef, A.; Azeez, M. A.; Asafa, T. B.; Yekeen, T. A.; Akinboro, A.; Oladipo, I. C.; Ajetomobi, F. E.; Gueguim-Kana, E. B.; Beukes, L. S. *Cola nitida*-mediated Biogenic Synthesis of Silver Nanoparticles Using Seed and Seed Shell Extracts and Evaluation of Antibacterial Activities. *BioNanoSci.* **2015,** *5* (4), 196–205.
82. Lateef, A.; Azeez, M. A.; Asafa, T. B.; Yekeen, T. A.; Akinboro, A.; Oladipo, I. C.; Azeez, L.; Ojo, S. A.; Gueguim-Kana, E. B.; Beukes, L. S. Cocoa Pod Husk Extract-mediated Biosynthesis of Silver Nanoparticles: Its Antimicrobial, Antioxidant and Larvicidal Activities. *J. Nanostruct. Chem.* **2016,** *6* (2), 159–169.
83. Lateef, A.; Azeez, M. A.; Asafa, T. B.; Yekeen, T. A.; Akinboro, A.; Oladipo, I. C.; Azeez, L.; Ajibade, S. E.; Ojo, S. A.; Gueguim-Kana, E. B.; Beukes, L. S. Biogenic Synthesis of Silver Nanoparticles Using a Pod Extract of *Cola nitida*: Antibacterial, Antioxidant Activities and Application as a Paint Additive. *J. Taibah Univ. Sci.* **2016,** *10*, 551–562.
84. Lateef, A.; Davies, T. E.; Adelekan, A.; Adelere, I. A.; Adedeji, A. A.; Fadahunsi, A. H. Akara Ogbomoso: Microbiological Examination and Identification of Hazards and Critical Control Points. *Food Sci. Technol. Int.* **2010,** *16*, 389–400.
85. Lateef, A.; Ojo, S. A.; Akinwale, A. S.; Azeez, L.; Gueguim-Kana, E. B.; Beukes, L. S. Biogenic Synthesis of Silver Nanoparticles Using Cell-free Extract of *Bacillus safensis* LAU 13: Antimicrobial, Free Radical Scavenging and Larvicidal Activities. *Biologia* **2015,** *70* (10), 1295–1306.
86. Lateef, A.; Ojo, S. A.; Azeez, M. A.; Asafa, T. B.; Yekeen, T. A.; Akinboro, A.; Oladipo, I. C.; Gueguim-Kana, E. B.; Beukes, L. S. Cobweb as Novel Biomaterial for the Green and Eco-friendly Synthesis of Silver Nanoparticles. *Appl. Nanosci.* **2016,** *6* (6), 863–874.
87. Lateef, A.; Ojo, S. A.; Elegbede, J. A. The Emerging Roles of Arthropods and Their Metabolites in the Green Synthesis of Metallic Nanoparticles. *Nanotechnol. Rev.* **2016,** *5* (6), 601–622.
88. Lateef, A.; Ojo, S. A.; Elegbede, J. A.; Azeez, M. A.; Yekeen, T. A.; Akinboro, A. Evaluation of Some Biosynthesized Silver Nanoparticles for Biomedical Applications:

Hydrogen Peroxide Scavenging, Anticoagulant and Thrombolytic Activities. *J. Clust. Sci.* **2017,** *28* (3), 1379–1392.

89. Lateef, A.; Ojo, S. A.; Folarin, B. I.; Gueguim-Kana, E. B.; Beukes, L. S. Kolanut (*Cola nitida*) Mediated Synthesis of Silver–Gold Alloy Nanoparticles: Antifungal, Catalytic, Larvicidal and Thrombolytic Applications. *J. Clust. Sci.* **2016,** *27* (5), 1561–1577.
90. Lateef, A.; Ojo, S. A.; Oladejo, S. M. Anti-candida, Anti-coagulant and Thrombolytic Activities of Biosynthesized Silver Nanoparticles Using Cell-free Extract of *Bacillus safensis* LAU 13. *ProcessBiochem.* **2016,** *51*, 1406–1412.
91. Lateef, A.; Oloke, J. K.; Gueguim-Kana, E. B. Antimicrobial Resistance of Bacterial Strains Isolated from Orange Juice Products. *Afr. J. Biotechnol.* **2004,** *3*, 334–338.
92. Lateef, A.; Oloke, J. K.; Gueguim-Kana, E. B. The Prevalence of Bacterial Resistance in Clinical, Food, Water and Some Environmental samples in Southwest Nigeria. *Environ. Monitor. Assess.* **2005,** *100*, 59–69.
93. Lateef, A.; Oloke, J. K.; Gueguim-Kana, E. B.; Pacheco, E. The Microbiological Quality of Ice Used to Cool Drinks and Foods in Ogbomoso Metropolis, Southwest, Nigeria. *Inter. J. Food Saf.* **2006,** *8*, 39–43.
94. Lateef, A.; Yekeen, T. A.; Ufuoma, P. E. Bacteriology and Genotoxicity of Some Pharmaceutical Wastewaters in Nigeria. *Int. J. Environ. Health* **2007,** *1* (4), 551–562.
95. Lateef, A.; Adelere, I. A.; Gueguim-Kana, E. B. The Biology and Potential Biotechnological Applications of *Bacillus safensis*. *Biologia* **2015,** *70*, 411–419.
96. Lateef. A.; Adeeyo, A. O. Green synthesis and Antibacterial Activities of Silver Nanoparticles Using Extracellular Laccase of *Lentinus edodes. Notulae Scientia Biologicae* **2015,** *7* (4), 405–411.
97. Lee, K. D.; Nagajyothi, P. C.; Sreekanth, T. V. M.; Park, S. Eco-friendly Synthesis of Gold Nanoparticles (AuNPs) Using *Inonotus obliquus* and Their Antibacterial, Antioxidant and Cytotoxic Activities. *J. Indus. Eng. Chem.* **2015,** *26,* 67–72.
98. Levi, M.; Schultz, M.; van der Poll, T. Disseminated Intravascular Coagulation in Infectious Disease. *Sem. Thromb. Hemost.* **2010,** *36* (4), 367–377.
99. Liu, Q.; Liu, H.; Yuan, Z.; Wei, D.; Ye, Y. Evaluation of Antioxidant Activity of Chrysanthemum Extracts and Tea Beverages by Gold Nanoparticles-based Assay. *Coll. Surf. B Biointerfaces* **2012,** *92*, 348–352.
100. Majdalawieh, A.; Kanan, M. C.; El-Kadri, O.; Kanan, S. M. Recent Advances in Gold and Silver Nanoparticles: Synthesis and Applications. *J. Nanosci. Nanotechnol.* **2014,** *14* (7), 4757–4780.
101. Malathi, S.; Ezhilarasu, T.; Abiraman, T.; Balasubramanian, S. One Pot Green Synthesis of Ag, Au and Au–Ag Alloy Nanoparticles Using Isonicotinic Acid Hydrazide and Starch. *Carbohydr. Polym.* **2014,** *111*, 734–743.
102. Manivasagan, P.; Alam, M. S.; Kang, K. H.; Kwak, M.; Kim, S. K. Extracellular Synthesis of Gold Bionanoparticles by *Nocardiopsis* sp. and Evaluation of its Antimicrobial, Antioxidant, and Cytotoxic Activities. *Bioproc. Biosyst. Eng.* **2015,** *38* (6), 1167–1177.
103. Marimuthu, S.; Rahuman, A. A.; Rajakumar, G.; Santhoshkumar, T.; Kirthi, A. V.; Jayaseelan, C.; Bagavan, A.; Zahir, A. A.' Elango, G.; Kamaraj, C. Evaluation of Green Synthesized Silver Nanoparticles Against Parasites. *Parasitol. Res.* **2011,** *108* (6), 1541–1549.
104. Markus, J.; Mathiyalagan, R.; Kim, Y. J.; Abbai, R.; Singh, P.; Ahn, S.; Perez, Z. E. J.; Hurh, J.; Yang, D. C. Intracellular Synthesis of Gold Nanoparticles with Antioxidant

Activity by Probiotic *Lactobacillus kimchicus* DCY51 T Isolated from Korean kimchi. *Enzyme Microb. Technol.* **2016,** *95*, 85–93.

105. Mata, R.; Nakkala, J. R.; Sadras, S. R. Catalytic and Biological Activities of Green Silver Nanoparticles Synthesized from *Plumeria alba* (frangipani) Flower Extract. *Mater. Sci. Eng. C* **2015,** *51*, 216–225.
106. Mehlhorn, H.; Al-Rasheid, K. A.; Al-Quraishy, S.; Abdel-Ghaffar, F. Research and Increase of Expertise in Arachno-entomology are Urgently Needed. *Parasitol. Res.* **2012,** *110* (1), 259–265.
107. Meyer, G.; Vicaut, E.; Danays, T.; Agnelli, G.; Becattini, C.; Beyer-Westendorf, J.; Bluhmki, E.; Bouvaist, H.; Brenner, B.; Couturaud, F.; Dellas, C. Fibrinolysis for Patients with Intermediate-risk Pulmonary Embolism. *New Eng. J. Med.* **2014,** *370* (15), 1402–1411.
108. Millán, M.; Dorado, L.; Dávalos, A. Fibrinolytic Therapy in Acute Stroke. *Curr. Cardiol. Rev.* **2010,** *6* (3), 218–226.
109. Mondal, S.; Roy, N.; Laskar, R. A.; Sk, I.; Basu, S.; Mandal, D.; Begum, N. A. Biogenic Synthesis of Ag, Au and Bimetallic Au/Ag Alloy Nanoparticles Using Aqueous Extract of Mahogany (*Swietenia mahogani* JACQ.) Leaves. *Coll. Surf. B Biointerfaces* **2011,** *82*, 497–504.
110. Moteriya, P.; Chanda, S. Synthesis and Characterization of Silver Nanoparticles Using *Caesalpinia pulcherrima* Flower Extract and Assessment of Their *In vitro* Antimicrobial, Antioxidant, Cytotoxic, and Genotoxic Activities. *Artif. Cells Nanomed. Biotechnol.* **2016,** *30*, 1–12.
111. MubarakAli, D.; Thajuddin, N.; Jeganathan, K.; Gunasekaran, M. Plant Extract Mediated Synthesis of Silver and Gold Nanoparticles and its Antibacterial Activity Against Clinically Isolated Pathogens. *Coll. Surf. B Biointerfaces* **2011,** *85* (2), 360–365.
112. Mukherjee, P. *Stenotrophomonas* and *Microbacterium*: Mediated Biogenesis of Copper, Silver and Iron Nanoparticles-Proteomic Insights and Antibacterial Properties Versus Biofilm Formation. *J. Clust. Sci.* **2017,** *28* (1), 331–358.
113. Murugan, K.; Benelli, G.; Panneerselvam, C.; Subramaniam, J.; Jeyalalitha, T.; Dinesh, D.; Nicoletti, M.; Hwang, J. S.; Suresh, U.; Madhiyazhagan, P. *Cymbopogon citratus*-Synthesized Gold Nanoparticles Boost the Predation Efficiency of Copepod *Mesocyclops aspericornis* Against Malaria and Dengue Mosquitoes. *Exp. Parasitol.* **2015,** *153*, 129–138.
114. Murugan, K.; Dinesh, D.; Paulpandi, M.; Althbyani, A. D. M.; Subramaniam, J.; Madhiyazhagan, P.; Wang, L.; Suresh, U.; Kumar, P. M.; Mohan, J.; Rajaganesh, R. Nanoparticles in the Fight Against Mosquito-borne Diseases: Bioactivity of *Bruguiera cylindrica*-synthesized Nanoparticles Against Dengue Virus DEN-2 (in vitro) and its Mosquito Vector *Aedes aegypti* (Diptera: Culicidae). *Parasitol. Res.* **2015,** *114* (12), 4349–4361.
115. Murugan, K.; Panneerselvam, C.; Samidoss, C. M.; Madhiyazhagan, P.; Suresh, U.; Roni, M.; Chandramohan, B.; Subramaniam, J.; Dinesh, D.; Rajaganesh, R.; Paulpandi, M. *In vivo* and *in vitro* Effectiveness of *Azadirachta indica*-synthesized Silver Nanocrystals Against *Plasmodium berghei* and *Plasmodium falciparum*, and Their Potential Against Malaria Mosquitoes. *Res. Vet. Sci.* **2016,** *106*, 14–22.
116. Murugan, K.; Panneerselvam, C.; Subramaniam, J.; Madhiyazhagan, P.; Hwang, J. S.; Wang, L.; Dinesh, D.; Udaiyan Suresh, U.; Roni, M.; Higuchi, A.; Nicoletti, M.; Benelli, G. Eco-friendly Drugs From the Marine Environment: Spongeweed-synthesized Silver

Nanoparticles are Highly Effective on *Plasmodium falciparum* and its Vector *Anopheles stephensi*, with Little Non-target Effects on Predatory Copepods. *Environ. Sci. Pollut. Res.*, **2016,** *23* (16), 16671–16685.
117. Muthukumar, T.; Sambandam, B.; Aravinthan, A.; Sastry, T. P.; Kim, J. H. Green Synthesis of Gold Nanoparticles and Their Enhanced Synergistic Antitumor Activity Using HepG2 and MCF7 Cells and its Antibacterial Effects. *Proc. Biochem.* **2016,** *51* (3), 384–391.
118. Nakkala, J. R.; Mata, R.; Bhagat, E.; Sadras, S. R. Green Synthesis of Silver and Gold Nanoparticles from *Gymnema sylvestre* Leaf Extract: Study of Antioxidant and Anticancer Activities. *J. Nanopart. Res.* **2015,** *17*, 151.
119. Nakkala, J. R.; Bhagat, E.; Suchiang, K.; Sadras, S. R. Comparative Study of Antioxidant and Catalytic Activity of Silver and Gold Nanoparticles Synthesized from *Costus pictus* Leaf Extract. *J. Mater. Sci. Technol.* **2015,** *31*, 986–994.
120. Nasrollahzadeh, M.; Sajadi, S. M.; Babaei, F.; Maham, M. *Euphorbia helioscopia* Linn as a Green Source for Synthesis of Silver Nanoparticles and Their Optical and Catalytic Properties. *J. Coll. Interface Sci.* **2015,** *450*, 374–380.
121. Naveena, B. E.; Prakash, S. Biological Synthesis of Gold Nanoparticles Using Marine Algae *Gracilaria corticata* and its Application as a Potent Antimicrobial and Antioxidant Agent. *Asian J. Pharm. Clin. Res.* **2013,** *6* (2), 179–182.
122. Nayak, D.; Ashe, S.; Rauta, P. R.; Kumari, M.; Nayak, B. Bark Extract Mediated Green Synthesis of Silver Nanoparticles: Evaluation of Antimicrobial Activity and Antiproliferative Response Against Osteosarcoma. *Mater. Sci. Eng. C* **2016,** *58*, 44–52.
123. Niraimathi, K. L.; Sudha, V.; Lavanya, R.; Brindha, P. Biosynthesis of Silver Nanoparticles Using *Alternanthera sessilis* (Linn.) Extract and Their Antimicrobial, Antioxidant Activities. *Coll. Surf. B Biointerfaces* **2013,** *102*, 288–291.
124. Ojo, S. A.; Lateef, A.; Azeez, M. A.; Oladejo, S. M.; Akinwale, A. S.; Asafa, T. B.; Yekeen, T. A.; Akinboro, A.; Oladipo, I. C.; Gueguim-Kana, E. B.; Beukes, L. S. Biomedical and Catalytic Applications of Gold and Silver–Gold Alloy Nanoparticles Biosynthesized Using Cell-free Extract of *Bacillus safensis* LAU 13: Antifungal, Dye Degradation, Anti-coagulant and Thrombolytic Activities. *IEEE Trans. Nanobiosci.* **2016,** *15* (5), 433–442.
125. Olajire, A. A.; Abidemi, J. J.; Lateef, A.; Benson, N. U. Adsorptive Desulfurization of Model Oil by Ag Nanoparticles-modified Activated Carbon Prepared From Brewer's Spent Grains. *J. Environ. Chem. Eng.* **2017,** *5* (1), 147–159.
126. Omer, N.; Rohilla, A.; Rohilla, S.; Kushnoor, A. Nitric Oxide: Role in Human Biology. *Int. J. Pharm. Sci. Drug Res.* **2012,** *4* (2), 105–109.
127. Panneerselvam, C.; Murugan, K.; Roni, M.; Suresh, U.; Rajaganesh, R.; Madhiyazhagan, P.; Subramaniam, J.; Dinesh, D.; Nicoletti, M.; Higuchi, A.; Alarfaj, A.A. Fern-synthesized Nanoparticles in the Fight Against Malaria: LC/MS Analysis of *Pteridium aquilinum* Leaf Extract and Biosynthesis of Silver Nanoparticles with High Mosquitocidal and Antiplasmodial Activity. *Parasitol. Res.* **2016,** *115* (3), 997–1013.
128. Park, T. J.; Lee, K. G.; Lee, S. Y. Advances in Microbial Biosynthesis of Metal Nanoparticles. *Appl. Microbiol. Biotechnol.* **2016,** *100*, 521-534.
129. Parveen, M.; Ahmad, F.; Malla, A. M.; Azaz, S. Microwave-assisted Green Synthesis of Silver Nanoparticles from *Fraxinus excelsior* Leaf Extract and its Antioxidant Assay. *Appl. Nanosci.* **2016,** *6* (2), 267–276.

130. Patra, J. K.; Baek, K. H. Novel Green Synthesis of Gold Nanoparticles Using *Citrullus lanatus* rind and Investigation of Proteasome Inhibitory Activity, Antibacterial, and Antioxidant Potential. *Int. J. Nanomed.* **2014,** *10*, 7253–7264.
131. Paul, B.; Bhuyan, B.; Purkayastha, D. D.; Dhar, S. S. Green Synthesis of Silver Nanoparticles Using Dried Biomass of *Diplazium esculentum* (retz.) sw. and Studies of Their Photocatalytic and Anticoagulative Activities. *J. Mol. Liq.* **2015,** *212*, 813–817.
132. Paul, B.; Bhuyan, B.; Purkayastha, D. D.; Vadivel, S.; Dhar, S. S. One-pot Green Synthesis of Gold Nanoparticles and Studies of Their Anticoagulative and Photocatalytic Activities. *Mater. Lett.* **2016,** *185*, 143–147.
133. Philip, D. *Mangifera indica* Leaf-assisted Biosynthesis of Well-Dispersed Silver Nanoparticles. *Spectrochim. Acta Part A Mol. Biomol. Spectrosc.* **2011,** *78* (1), 327–331.
134. Phull, A. R.; Abbas, Q.; Ali, A.; Raza, H.; Zia, M.; Haq, I. U. Antioxidant, Cytotoxic and Antimicrobial Activities of Green Synthesized Silver Nanoparticles From Crude Extract of *Bergenia ciliata. Future J. Pharm. Sci.* **2016,** *2* (1), 31–36.
135. Prandoni, P.; Falanga, A.; Piccioli, A. Cancer, Thrombosis and Heparin-induced Thrombocytopenia. *Thromb. Res.* **2007,** *120*, S137–S140.
136. Prasannaraj, G.; Venkatachalam, P. Enhanced Antibacterial, Anti-biofilm and Antioxidant (ROS) Activities of Biomolecules Engineered Silver Nanoparticles Against Clinically Isolated Gram Positive and Gram Negative Microbial Pathogens. *J. Clust. Sci.* **2017,** *28* (1), 645–664.
137. Priyadarshini, K. A.; Murugan, K.; Panneerselvam, C.; Ponarulselvam, S.; Hwang, J. S.; Nicoletti, M. Biolarvicidal and Pupicidal Potential of Silver Nanoparticles Synthesized Using *Euphorbia hirta* against *Anopheles stephensi* Liston (Diptera: Culicidae). *Parasitol. Res.* **2012,** *111* (3), 997–1006.
138. Qayyum, S.; Khan, A. U. Biofabrication of Broad Range Antibacterial and Antibiofilm Silver Nanoparticles. *IET Nanobiotechnol.* **2016,** *10* (5), 349–357.
139. Rahman, L.; Shah, A.; Khan, S. B.; Asiri, A. M.; Hussain, H.; Han, C.; Qureshi, R.; Ashiq, M. N.; Zia, M. A.; Ishaq, M.; Kraatz, H. B. Synthesis, Characterization, and Application of Au–Ag Alloy Nanoparticles for the Sensing of an Environmental Toxin, Pyrene. *J. Appl. Electrochem.* **2015,** *45*, 463–472.
140. Rai A.; Prabhune, A.; Perry, C. Antibiotic Mediated Synthesis of Gold Nanoparticles with Potent Antimicrobial Activity and Their Application in Antimicrobial Coatings. *J. Mater. Chem.* **2010,** *20*, 6789–6798.
141. Raja, S.; Ramesh, V.; Thivaharan, V. Antibacterial and Anticoagulant Activity of Silver Nanoparticles Synthesised from a Novel Source–pods of *Peltophorum pterocarpum. J. Indus. Eng. Chem.* **2015,** *29*, 257–264.
142. Rajakumar, G.; Rahuman, A. A. Larvicidal Activity of Synthesized Silver Nanoparticles Using *Eclipta prostrata* Leaf Extract Against Filariasis and Malaria Vectors. *Acta Trop.* **2011,** *118* (3), 196–203.
143. Rajan, A.; Rajan, A. R.; Philip, D. *Elettaria cardamomum* Seed Mediated Rapid Synthesis of Gold Nanoparticles and its Biological Activities. *OpenNano* **2017,** *2*, 1–8.
144. Raman, N.; Sudharsan, S.; Veerakumar, V.; Pravin, N.; Vithiya, K. *Pithecellobium dulce* Mediated Extra-cellular Green Synthesis of Larvicidal Silver Nanoparticles. *Spectrochim. Acta Part A Mol. Biomol. Spectros.* **2012,** *96,* 1031–1037.
145. Ramanibai, R.; Velayutham, K. Bioactive Compound Synthesis of Ag Nanoparticles from Leaves of *Melia azedarach* and its Control for Mosquito Larvae. *Res. Vet. Sci.* **2015,** *98*, 82–88.

146. Ramkumar, V. S.; Pugazhendhi, A.; Gopalakrishnan, K.; Sivagurunathan, P.; Saratale, G. D.; Dung, T. N. B.; Kannapiran, E. Biofabrication and Characterization of Silver Nanoparticles Using Aqueous Extract of Seaweed *Enteromorpha compressa* and its Biomedical Properties. *Biotechnol. Rep.* **2017,** *14*, 1–7.
147. Rath, G.; Hussain, T.; Chauhan, G.; Garg, T.; Goyal, A. K. Collagen Nanofiber Containing Silver Nanoparticles for Improved Wound-healing Applications. *J. Drug Target.* **2016,** *24*, 520–529.
148. Ravichandran, V.; Vasanthi, S.; Shalini, S.; Shah, S. A. A.; Harish, R. Green Synthesis of Silver Nanoparticles Using *Atrocarpus altilis* Leaf Extract and the Study of Their Antimicrobial and Antioxidant Activity. *Mater. Lett.* **2016,** *180*, 264–267.
149. Ravikumar, Y. S.; Mahadevan, K. M.; Kumaraswamy, M. N.; Vaidya, V. P.; Manjunatha, H.; Kumar, V.; Satyanarayana, N. D. Antioxidant, Cytotoxic and Genotoxic Evaluation of Alcoholic Extract of *Polyalthia cerasoides* (Roxb.) Bedd. *Environ. Toxicol. Pharmacol.* **2008,** *26* (2), 142–146.
150. Rawani, A.; Ghosh, A.; Chandra, G. Mosquito Larvicidal and Antimicrobial Activity of Synthesized Nano-crystalline Silver Particles Using Leaves and Green Berry Extract of *Solanum nigrum* L.(Solanaceae: Solanales). *Acta Trop.* **2013,** *128* (3), 613–622.
151. Reddy, N. J.; Vali, D. N.; Rani, M.; Rani, S. S. Evaluation of Antioxidant, Antibacterial and Cytotoxic Effects of Green Synthesized Silver Nanoparticles by *Piper longum* Fruit. *Mater. Sci. Eng. C* **2014,** *34,* 115–122.
152. Ribeiro, A. B.; Chisté, R. C.; Freitas, M.; da Silva, A. F.; Visentainer, J. V.; Fernandes, E. *Psidium cattleianum* Fruit Extracts are Efficient *In vitro* Scavengers of Physiologically Relevant Reactive Oxygen and Nitrogen Species. *Food Chem.* **2014,** *165*, 140–148.
153. Roni, M.; Murugan, K.; Panneerselvam, C.; Subramaniam, J.; Hwang, J. S. Evaluation of Leaf Aqueous Extract and Synthesized Silver Nanoparticles Using *Nerium oleander* Against *Anopheles stephensi* (Diptera: Culicidae). *Parasitol. Res.* **2013,** *112* (3), 981–990.
154. Roopan, S. M.; Madhumitha, G.; Rahuman, A. A.; Kamaraj, C.; Bharathi, A.; Surendra, T. V. Low-cost and Eco-friendly Phyto-synthesis of Silver Nanoparticles Using *Cocos nucifera* coir Extract and its Larvicidal Activity. *Ind. Crops Prod.* **2013,** *43*, 631–635.
155. Salunke, G. R.; Ghosh, S.; Kumar, R. S.; Khade, S.; Vashisth, P.; Kale, T.; Chopade, S.; Pruthi, V.; Kundu. G.; Bellare, J.R.; Chopade, B. A. Rapid Efficient Synthesis and Characterization of Silver, Gold, and Bimetallic Nanoparticles From the Medicinal Plant *Plumbago zeylanica* and Their Application in Biofilm Control. *Int. J. Nanomed.* **2014,** *9*, 2635.
156. Sathishkumar, G.; Jha, P. K.; Vignesh, V.; Rajkuberan, C.; Jeyaraj, M.; Selvakumar, M.; Jha, R.; Sivaramakrishnan, S. Cannonball Fruit (*Couroupita guianensis*, Aubl.) Extract Mediated Synthesis of Gold Nanoparticles and Evaluation of its Antioxidant Activity. *J. Mol. Liquids* **2016,** *215*, 229–236.
157. Savan, E. K.; Kucukbay, F. Z. Essential Oil Composition of *Elettaria cardamomum* Maton. *J. Appl. Biol. Sci.* **2013,** *7* (3), 42–44.
158. Shankar, S.; Jaiswal, L.; Aparna, R. S. L.; Prasad, R. G. S. V. Synthesis, Characterization, *In vitro* Biocompatibility, and Antimicrobial Activity of Gold, Silver and Gold–Silver Alloy Nanoparticles Prepared From *Lansium domesticum* Fruit Peel Extract. *Mater. Lett.* **2014,** *137*, 75–78.
159. Shanmugam, C.; Sivasubramanian, G.; Parthasarathi, B.; Baskaran, K.; Balachander, R.; Parameswaran, V. R. Antimicrobial, Free Radical Scavenging Activities and Catalytic

Oxidation of Benzyl Alcohol by Nano-Silver Synthesized from the Leaf Extract of *Aristolochia indica*. *Appl. Nanosci.* **2016,** *6* (5), 711–723.

160. Shanmugasundaram, T.; Radhakrishnan, M.; Gopikrishnan, V.; Pazhanimurugan, R.; Balagurunathan, R. A Study of the Bactericidal, Anti-biofouling, Cytotoxic and Antioxidant Properties of Actinobacterially Synthesised Silver Nanoparticles. *Coll. Surf. B Biointerfaces* **2013,** *111*, 680–687.
161. Shedbalkar, U.; Singh, R.; Wadhwani, S.; Gaidhani, S.; Chopade, B. A. Microbial Synthesis of Gold Nanoparticles: Current Status and Future Prospects. *Adv. Coll. Interface Sci.* **2014,** *209*, 40–48.
162. Sheny, D. S.; Mathew, J.; Philip, D. Phytosynthesis of Au, Ag and Au–Ag Bimetallic Nanoparticles Using Aqueous Extract and Dried Leaf of *Anacardium occidentale*. *Spectrochim. Acta Part A Mol. Biomol. Spectros.* **2011,** *79*, 254–262.
163. Shiang, Y. C.; Hsu, C. L.; Huang, C. C.; Chang, H. T. Gold Nanoparticles Presenting Hybridized Self-Assembled Aptamers That Exhibit Enhanced Inhibition of Thrombin. *Angewandte Chemie* **2011,** *123*, 7802–7807.
164. Shrivastava, S.; Bera, T.; Singh, S.K.; Singh, G.; Ramachandrarao, P.; Dash, D. Characterization of Antiplatelet Properties Of Silver Nanoparticles. *ACS Nano* **2009,** *3*, 1357–1364.
165. Singh, P.; Kim, Y. J.; Wang, C.; Mathiyalagan, R.; El-Agamy Farh, M.; Yang, D. C. Biogenic Silver and Gold Nanoparticles Synthesized Using Red Ginseng Root Extract, and Their Applications. *Artif. Cells Nanomed. Biotechnol.* **2016,** *44* (3), 811–816.
166. Singh, P.; Kim, Y. J.; Wang, C.; Mathiyalagan, R.; Yang, D. C. The Development of a Green Approach for the Biosynthesis of Silver and Gold Nanoparticles by Using *Panax ginseng* Root Extract, and Their Biological Applications. *Artif. Cells Nanomed. Biotechnol.* **2016,** *44* (4), 1150–1157.
167. Singh, P.; Kim, Y. J.; Yang, D. C. A Strategic Approach for Rapid Synthesis of Gold and Silver Nanoparticles by *Panax ginseng* Leaves. *Artif. Cells Nanomed. Biotechnol.* **2016,** *44* (8), 1949–1957.
168. Singh, P.; Kim, Y. J.; Zhang, D.; Yang, D. C. Biological Synthesis of Nanoparticles from Plants and Microorganisms. *Trends Biotechnol.* **2016,** *34*, 588–599.
169. Singh, P.; Singh, H.; Kim, Y. J.; Mathiyalagan, R.; Wang, C.; Yang, D. C. Extracellular Synthesis of Silver and Gold Nanoparticles by *Sporosarcina koreensis* DC4 and Their Biological Applications. *Enzyme Microb. Technol.* **2016,** *86,* 75–83.
170. Singh, R.; Nalwa, H. S. Medical Applications of Nanoparticles in Biological Imaging, Cell Labeling, Antimicrobial Agents, and Anticancer Nanodrugs. *J. Biomed. Nanotechnol.* **2011,** *7* (4), 489–503.
171. Sosa, I. O.; Noguez, C.; Barrera, R. G. Optical Properties of Metal Nanoparticles with Arbitrary Shapes. *J. Phys. Chem. B* **2003,** *107* (26), 6269–6275.
172. Sriranjani, R.; Srinithya, B.; Vellingiri, V.; Brindha, P.; Anthony, S. P.; Sivasubramanian, A.; Muthuraman, M. S. Silver Nanoparticle Synthesis Using *Clerodendrumphlomidis* Leaf Extract and Preliminary Investigation of its Antioxidant and Anticancer Activities. *J. Mol. Liq.* **2016,** *220*, 926–930.
173. Subramaniam, J.; Murugan, K.; Panneerselvam, C.; Kovendan, K.; Madhiyazhagan, P.; Dinesh, D.; Kumar, P. M.; Chandramohan, B.; Suresh, U.; Rajaganesh, R.; Alsalhi, M. S. Multipurpose Effectiveness of *Couroupita guianensis*-synthesized Gold Nanoparticles: High Antiplasmodial Potential, Field Efficacy Against Malaria Vectors and Synergy with *Aplocheilus lineatus* Predators. *Environ. Sci. Pollut. Res.* **2016,** *23* (8), 7543–7558.

174. Sun, L.; Zhang, J.; Lu, X.; Zhang, L.; Zhang, Y. Evaluation to the Antioxidant Activity of Total Flavonoids Extract From Persimmon (*Diospyros kaki* L.) Leaves. *Food Chem. Toxicol.* **2011,** *49* (10), 2689–2696.
175. Suresh, U.; Murugan, K.; Benelli, G.; Nicoletti, M.; Barnard, D. R.; Panneerselvam, C.; Kumar, P. M.; Subramaniam, J.; Dinesh, D.; Chandramohan, B. Tackling the Growing Threat of Dengue: *Phyllanthus niruri*-mediated Synthesis of Silver Nanoparticles and Their Mosquitocidal Properties Against the Dengue Vector *Aedes aegypti* (Diptera: Culicidae). *Parasitol. Res.* **2015,** *114* (4), 1551–1562.
176. Swain, S.; Barik, S. K.; Behera, T.; Nayak, S. K.; Sahoo, S. K.; Mishra, S. S.; Swain, P. Green synthesis of Gold Nanoparticles Using Root and Leaf Extracts of *Vetiveria zizanioides* and *Cannabis sativa* and its Antifungal Activities. *BioNanoScie.* **2016,** *6* (3), 205–213.
177. Tahir, K.; Nazir, S.; Li, B.; Khan, A. U.; Khan, Z. U. H.; Gong, P. Y.; Khan, S. U.; Ahmad, A. *Nerium oleander* Leaves Extract Mediated Synthesis of Gold Nanoparticles and its Antioxidant Activity. *Mater. Lett.* **2015,** *156*, 198–201.
178. Thirumurugan, A.; Tomy, N. A.; Kumar, H. P.; Prakash, P. Biological Synthesis of Silver Nanoparticles by *Lantana camara* Leaf Extracts. *Int. J. Nanomater. Biostruct.* **2011,** *1* (2), 22–24.
179. Thunugunta, T.; Reddy, A. C.; Reddy, D. C. Green Synthesis of Nanoparticles: Current Prospectus. *Nanotechnol. Rev.* **2015,** *4*, 303–323.
180. Velu, K.; Elumalai, D.; Hemalatha, P.; Janaki, A.; Babu, M.; Hemavathi, M.; Kaleena, P. K. Evaluation of Silver Nanoparticles Toxicity of *Arachis hypogaea* Peel Extracts and its Larvicidal Activity Against Malaria and Dengue Vectors. *Environ. Sci. Pollut. Res.* **2015,** *22* (22), 17769–17779.
181. Vizuete, K. S.; Kumar, B.; Guzmán, K.; Debut, A.; Cumbal, L. Shora (*Capparis petiolaris*) Fruit Mediated Green Synthesis and Application of Silver Nanoparticles. *Green Process. Synth.* **2016,** *6* (1), 23–30.
182. Wani, I. A.; Ahmad, T. Size and Shape Dependant Antifungal Activity of Gold Nanoparticles: A Case Study of *Candida*. *Coll. Surf. B Biointerfaces* **2013,** *101*, 162–170.
183. Yallappa, S.; Manjanna, J.; Dhananjaya, B. L. Phytosynthesis of Stable Au, Ag and Au–Ag Alloy Nanoparticles Using *J. sambac* Leaves Extract and Their Enhanced Antimicrobial Activity in Presence of Organic Antimicrobials. *Spectrochim. Acta Part A Mol. Biomol. Spectrosc.* **2015,** *137*, 236–243.
184. Zaki, S.; El Kady, M. F.; Abd-El-Haleem, D. Biosynthesis and Structural Characterization of Silver Nanoparticles from Bacterial Isolates. *Mater. Res. Bull.* **2011,** *46* (10), 1571–1576.
185. Zheng, B.; Qian, L.; Yuan, H.; Xiao, D.; Yang, X.; Paau, M. C.; Choi, M. M. Preparation of Gold Nanoparticles on Eggshell Membrane and Their Biosensing Application. *Talanta* **2010,** *82*, 177–183.
186. Zielińska-Górska, M. K.; Sawosz, E.; Górski, K.; Chwalibog, A. Does Nanobiotechnology Create New Tools to Combat Microorganisms? *Nanotechnol. Rev.* **2016,** *6* (2), 171–189.

CHAPTER 2

MICROEMULSIONS: PRINCIPLES, SCOPE, METHODS, AND APPLICATIONS IN TRANSDERMAL DRUG DELIVERY

IRINA PEREIRA, SARA ANTUNES, ANA C. SANTOS, FRANCISCO J. VEIGA, AMÉLIA M. SILVA, PRAPAPORN BOONME, and ELIANA B. SOUTO*

*Corresponding author. E-mail: ebsouto@ebsouto.pt

ABSTRACT

A large variety of oils, surfactants, cosurfactants, and aqueous phases are available to obtain microemulsions. The percentage (or ratio) of the different components, as well as the interaction between them, should be insightfully controlled due to their capacity to induce alterations in microemulsion properties, namely structure, electric conductivity, viscosity, and drug solubilization. Microemulsions are endowed of a small droplet size and the capacity of solubilizing lipophilic and hydrophilic drugs, allowing for a superior drug accumulation in the active site. These drug delivery systems act as penetration enhancers, due to the presence of surfactants and oily components, increasing the transdermal absorption of the drug. The addition of a cosurfactant to the surfactant mixture enhances this effect synergistically. Nevertheless, some of the components in microemulsions, particularly, surfactants, cosurfactants, and oily components, may be responsible for the occurrence of skin irritation reactions and comedogenic effects, although these adverse effects can be minimized. Currently, the search worldwide for new drug delivery systems is one of the greatest scientific purposes. Current technological developments enable the improvement and the production of new and advanced analytical devices. Improvements in the analytical devices allow consequently for a more efficient and reliable characterization of microemulsions. Furthermore, to maximize clinical and cosmetic potentials of microemulsions, not only the characterization of the formulations but also the safety concerns must be addressed during the optimization process.

2.1 INTRODUCTION

Skin is an accessible organ that possesses a large surface area, which renders it a focus of much research and product development for drug administration. In the last few years, transdermal drug delivery has aroused considerable interest among the scientific community due to benefits compared to oral or intravenous routes. This route downsizes the gastrointestinal and systemic toxicity and adverse effects (pH changes, enzymatic deactivation, and gastric retention), prevents the first-pass metabolism and allows for an improved control of drug's blood concentrations. However, skin impermeability to the passage of exogenous substances limits significantly the number of drugs that are able to exert systemic effects and wield therapeutic efficacy at a sufficiently high rate. This route of administration is also not feasible for drugs that irritate or sensitize the skin.[53,94] Indeed, skin constitutes a physical barrier between the human organism and the external environment, conferring protection against excessive water loss and blocking the passage of infectious agents into the body, such as bacteria and viruses, or injurious substances.[81]

Stratum corneum (SC) is the epidermis outermost *strata* and contributes primarily to the barrier function of the skin. This *stratum* is characterized by a thin layer of squamous cells organized in a dense configuration with specialized lipids in the intercellular spaces, and with an acid pH character. The SC is constituted by layers of flattened cells designated by corneocytes. The life cycle of these cells starts as keratinocytes in the basal layer of epidermis, followed by their displacement along time through the numerous layers of epidermis until reaching the skin surface. Once at this stage, they have been differentiated to nonliving pancake-flat cells called corneocytes, which are tightly bound into sheets. When they reach the outermost layer of the SC, corneocytes shed and their life cycle starts over again. The dense and thick layer with 15–20 μm of corneocytes in the skin is responsible for providing the physical barrier protection.[66,88] Under the SC, there is an impermeable lipid-based layer that is tightly stacked to corneocytes. These lipids, originating from unique epidermis structures called lamellar bodies, are structured in bilayers and consist of ceramides; cholesterol, and essential fatty acids. This lipid mixture layer plays a key role on the proper skin barrier function by the prevention of water loss, and also by avoiding the entrance of undesirable substances.[28,81] Where the lipid layer integrity is affected, the skin barrier function is disrupted causing dehydration. This compromises the normal shedding of cells from the SC surface, and, consequently, the skin becomes dry and flaky.

The maintenance of the normally acidic pH of the SC is likewise fundamental for the proper establishment of the corneocytes surrounding lipid layer. When the skin pH is enhanced, the regular shedding of cells from the SC is affected, triggering also skin flaking. Where no underlying genetic ground for a damaged skin barrier exists, the utilization of skin care products capable of preserving an acid skin pH can improve this function. This occurs by restricting the lipid layer disruption, improving cell shedding and by an adequate skin surface moisturizing.[3,37]

Topical or dermatological drugs consist of a set of products that are applied to the skin or to the mucous membranes and potentiate or retrieve the basic function of the skin, or pharmacologically alter the action of certain tissue. Drugs applied topically can target the superficial damaged by disease skin and be absorbed through topical/cutaneous route. Alternatively, drug absorption can occur through the transdermal route. In this case, the drug diffuses through the SC and inner skin layers which allow a prolonged drug release, maintaining drug plasma concentrations constant over time, and decreasing drug dosing frequency.[68] The rate of drug absorption depends on drug molecular size and lipophilicity. This is intrinsically related to the evidenced tendency of skin to exclude drug molecules higher than 500 Da, particularly hydrophilic molecules.[19]

Cutaneous drug permeation can be accomplished by several routes the transcellular, intercellular, or transappendageal (via the eccrine glands or hair follicles). The administration by this route is facilitated by the use of techniques such as iontophoresis, due to the very small surface area of appendages.[61] After drug dissolution and diffusion of the carrier system along the skin surface, it passes across the SC and it is distributed along the epidermis, a more hydrophilic layer. Lastly, the drug diffuses by the inner layers of skin (dermis) to be captured by blood capillary cells, allowing for drug release into the systemic circulation. Transdermal permeation depends on skin's physiological factors such as the thickness of the skin; the lipid content; the density of hair follicles; the density of eccrine glands; skin pH; blood flow; the skin moisturizing state and its inflammatory conditions; and on drug's physicochemical factors, like its partition coefficient, molecular weight, degree of ionization, and ability to be incorporated and released from the drug delivery system.[58,87]

Large and hydrophilic compounds, like peptides or proteins, do not easily diffuse through the skin. Therefore to promote skin permeation, direct effect on the skin or some type of physical and/or chemical modification of the formulation is usually necessary, in order to change partition, diffusion or solubility of these molecules. Mechanisms of direct effect on the skin consist of: denaturation or changes in conformation of intracellular keratin that can cause swelling and increase hydration; destruction of desmosomes (or *macula adherens*), specialized in the cell–cell adhesion, maintaining the cohesion between corneocytes; modification of lipid bilayers, for example, by using removing solvents that can reduce the resistance to permeation; and changes in properties of SC solvents to modify their partitioning capacity, for example, surfactants/cosurfactants.[61]

Modifications on the formulations consist of different approaches, such as adding a volatile solvent that allows the achievement of a drug supersaturation state on the formulation with higher thermodynamic stability; adding penetration (or permeation enhancer molecules, like good solvents for the drug, or fatty acids, like oleic acid that acts as a surfactant and create a lipidic pool between bilayers and disrupt them uniformly, improving drug partitioning in the SC.[57,104] Some mechanical processes—like iontophoresis,[69] microneedle technology,[6] electroporation,[43,51] sonophoresis[76] or thermal ablation[65]—are also used to enhance skin permeation. For example, verified that iontophoresis in combination with fatty acids, like R(+)-limonene, palmitic, palmitoleic, stearic, oleic, linoleic, or linolenic acids, enhance the in vitro permeation of insulin through the skin.[90] Among all, the authors[90] found that the most efficient penetration enhancer was linolenic acid, either by passive or iontophoretic transport.

In light of the foregoing, generally, a drug molecule should be sufficiently lipophilic to diffuse through the SC lipophilic layer, and sufficiently hydrophilic that it can also cross through the hydrophilic layer of the epidermis. Very few drugs in their free form can overcome the various skin layers, thus, their administration can be strengthened by using percutaneous routes or penetration enhancers. Taking into account all the advantages of transdermal and topical drug delivery, it is essential to overcome drug requirements and, for that, different carrier systems have been developed, mostly of lipophilic nature. Microemulsions emerge as drug delivery systems that combine hydrophilic and hydrophobic properties, allowing for transdermal drug permeation through the skin.[40,61,64]

This chapter discusses the facts and features of microemulsions as drug delivery systems, the formation theory, and their advantages as innovative carriers for transdermal administration of drugs.

2.2 BASICS AND FUNDAMENTAL CONCEPTS OF MICROEMULSIONS

The term "microemulsion" was firstly introduced by researchers Hoar and Schulman in 1943, to describe thermodynamically stable liquids, simple and optically isotropic systems, constituted by surfactants, water, and oil.[45]

Microemulsion is derived from the term emulsion. However, microemulsions and emulsions constitute distinct systems. Microemulsions display a translucent appearance, a fluid consistency (without the tendency to coalesce) and have a nanometric droplet size (usually, from 10 to 100 nm); while emulsions are milky-like dispersions with droplet sizes above 500 nm.[64] The stability is an additional difference among the aforementioned systems. In terms of thermodynamic, emulsions are less stable than microemulsions and require energy for their formation process, under penalty of eventual phase-separation. In opposition to emulsions, microemulsions are simply formed by gentle mixing, stirring or heating under the appropriate experimental conditions requiring a minor energy input.

Thermodynamic stability means that when separated the water and oil (immiscible liquids) have lower free energy than the mixture of both components which form an emulsion.

In the case of microemulsions, it requires a great amount of surfactant and cosurfactant mixture as an interfacial film to promote the spontaneous formation, which does not happen in the case of emulsions. However, the presence of surfactant combinations in microemulsions presents an enhanced cutaneous irritation potential.[64,68]

The term "nanoemulsion" has emerged most recently. The description of nanoemulsions is identical to that of microemulsions—both are fluid dispersions with a droplet size inferior to 250 nm.

Similarly to emulsions, thermodynamically, nanoemulsions are unstable, so they require high-energy input methods to be formed and thus are costly. Nevertheless, nanoemulsions have a beneficial feature since these systems might be produced with lower quantities of tensioactive agents (so-called surfactants), suggesting lower irritation effects. Nanoemulsions and microemulsions differ in droplet size. The prefix "nano" (10^{-9}) and "micro" (10^{-6})

implies that the dispersed phase in nanoemulsions is inferior in size when compared to microemulsions. Yet, in practical terms, the opposite assumption is commonly observed.[70]

A microemulsion is commonly known as an O/W (oil-in-water) or a W/O (water-in-oil) dispersion. However, it can also be classified as a bicontinuous type, in which the oil and water phases are dispersed within each other and the system is stabilized by a surfactant film, but without forming droplets. Thus, the concept of "microemulsion" combines[27]:

- Aqueous micellar surfactant solutions containing a solubilized lipid, in an "O/W microemulsion";
- Lipophilic micellar surfactant solutions containing solubilized water, the so-called reversed micellar solutions or "W/O microemulsions";
- Systems in which a continuous transition from an aqueous to lipophilic solution exists;
- A "surfactant phase" in nonionic surfactant systems.

Microemulsions are ideal vehicles for the transport of drugs since they present the main characteristics of fluid systems, like long-term storage stability and undemanding production (are produced spontaneously with zero interfacial tension). Moreover, microemulsions exhibit a Newtonian behavior due to their fluidity, have a large surface area and extremely low droplet size that facilitates membrane adherence enabling a controlled transport of bioactive agents.

Incorporated drugs in microemulsions may be administrated into the body by the oral route, by the topical route through the skin or nose, or via a direct entry through an aerosol in the lungs. The use of microemulsions presents an advantage due to the ease and low-cost process preparation—since no specialized equipment is necessary—and also their enhanced bioavailability. Microemulsions constitute, thus, a valuable technological tool capable of protecting labile drugs, increasing drug solubility, regulating drug release kinetics, and decreasing patient variability.[61,79]

Microemulsions can be characterized by their physical characteristics such as low viscosity, transparency, optical isotropy, and thermodynamic stability.[10,110] Notwithstanding in order to understand microemulsions real properties, it is necessary to characterize them, including an acquaintance of the drug localization in loaded microemulsions. The microstructure of microemulsions can be changed due to intermolecular interactions between the drug and the microemulsion components. Additionally, the reduced droplet size and the oscillating interfaces might promote variations in the microemulsions microstructure. In order to properly characterize microemulsions is crucial to combine different characterization methods

Some of the methods used are[44,48,50,52,61]:

- Polarizing light microscopy to analyze the optical isotropy;
- Rheological and textural measurements to study the viscosity and spreadability (properties related to the percentage of oil and surfactant phases);
- Electron microscopy techniques (e.g., Cryo-transmission electron microscopy (cryo-tem)) to analyze structural differences between bicontinuous and droplet-like phases of microemulsions;
- Conductivity measurements to ascertain the nature of the continuous phase and to estimate the percolation threshold—transformation process into a bicontinuous microemulsion—which is, in turn, related to the formulation viscosity and influences skin permeation ability;
- Pulsed-field gradient nuclear magnetic resonance (PFG-NMR) to assess the interdependence between the transdermal permeation rate and structural characteristics of microemulsions;
- Dynamic light scattering (DLS) to determine the droplet size and the polydispersity index;
- Differential scanning calorimetry (DSC) to track the drugs in the microemulsions;
- Small angle X-ray scattering (SAXS);
- Small angle neutron scattering (SANS) which can be used to study the droplet size, shape and morphology.

2.3 COMPOSITION AND PRODUCTION METHODS

2.3.1 FORMATION THEORY OF MICROEMULSIONS

The mixture of two immiscible liquids by constant stirring forms dispersed droplets. The stirring cessation causes droplet coalescence, which, in turn, leads to the separation of the two phases. The shelf time of an emulsion includes the time passed since the moment of the homogenization of the two phases till the separation of the two immiscible phases. The stability of the system has a straightforward relationship with the shelf time, that is, a stable emulsion has a longer shelf time.

When one liquid is mixed with the other, it is formed an internal phase (dispersed or discontinuous phase) surrounded by an external phase (dispersant or continuous phase). The process of emulsification implies a large increase in interfacial area ($\Delta A > 0$), which leads to a sharp increase of the surface free energy ($\Delta G > 0$). This process, at constant temperature, volume, and number of moles, can be described by the Gibbs–Helmholtz equation:

$$\Delta G = \gamma_i \cdot \Delta A - T \cdot \Delta S \tag{2.1}$$

In eq (2.1), γ_i corresponds to the interfacial tension between the oil and the aqueous phases, ΔA corresponds to the adjustment in the interfacial area that occurs throughout the formation process, ΔS represents the entropy variation, and T corresponds to the temperature.[44,64] Here, it is evident that ΔG is directly influenced by the interfacial area and that the spontaneous emulsification is possible if γ_i is negative, which means that ΔG will be < 0. If increasing the interfacial area is essential in the technological point of view, one of the alternatives for stabilization of an emulsion would be to provide mechanical energy continuously to maintain the increased interfacial area.[96] Although this factor is necessary for dispersion, yet the effect is temporary because it only occurs during stirring. System stabilization may be feasibly accomplished by the reduction of the tension between the two immiscible phases, which as a result of the increase of interfacial area, conduces to a reduction of the free energy of the system, contributing ultimately to a thermodynamically stable system (at a constant temperature).[18]

Surfactants decrease the interfacial tension between oil and water and, that way, contribute to emulsions and microemulsions stability. However, not all of them can entirely decrease the free energy by increasing the interfacial area. In these cases, to obtain separated phases as long as possible, a cosurfactant can be added with the view of decreasing the interfacial tension to lower values than the common surfactants. Microemulsions differ from emulsions by the presence of a cosurfactant, which impart a reduction of the internal phase dimensions, increasing their thermodynamic stability and optical transparency. By virtue of reduced dimensions of the internal phase, microemulsions droplets show higher bending angles and diffusion coefficients compared to emulsions.[68,89]

In a four-component system (microemulsions), a mixed film of surfactant/cosurfactant is adsorbed on the oil–water interface. These molecules are oriented at the interface with polar heads facing the aqueous phase and polar chains in contact to the oil phase. An increased interfacial tension between oil and water promotes a higher number of molecules per unit area. They begin to compress, alongside each other, developing a two-dimensional lateral pressure (π). An analysis of the surfactant film shows that the surface/interfacial tension (γ_i) decreases proportionally with increasing pressure. This phenomenon can be expressed by eq (2.2), which shows that when the repulsion of the species of film (π pressure) exceeds γ O/W, γ_i is negative:

$$\gamma_i = \gamma_o/w - \pi \tag{2.2}$$

The surface free energy (ΔG) makes possible the spontaneous expansion of the interface because the temporary existence of $\pi > \gamma_o/w$ orients the force that reduces the size of droplets, until no more energy is needed to increase the interfacial area. Equilibrium is reached when the negative interfacial tension returns to zero, due to the decompression of the molecules with a consequent decrease in π pressure.

A more detailed analysis of the system shows that the negative tension is a consequence not only of the high initial pressure of the film (π) but also of the large decrease of the original tension between oil and water (γ_o/w). This reduction occurs because cosurfactant (medium chain alcohols), being soluble in both oil phase and interface, will be partitioning in both phases; therefore, the fraction dissolved in the oil reduces the initial tension between the oil and aqueous phases. At the water–oil interface, the surfactant/cosurfactant film shows different tensions on each side. These different forces generate the curvature of the microemulsions droplets. The side with higher tension is concave and, therefore, will include the other liquid, making it the internal phase of the system.[36,89]

2.3.2 COMPOSITION OF MICROEMULSIONS

Microemulsions are constituted by: an oil phase, surfactants, cosurfactants, and an aqueous phase. Each of these constituents is discussed in this section.

2.3.2.1 OIL PHASE

Numerous compounds of either natural or synthetic origin, with different features, were employed as oil phase in the production of microemulsions.

They generally present low water solubility and are selected based on their penetration-enhancing characteristics. These components may also influence parameters such as the drug loading capacity and kinetic behavior during release, as well as the skin permeation. Compounds—as saturated and unsaturated fatty acids, alcohols, sulfoxides, surfactants, or amides—can be added to the microemulsion to promote drug skin permeation. The drug may permeate the skin due to the disruption of the SC lipid structure. Furthermore, the partition coefficient of the drug may be altered in order to favor skin permeation.[32,63,68]

In the study of Aungst et al.,[7] the authors advanced that compounds like oleic acid which has a C10 unsaturated alkyl chain or molecules with a saturated C10–C12 alkyl chain and a polar head are those with the best capacity as skin penetration enhancers.

Besides in the case of oleic acid, *cis* configuration disturbs more actively the lipid layer of the skin that *trans* unsaturated configuration. Since the SC is hydrophobic, fatty acids are able to enter into the lipid layer, creating separate domains and inducing highly permeable paths.[61] The effect of oleic acid as penetration enhancer is also influenced by the drug physicochemical properties. While Rhee et al.[91] observed that oleic acid enhances ketoprofen transport (log $P > 2$), Kim et al.[59] contrarily reported that oleic acid decreases transdermal delivery of triptolide (hydrophilic, log $P < 0$). It is also proved that oleic acid acts better as penetration enhancer in hydrophilic microemulsions than in lipophilic ones.[99]

Other fatty acids such as linolenic acid, in combination with iontophoresis technique, are capable of stimulating lipid disorganization of SC and potentiate skin permeation of drugs adsorbed on microemulsions.[4,90] Other examples used as oil phases, due to their permeation enhancing ability, are isopropyl myristate, isostearilic isostearate, isopropyl palmitate, medium chain triglycerides, R(+)-limonene, and triacetin.[61,72] Some compounds (like vegetable oils) hold occlusive properties, which alter the water gradient in the most external skin layers by preventing water evaporation and, thus, contributing to skin moisturizing enhancement.[85]

Although the selection of oil phase components is performed to favor the solubilization of the drug, yet an enhancement of drug solubility can affect the oil phase drug partition and consequently lower thermodynamic drug activity. Therefore, enhancing drug solubility is of less pertinence being more relevant to the biocompatibility, as well as, the effects of the oil phase on the skin.[46,68]

2.3.2.2 SURFACTANTS

Several surfactants, surfactant mixtures, and cosurfactants are used in the production of microemulsions intended for topical and transdermal administration. The formation of microemulsions, the droplet size, and shape of aggregates are affected by the concentration and type of tensioactives (surfactant and cosurfactants) used, as well by the percentage of each tensioactive component in the mixture.[30,44] These last mentioned factors also affect the water solubilization.

Anionic, cationic, and amphoteric surfactants can be used in transdermal microemulsions; however, nonionic ones are more frequently used due to lower risk of skin irritancy.

It is possible to classify microemulsions according to the main surfactant:

- Aerosol® OT-based microemulsions;
- PEGylated fatty alcohol-based microemulsions;
- Polysorbate-based microemulsions;
- PEGylated fatty acid esters-based microemulsions;
- Phosphatidylcholine-based microemulsions, among others.[44]
- Aerosol® OT salt (bis(2-ethylhexyl) sulfosuccinate sodium),[17,54,55] cationic CTAB (hexadecyltrimethyl ammonium bromide)[26] and lecithin[23,24,80] can be also used as surfactants in transdermal microemulsions.

Lecithin is the major source of phosphatidylcholine; and in parallel, is one of the most common phospholipid constituents of the lipid bilayers of the body. It is presented as a nontoxic and biocompatible surfactant that acts as a promoter of cutaneous permeation, which, when it is in a fluid state and due to their physicochemical properties, can merge with the of SC lipids, disrupt its structure and facilitate the dermal and transdermal drug permeation.[60]

Caprylocaproyl macroglycerides (Labrasol®), Plurol Isostearique®, or Plurol Oleique® are examples of surfactants that barely irritate the skin and are widely used in the production of microemulsions for topical administration.[61] Nonionic surfactants such as Brij® (a PEGylated fatty alcohol) and Tween® (Polysorbates) are used in microemulsion formation processes. For example, Span® 20 or Tween® 20 affect SC lipids and make them more fluid, which is ideal for the diffusion of lipophilic molecules through the SC.[34]

Since surfactants have the ability to enhance skin permeation, some authors suggest that increasing the concentration of these compounds in microemulsions can trigger an enhancement of their ability to disrupt SC lipid layers and thus an improvement of the drug diffusion through the skin.[49] However, in practice, this is not a straight rule, and this conclusion depends on the thermodynamic stability of the drug. For example, increasing the mixture ratio of surfactant/oil phase in microemulsions increases the viscosity of the system, which results in the formation of compacted and dense structures and a lower dilution capacity of the drugs in the formulation.[62]

2.3.2.3 COSURFACTANTS

Microemulsions contain high amounts of the surfactant, usually in association with a cosurfactant, in most cases alcohol, solvents, and cosolvents that are not active compounds but may involve a risk in the pharmaceutical field.[56,107] Alkyl alcohols with simple chains are widely used to promote microemulsion formation and are also applied as penetration enhancers due to the high capacity to solubilize oils.[32,103] Ethanol is one of the most widely used cosurfactants owing to its capacity to increase the solubility of hydrosoluble drugs in formulation and to concomitantly solubilize some of the SC lipids, enhancing the drug flux and permeability through the skin.[71]

Ethanol volatility renders another mechanism since the evaporation of ethanol from the dosage form intensifies drug concentration which leads to a supersaturated state thus facilitating permeation. The penetration enhancer effect is dependent on its concentration in the formulation.[61] Other examples of cosurfactants that can facilitate skin permeation are 1-butanol, decanol (as saturated fatty alcohol), propylene glycol, that act similarly to ethanol, or Transcutol®.[61,74]

In opposition to surfactants, which only experiment partition at the oil/water interface, cosurfactants experiment partition between the oil phase and the oil/water interface. Shorter chain alcohols (methanol, ethanol, or 1-propanol) are hydrosoluble, therefore, their use as cosurfactants is inadequate.[11] Increasing the chain length from *n*-butanol to *n*-octanol leads to a decrease in the cosurfactant polarity and to an increase of the system viscosity, making the homogeneous system more unstable, and becoming the clear microemulsion formation less favorable. It also enhances the oil/water interface affinity of *n*-alkanol which corresponds to a greater incorporation of water per oil gram (water solubilization) which contributes to form microemulsions.[77,86]

Relative to drug release, Liu et al.[67] demonstrated that increasing the cosurfactant chain length (in this case, from ethanol to isopropanol) leads to a reduction in curcumin transdermal flux. Antagonistically, increasing the number of hydroxyl groups (replacement of isopropanol with propylene glycol) amplifies curcumin transdermal flux. Besides alcohols, polyhydroxy compounds (such as glycerin, propylene glycol, and 1,3-butylene glycol[16]) can be used as cosurfactants in a microemulsion system. Additionally, these compounds can be used as cosolvents in the aqueous phase.[106]

According to several research studies, it is also necessary to take into account the used cosurfactants concentrations, and the membranes in which the study is carried out. Drug permeation mechanism varies in a cellulose membrane, in a membrane pre-treated with additives, in a dislipid membrane or even in an extract of SC. Cosurfactants (such as ethanol, phospholipids, or propylene glycol known as penetration enhancers) can act as drug permeation retardants depending on its concentration and the membrane used in the studies.[33,61]

To investigate the importance of the ratio between the surfactant and the cosurfactant on microemulsions using isopropyl palmitate as the oil phase, Basheer et al.[11] used Tween® 20 to 80 as surfactant and methanol to 1-pentanol as cosurfactant. All obtained systems were further characterized by stability analysis and microemulsion formation. They found that to form microemulsions, 3:1 was the optimal surfactant/cosurfactant ratio. Besides, the order of surfactant efficacy was Tween® 80 > 60 > 40 > 20; while in the case of alcohols (cosurfactants), the efficacy order was 1-butanol > 1-pentanol > 1-propanol > ethanol = methanol. The microemulsion composed by the surfactant mixture Tween® 80/1-butanol (3:1) exhibited long-term stability, an optimal pH, droplet size, and a negative charge.

2.3.2.4 *AQUEOUS PHASE*

According to several scientific studies on microemulsions, water constitutes typically the aqueous phase. The presence of water favors the transdermal permeation of microemulsions containing lipophilic and hydrophilic drugs.[104] The ratio of aqueous phase, in turn, affects microemulsions structure, which has influence in the transport of the drug across the skin.[5] The enhancement of the water ratio in the microemulsion changes the drug solubility in the external phase, enhances skin moisturizing, and consequently skin permeation.[68]

Microemulsions in certain critical water ratios can display percolation phenomena.[20,52,68] This means that an increase in microemulsion water ratio to the percolation threshold can lead to transition of W/O microemulsions (with low water ratio) to O/W (with low oil phase ratio) leading to the development of a bicontinuous system and aggregates. The aforementioned transition causes the aqueous droplets to interconnect and give rise to larger structures, leading to an increase in electrical conductivity and viscosity, which, in turn, affects skin permeation.

In addition, increasing microemulsion water ratio can also alter the stability of the system, since it can lead to drug removal from the delivery system, growth of microorganisms, and also trigger hydrolysis of the components that constitute oil phase.[73] Microemulsions may undergo a lyophilization process (also known as freeze-drying process) which enhances the formulation storage stability by removing the water content. In this lyophilization process, it is crucial to use cryoprotectans in order to maintain intact the composition of the microemulsion.

Furthermore, aqueous phase in microemulsions can be replaced by other liquids besides water, such as: phosphate buffer. The pH of the phosphate buffer can be 7.4 (body fluid pH) or 5.5 (skin pH) depending on the administration route.[26,61] Also glycerol, polyethylene glycol, and propylene glycol can also replace water in some microemulsion systems.[20,108] These solutions are denominated in literature as "nonaqueous systems" although they contain some percentage of water in their composition.[20] In their study, Carvalho et al.[20] used three different drugs (α-tocopherol, lycopene, and progesterone) to compare "nonaqueous microemulsions" (with propylene glycol as a water substitute) with "aqueous microemulsions" with identical composition. The authors concluded that "aqueous microemulsions" promote transdermal penetration of the drugs with low partition coefficient (log P), like progesterone (log $P = 4.04$), and α-tocopherol (log $P = 7.8$). "Nonaqueous microemulsions" increase the transdermal penetration of drugs with values of log P near 17 such as lycopene. Therefore, drug log P value is an important drug characteristic to take into account when choosing the aqueous phase in microemulsion systems.

2.3.3 PRODUCTION METHODS

Microemulsions, in practice, can be easily prepared by simply stirring all components in suitable ratios.[68] In the formation process, no high energy input or specific technique or equipment is required making production more affordable. Therefore, it is possible to scale up microemulsion production.

Notwithstanding, microemulsions can only be produced in a specific interval of concentration of the components oil phase, aqueous phase, and surfactant mixture (includes surfactant and cosurfactant).[52] The interval, where it is possible to obtain stable microemulsions, is graphically represented by two types of diagrams, designated by "*ternary phase diagram*" and "*pseudo-ternary phase diagram*". "*Ternary phase diagram*" consists in a triangle, where each of the vertices corresponds to 100% of oil phase, aqueous phase, and surfactant. In "*pseudo-ternary phase diagram*", additional components such as cosurfactants and/or drugs can be introduced. The mixture between two microemulsions components (surfactant/cosurfactant, water/drug, or oil/drug) represents one of the vertices of the "pseudo-ternary phase diagram."[52]

A phase diagram for the determination of the microemulsion region can be constructed by the titration method.[82] In the titration method, a surfactant and a cosurfactant are mixed to obtain the surfactant/cosurfactant mixture. Oil and surfactant/cosurfactant are mixed at various weight ratios and then diluted dropwise with the addition of the aqueous phase under agitation.[18,68,82]

The different regions in the diagram (Fig. 2.1) represent appropriate experimental conditions for the existence of different types of systems such as micelles, lamellar bicontinuous structures, macroemulsions, and microemulsions.[82] Normal or reverse micelles solutions are produced in the absence of a surfactant mixture and with a quite low volume of the internal phase. In turn, lamellar bicontinuous structures are formed when a high ratio of surfactant and cosurfactant mixture is organized in the continuous oil/water interface, separating both phases. Macroemulsions are very unstable and are formed when occurs separation between aqueous and oil phases.

In a "*pseudo-ternary phase diagram*," the two types of microemulsions—O/W (normal-type structures) and W/O (reverse structures)–correspond to specific regions. The formation of O/W microemulsions occurs, in the presence of a surfactant mixture, due to an increase in the ratio of the aqueous phase concentration, so-called, in this case, continuous phase.[68,82] In contrast, W/O microemulsions are formed due to the aqueous droplets intern dispersion into the continuous oil phase.

Microemulsions can also be produced by a second method designated as "*phase inversion method*," which consists of the addition of excess dispersed phase.[82] Microemulsions are under constant fluctuations. Therefore, in the "phase inversion method," some parameters (such as temperature, salt concentration, or pH value) can be altered in order to force the transition between O/W microemulsions to W/O or the opposite. As mentioned above, increasing water volume can invert W/O to O/W microemulsions. At the inversion point, bicontinuous microemulsions can be formed. Obtained samples, by either one of the methods, are classified as microemulsions when they appear as transparent isotropic liquids.

For microemulsions characterization, a set of different devices and assays are applied. As previously mentioned, SAXS, SANS, DLS, and DSC are the main techniques used to characterize microemulsions. Additionally, conductivity measurements, stability, and in vitro release studies are also used to complement microemulsions characterization.

2.4 APPLICATIONS IN TRANSDERMAL DRUG DELIVERY

The transdermal delivery of drugs encapsulated in microemulsions can be influenced by the structure, electric conductivity, and viscosity of the microemulsion system. According to a large body of evidence, microemulsions are now being optimized as transdermal and dermal permeation drug delivery

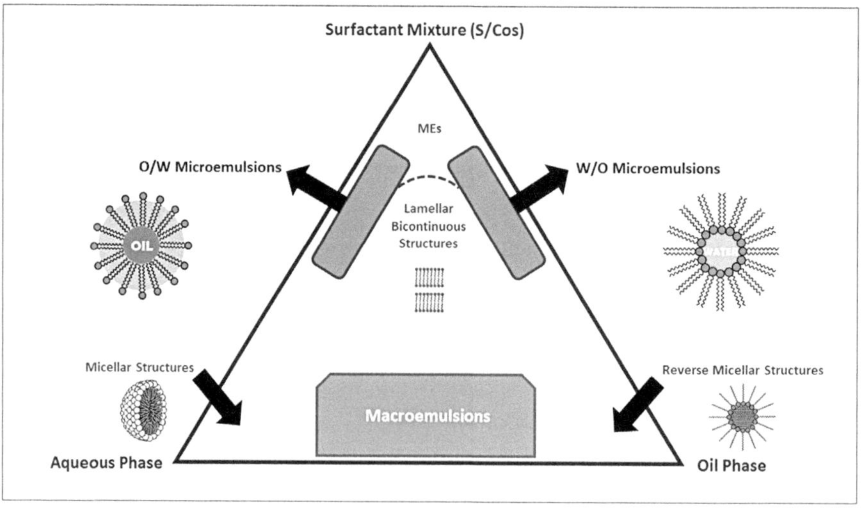

FIGURE 2.1 (See color insert.) Diagram showing different drug delivery systems regions in a "*pseudo-ternary phase diagram*". The vertices of the triangle represent: oil phase, aqueous phase, and surfactant/cosurfactant mixture. All area above the curved dotted line corresponds to the possible formation of microemulsions.[42,84]

carriers. Table 2.1 summarizes microemulsion applications in transdermal drug delivery and also describes the wide range of alternative components used in these drug delivery systems.

Microemulsions are advantageous encapsulation systems of lipophilic and hydrophilic drugs for topical administration. The incorporation of drugs occurs in the continuous phase, dispersed phase or in the microemulsion interface. In general, O/W systems are used to solubilize lipophilic drugs whereas W/O systems seem to be a better choice for hydrophilic drugs.[101]

Microemulsions skin penetration effect is due to the presence of a surfactant-containing interface and the organized structure of oil and aqueous phases. The interaction between oil and aqueous phases promotes a small droplet size and increases microemulsion surface area. Those characteristics create additional drug solubility regions and enhance drug concentration gradients.[63]

It was demonstrated that microemulsions with droplet size within 100–300 nm enhance transdermal delivery by increasing surface area and skin adhesiveness.[68] The components of oil phase are relevant to microemulsion interactions with the SC since this phase shows direct influence in drug release. The mixture of surfactants/cosurfactants stabilizes microemulsion and may increase drug solubility in the system.[44,68] The surfactants mixture in microemulsion is continuously fluctuating between interfaces which confer poor viscosity and spreadability to the formulation. Indeed, low viscosity values can limit the application of microemulsions in the skin.[9] The addition of polymers changes microemulsions texture by increasing the viscosity, owed to the ability that polymers exhibit in promoting stronger interactions among the formulation and the skin.[25,29,82,113] It should be effectively achieved an ideal viscosity level of the formulation (neither too fluid nor too viscous) in order to increase retention time and drug diffusion capacity through the skin.

In order to potentiate skin permeation, it is crucial to use drug delivery systems with cationic charge.[12,68] The microemulsion charge is mainly related with the type of surfactant used (anionic, cationic, amphoteric, or nonionic).[82] Anionic protein residues are present in the external membrane surface of skin epithelial cells contributing to the negative surface charge of this skin cells. For that reason, the delivery of positive charge microemulsions enhances the skin interaction and favors membrane adherence.

Oil, surfactants mixture and water (or other substitutes for the aqueous phase) penetrate into the SC at different extents. Nevertheless, it is the synergy of all microemulsion components that is responsible for SC relative drug uptake.[40]

TABLE 2.1 Examples of Transdermal Drug Delivery Systems Based on Microemulsions.

Drug	Components of microemulsions			Reference
	Oil phase	**Surfactants/cosurfactants**	**Aqueous phase**	
5-Fluorouracil	Isopropyl myristate	AOT (Aerosol® OT)	Water	[39]
8-Methoxsalen	Isopropyl myristate	Tween® 80 + Span® 80/1,2-octanediol	Water	[8], [10]
Anticholinesterase alkaloidal extract from *Tabernaemontana divaricata*	*Zingiber cassumunar* oil	Triton™ X-114/ethanol	Water	[21]
Anticholinesterase plant oil	*Cymbopogon citratus* oil	Tween® 20/ethanol	Water	[22]
Artemether	Peceol®	Labrasol®/Transcutol® P	Water	[1]
Clobetasol propionate	Isopropyl myristate	Cremophor® EL/isopropyl alcohol	Water, carbopol 934P	[82], [83]
Clonazepam	Labrafac™ Hydro WL1219	Tween® 20/Transcutol®	Water	[75]
Curcumin	Isopropyl palmitate	Labrasol® + Glyceryl oleate/propylene carbonate	Water	[98]
Diclofenac sodium	Soybean oil	Brij® 58 + Span® 80/ ethanol, isopropyl alcohol, or propanol	Water	[56]
Estradiol	Epicuron™ 200; Oleic acid; isopropyl myristate	Tween® 20, Tween® 80 or Span® 80/ethanol, or isopropanol	Phosphate buffer (pH 7.4)	[87]
17-estradiol; progesterone; cyproterone acetate; finasteride	Eucalyptus oil	Brij® 30/ethanol	Water	[8], [15]
Hydrocortisone	Eucalyptus oil	Tween® 80/ethanol, isopropanol, or propylene glycol	Water	[8], [32]
Insulin	Olive oil	Tween® 80 + Span® 80	Water	[8], [105]
Ketoprofen	Miglyol® 812 N	Soybean lecithin/*n*-butanol	Water	[80], [97]
Lacidipine	Isopropyl myristate + dimethyl formamide	Tween® 80/Labrasol®	Water	[38]
Nicotinic acid and prodrugs	Isopropyl myristate	Labrasol®/Peceol™	Water	[100]
Ondansetron	Oleic acid	Tween® 20/PEG 400	Water	[2]

TABLE 2.1 *(Continued)*

Drug	Components of microemulsions			Reference
	Oil phase	Surfactants/cosurfactants	Aqueous phase	
Penciclovir	Oleic acid	Cremorphor® EL/ethanol	Water, Carbomer 940	[113]
	Oleic acid	Cremorphor® EL/ethanol	Water	[109], [112]
Phenobarbital	Isopropyl myristate	Labrasol®/ ethanol	Water	[35]
Piroxicam	Isopropyl myristate	Cetyltrimethyl ammonium bromide (CTAB)	Aqueous buffer (pH 5.5)	[8], [26]
	Oleic acid	Tween® 80/ethanol	Water	[8], [29]
Progesterone	Tributyrin	Polyoxyethylene (10) lauryl ether	Water	[14]
Sertaconazole	Oleic acid	Tween® 80/propylene glycol	Water	[8], [93], [92]
Sodium fluorescein	Mygliol 812®; soybean oil or tributyrin	Brij® 97	Water	[8], [102]
Testosterone	Oleic acid	Tween® 20/Transcutol®	Water	[41]
Theophylline	Oleic acid	Cremophor® RH40/ Labrasol®	Water	[111]
Triptolide	Oleic acid	Tween® 80/ propylene glycol	Water	[59]
	Oleic acid + Gemseal® 40	Labrasol® + Tween® 80/ethanol	Water	[25]
Vinpocetine	Oleic acid	Labrasol®/ Transcutol® P	Water	[8], [48]

Only after optimization and characterization of a microemulsion, it is possible to predict if the formulation will be used as a cutaneous or transdermal delivery system.[47] In cutaneous delivery, the encapsulated drug in microemulsions is supposed to be released in a specific location in the outer layer of the skin (epidermis). In the case of transdermal delivery, the drug nanocarrier is supposed to penetrate SC and enter in contact with the capillaries system in the dermis for a prolonged release and a systemic effect. Some microemulsions can be used for cutaneous or transdermal delivery. Recent studies indicate that microemulsions can also be used as drug carriers for transfollicular pathway.[13]

The topical administration of microemulsions is particularly captivating for the cosmetic industry. Furthermore, recent studies demonstrate a growing and substantial interest in the encapsulation of natural oils in microemulsions for topical administration. This is due to the wide range of biological effects that these natural oils and its components offer.[106]

2.4.1 DISADVANTAGES OF MICROEMULSIONS AS TRANSDERMAL FORMULATIONS

Despite all the advantages mentioned in this chapter, microemulsion topical, and transdermal applications are not exempt of potential skin irritant and comedogenic effects.[68] However, these adverse side effects are essentially related to the exposure time and composition of the microemulsion, particularly with the type of surfactant and oil phase chosen.

In fact, the need for a large amount of surfactant and cosurfactants for the stabilization of nanodroplets contributes to the risk of skin irritation.[56,78]

Comedogenic effects are associated with hyperkeratosis. A relation between the use of microemulsion formulations and comedogenic effects is not scientifically established. Nevertheless, comedogenic effects can be triggered by microemulsion individual compounds, principally alcohols (e.g., lauryl alcohol) added into the surfactant mixture.[31,68] Thus, a microemulsion can be considered noncomedogenic if the comedogenic compound ratio is low.

The duration and frequency of exposure and also the ratio of each compound in microemulsions is intimately related to the appearance of adverse effects. Natural surfactants, like lecithin or sugars (like polyglucosides), are frequently used because of their biocompatibility, which significantly reduces the risk of adverse effects.[46,95] It is, thus, crucial to use noncomedogenic components to create noncomedogenic microemulsions.

2.5 SUMMARY

Microemulsions are optically isotropic liquid dispersions, comprising an aqueous phase and oil phase stabilized by a single tensioactive or a tensioactive mixture. In comparison with emulsions, microemulsions are transparent colloidal dispersions with smaller droplet size—less than 100 nm—thermodynamically stable, with no tendency to coalescence. Depending on the ratio of the different components, microemulsions can be designated as O/W, W/O, or bicontinuous dispersions.

The Newtonian behavior, fluidity, drug solubilizing capacity, long-term shelf-life, small droplet size, and large surface area allied to the spontaneous formation are some of the appealing microemulsions properties. Microemulsions show the desired particle size for membranes adherence and for a controllable transport of different kinds of drugs and molecules. Still, microemulsions can be administered through different administration routes, predominantly nasal, oral and transdermal routes. In relation to the production process of microemulsions, it is characterized to be efficient, does not entail high costs and it is possible to scale up.

In terms of application, microemulsions are particularly interesting for the cosmetic industry, because of their properties and characteristics. Some adverse effects can be induced following microemulsions administration; however, those can be easily surpassed. A state of art discussion of the suitable use of microemulsions as drug delivery systems is presented throughout this chapter.

ACKNOWLEDGMENTS

The authors wish to thank *Nanotec-PSU Center of Excellence on Drug Delivery System, Prince of Songkla University, Thailand.* The financial support was also received from *Portuguese Science and Technology Foundation (FCT/MCT),* and *European Funds (PRODER/COMPETE)* under the projects M-ERA-NET-0004/2015-PAIRED and UID/AGR/04033/2019 (CITAB) and co-financed by FEDER, under the Partnership Agreement PT2020. The authors wish to acknowledge the contribution of the Master Student Ms. Irina Pereira in the reading of the manuscript.

KEYWORDS

- **comedogenic effects**
- **cryoprotectans**
- **lyophilization process**
- **nonaqueous microemulsion**
- **small angle X-ray scattering**
- **Stratum corneum**

REFERENCES

1. Agubata, C. O.; Nzekwe, I. T.; Obitte, N. C.; Ugwu, C. E.; Attama, A. A.; Onunkwo, G. C. Effect of Oil, Surfactant and Co-surfactant Concentrations on the Phase Behavior, Physicochemical Properties and Drug Release from Self-emulsifying Drug Delivery Systems. *J. Drug Discov. Dev. Deli.* **2014,** *1*, 7–14.
2. Al Abood, R. M.; Talegaonkar, S.; Tariq, M.; Ahmad, F. J. Microemulsion as a Tool for the Transdermal delivery of Ondansetron for the Treatment of Chemotherapy Induced Nausea and Vomiting. *Colloids Surf. B Biointerfaces* **2013,** *101*, 143–151.
3. Ali, S. M.; Yosipovitch, G. Skin pH: From Basic Science to Basic Skin Care. *Acta Dermato-Venereologica* **2013,** *93*, 261–267.
4. Alvarez-Figueroa, M. J.; Blanco-Mendez, J. Transdermal Delivery of Methotrexate: Iontophoretic Delivery from Hydrogels and Passive Delivery from Microemulsions. *Int. J. Pharm.* **2001,** *215*, 57–65.
5. Araujo, L. M.; Thomazine, J. A.; Lopez, R. F. Development of Microemulsions to Topically Deliver 5-Aminolevulinic Acid in Photodynamic Therapy. *Eur. J. Pharm. Biopharm.* **2010,** *75*, 48–55.
6. Arora, A.; Prausnitz, M. R.; Mitragotri, S. Micro-scale Devices for Transdermal Drug Delivery. *Int. J. Pharm.* **2008,** *364*, 227–236.
7. Aungst, B. J.; Rogers, N. J.; Shefter, E. Enhancement of Naloxone Penetration Through Human Skin in vitro Using Fatty Acids, Fatty Alcohols, Surfactants, Sulfoxides and Amides. *Int. J. Pharm.* **1986,** *33*, 225–234.
8. Azeem, A; Khan, Z. I; Aqil, M.; Ahmad, F. J.; Khar, R. K.; Talegaonkar, S. Microemulsions as a Surrogate Carrier for Dermal Drug Delivery. *Drug Dev. Indus. Pharm.* **2009,** *35*, 525–547.
9. Balasubramanian, R.; Sughir, A.; Damodar, G. Oleogel: A Promising Base for Transdermal Formulations. *Asian J. Pharm.* **2012,** *6*, 1.
10. Baroli, B.; López-Quintela, M. A; Delgado-Charro, M. B; Fadda, A. M; Blanco-Méndez, J. Microemulsions for Topical Delivery of 8-Methoxsalen. *J. Control. Release*, **2000,** *69*, 209–218.
11. Basheer, H. S.; Noordin, M. I.; Ghareeb, M. M. Characterization of Microemulsions Prepared Using Isopropyl Palmitate with Various surfactants and Cosurfactants. *Trop. J. Pharm. Res.* **2013,** *12*, 305–310.
12. Baspinar, Y.; Borchert, H. H. Penetration and Release Studies of Positively and Negatively Charged Nanoemulsions--Is There a Benefit of the Positive Charge? *Int. J. Pharm.* **2012,** *430*, 247–252.
13. Bhatia, G.; Zhou, Y.; Banga, A. K. Adapalene Microemulsion for Transfollicular Drug Delivery. *J. Pharm. Sci.* **2013,** *102*, 2622–2631.
14. Biruss, B.; Valenta, C. The Advantage of Polymer Addition to a Non-ionic Oil in Water Microemulsion for the Dermal Delivery of Progesterone. *Int. J. Pharm.* **2008,** *349*, 269–273.
15. Biruss, B.; Kahlig, H.; Valenta, C. Evaluation of an Eucalyptus Oil Containing Topical Drug Delivery System for Selected Steroid Hormones. *Int. J. Pharm.* **2007,** *328*, 142–151.

16. Boonme, P.; Songkro, S.; Junyaprasert, V. B. Effects of Polyhydroxy Compounds on the Formation of Microemulsions of Isopropyl Myristate, Water and Brij® 97. *Mahidol Univ. J. Pharm. Sci.* **2004,** *31*, 8–13.
17. Boonme, P.; Krauel, K.; Graf, A.; Rades, T.; Junyaprasert, V. B. Characterisation of Microstructures Formed in Isopropyl Palmitate/Water/Aerosol OT:1-Butanol (2:1) System. *Die Pharmazie Int. J. Pharm. Sci.* **2006,** *61*, 927–932.
18. Boonme, P.; Souto, E. B.; Wuttisantikul, N.; Jongjit, T.; Pichayakorn, W. Influence of Lipids on the Properties of Solid Lipid Nanoparticles from Microemulsion Technique. *Eur. J. Lipid Sci. Technol.* **2013,** *115*, 820–824.
19. Bruno, B. J.; Miller, G. D.; Lim, C. S. Basics and Recent Advances in Peptide and Protein Drug Delivery. *Ther. Deliv.* **2013,** *4*, 1443–1467.
20. Carvalho, V. F.; Lemos, D. P. ; Vieira, C. S.; Migotto, A.; Lopes, L. B. Potential of Non-aqueous Microemulsions to Improve the Delivery of Lipophilic Drugs to the Skin. *AAPS PharmSciTech*, **2016,** 1–11.
21. Chaiyana, W.; Rades, T.; Okonogi, S. Characterization and in vitro Permeation Study of Microemulsions and Liquid Crystalline Systems Containing the Anticholinesterase Alkaloidal Extract from Tabernaemontana divaricata. *Int. J. Pharm.* **2013,** *452*, 201–210.
22. Chaiyana, W.; Saeio, K.; Hennink, W. E.; Okonogi, S. Characterization of Potent Anticholinesterase Plant Oil Based Microemulsion. *Int. J. Pharm.* **2010,** *401*, 32–40
23. Changez, M.; Chander, J.; Dinda, A. K. Transdermal Permeation of Tetracaine Hydrochloride by Lecithin Microemulsion: in vivo. *Colloids Surf. B Biointerfaces* **2006,** *48*, 58–66.
24. Changez, M.; Varshney, M.; Chander, J.; Dinda, A. K. Effect of the Composition of Lecithin/n-Propanol/Isopropyl Myristate/Water Microemulsions on Barrier Properties of Mice Skin for Transdermal Permeation of Tetracaine Hydrochloride: in vitro. *Colloids Surf. B Biointerfaces* **2006,** *50*, 18–25.
25. Chen, L.; Zhao, X.; Cai, J.; Guan, Y.; Wang, S.; Liu, H.; Zhu, W.; Li, J. Triptolide-loaded Microemulsion-based Hydrogels: Physical Properties and Percutaneous Permeability. *Acta Pharm. Sinica B* **2013,** *3*, 185–192.
26. Dalmora, M. E.; Dalmora, S. L.; Oliveira, A. G. Inclusion Complex of Piroxicam with Beta-cyclodextrin and Incorporation in Cationic Microemulsion. In vitro Drug Release and in vivo Topical Anti-inflammatory Effect. *Int. J. Pharm.* **2001,** *222*, 45–55.
27. Danielsson, I.; Lindman, B. . The Definition of Microemulsion. *Colloids Surf.* **1981,** *3*, 391–392.
28. Darlenski, R.; Kazandjieva, J.; Tsankov, N. Skin Barrier Function: Morphological Basis and Regulatory Mechanisms. *J. Clin. Med.* **2011,** *4*, 36–45.
29. Dhawan, B.; Aggarwal, G.; Harikumar, S. L. Enhanced Transdermal Permeability of Piroxicam Through Novel Nanoemulgel Formulation. *Int. J. Pharm. Invest.* **2014,** *4*, 65–76.
30. Djekic, L.; Primorac, M. The Influence of Cosurfactants and Oils on the Formation of Pharmaceutical Microemulsions Based on PEG-8 Caprylic/Capric Glycerides. *Int. J. Pharm.* **2008,** *352*, 231–239.
31. Draelos, Z. D.; Dinardo, J. C. A Re-evaluation of the Comedogenicity Concept. *J. Am. Acad. Dermatol.* **2006,** *54*, 507–512.
32. El Maghraby, G. M. Transdermal Delivery of Hydrocortisone from Eucalyptus Oil Microemulsion: Effects of Cosurfactants. *Int. J. Pharm.* **2008,** *355*, 285–292.

33. Fang, J. Y.; Hwang, T. L.; Leu, Y. L. Effect of Enhancers and Retarders on Percutaneous Absorption of Flurbiprofen from Hydrogels. *Int. J. Pharm.* **2003,** *250*, 313–325.
34. Fang, J. Y.; Yu, S. Y.; Wu, P. C.; Huang, Y. B.; Tsai, Y. H. In vitro Skin Permeation of Estradiol from Various Proniosome Formulations. *Int. J. Pharm.* **2001,** *215*, 91–99.
35. Figueiredo, K. A.; Medeiros, S. C.; Neves, J. K.; Silva, J. A. Da; Rocha Tome, A. Da; Carvalho, A. L.; Freitas, R. M. De. In vivo Evaluation of Anticonvulsant and Antioxidant Effects of Phenobarbital Microemulsion for Transdermal Administration in Pilocarpine Seizure Rat Model. *Pharmacol. Biochem. Behav.* **2015,** *131*, 6–12.
36. Friberg, S. E.; Bothorel, P., *Microemulsions: Structure and Dynamics*; CRC Press Inc.: Boca Raton, FL, None, United States, 1987.
37. Ganceviciene, R.; Liakou, A. I.; Theodoridis, A.; Makrantonaki, E.; Zouboulis, C. C. Skin Anti-aging strategies. *Dermato-Endocrinol.* **2012,** *4*, 308–319.
38. Gannu, R.; Palem, C. R.; Yamsani, V. V.; Yamsani, S. K.; Yamsani, M. R. Enhanced Bioavailability of Lacidipine via Microemulsion Based Transdermal Gels: Formulation Optimization, Ex Vivo and In Vivo Characterization. *Int. J. Pharm.* **2010,** *388*, 231–241.
39. Gupta, R. R.; Jain, S. K.; Varshney, M. AOT Water-in-oil Microemulsions as a Penetration Enhancer in Transdermal Drug Delivery of 5-Fluorouracil. *Colloids Surf. B Biointerfaces* **2005,** *41*, 25–32.
40. Hathout, R. M.; Mansour, S.; Mortada, N. D.; Geneidi, A. S.; Guy, R. H. Uptake of Microemulsion Components into the Stratum Corneum and their Molecular Effects on Skin Barrier Function. *Mol. Pharm.* **2010,** *7*, 1266–1273.
41. Hathout, R. M.; Woodman, T. J.; Mansour, S.; Mortada, N. D.; Geneidi, A. S.; Guy, R. H. Microemulsion Formulations for the Transdermal Delivery of Testosterone. *Eur. J. Pharm. Sci.* **2010,** *40*, 188–196.
42. Hegde, R. R.; Verma, A.; Ghosh, A. Microemulsion: New Insights into the Ocular Drug Delivery. *ISRN Pharm* **2013,** *2013*, 1–11.
43. Herwadkar, A.; Banga, A. K. Peptide and Protein Transdermal Drug Delivery. *Drug Discov. Today Technol.* **2012,** *9*, e71–e174.
44. Heuschkel, S.; Goebel, A.; Neubert, R. H. Microemulsions - Modern Colloidal Carrier for Dermal and Transdermal Drug Delivery. *J. Pharm. Sci.* **2008,** *97*, 603–631.
45. Hoar, T. P.; Schulman, J. H. Transparent Water-in-oil Dispersions: The Oleopathic Hydromicelle. *Nature* **1943,** *152*, 102–103.
46. Hoeller, S.; Klang, V.; Valenta, C. Skin-compatible Lecithin Drug Delivery Systems for Fluconazole: Effect of Phosphatidylethanolamine and Oleic Acid on Skin Permeation. *J. Pharm. Pharmacol.* **2008,** *60*, 587–591.
47. Hosmer, J.; Reed, R.; Bentley, M. V.; Nornoo, A.; Lopes, L. B. Microemulsions Containing Medium-chain Glycerides as Transdermal Delivery Systems for Hydrophilic and Hydrophobic Drugs. *AAPS PharmSciTech* **2009,** *10*, 589–596.
48. Hua, L.; Weisan, P.; Jiayu, L.; Ying, Z. Preparation, Evaluation, and NMR Characterization of Vinpocetine Microemulsion for Transdermal Delivery. *Drug Dev. Indus. Pharm.* **2004,** *30*, 657–666.
49. Huang, Y. B.; Lin, Y. H.; Lu, T. M.; Wang, R. J.; Tsai, Y. H.; Wu, P. C. Transdermal Delivery of Capsaicin Derivative-sodium Nonivamide Acetate Using Microemulsions as Vehicles. *Int. J. Pharm.* **2008,** *349*, 206–211.
50. Ingham, B.; Smialowska, A.; Erlangga, G. D.; Matia-Merino, L.; Kirby, N. M.; Wang, C.; Haverkamp, R. G.; Carr, A. J. Revisiting the Interpretation of Casein Micelle SAXS Data. *Soft Matter* **2016,** *12*, 6937–6953.

51. Ita, K. Perspectives on Transdermal Electroporation. *Pharmaceutics* **2016,** *8*.
52. Ita, K. Progress in the Use of Microemulsions for Transdermal and Dermal Drug Delivery. *Pharm. Dev. Technol.* **2016,** 1–9.
53. Jain, S.; Patel, N.; Shah, M. K.; Khatri, P.; Vora, N. Recent Advances in Lipid-based Vesicles and Particulate Carriers for Topical and Transdermal Application. *J. Pharm. Sci.* **2016**.
54. Junyaprasert, V. B.; Boonme, P.; Wurster, D. E.; Rades, T. Aerosol OT Microemulsions as Carriers for Transdermal Delivery of Hydrophobic and Hydrophilic Local Anesthetics. *Drug Deliv.* **2008,** *15*, 323–330.
55. Junyaprasert, V. B.; Boonsaner, P.; Leatwimonlak, S.; Boonme, P. Enhancement of the Skin Permeation of Clindamycin Phosphate by Aerosol OT/1-butanol Microemulsions. *Drug Dev. Ind. Pharm.* **2007,** *33*, 874–880.
56. Kantarci, G.; Ozguney, I.; Karasulu, H. Y.; Arzik, S.; Guneri, T. Comparison of Different Water/oil Microemulsions Containing Diclofenac Sodium: Preparation, Characterization, Release Rate, and Skin Irritation Studies. *AAPS PharmSciTech* **2007,** *8*, E91.
57. Karande, P.; Mitragotri, S. Enhancement of Transdermal Drug Delivery via Synergistic Action of Chemicals. *Biochimica et Biophys. Acta Biomembr.* **2009,** *1788*, 2362–2373.
58. Kaur, J.; Kaur, J.; Jaiswal, S.; Das Gupta, G. A Review on Novel Approach of Antifungal Emulgel for Topical Delivery in Fungal Infections. *Indo Am. J. Pharm. Res.* **2016,** *6*, 6312–6324.
59. Kim, H. J.; Chen, F.; Wu, C.; Wang, X.; Chung, H. Y.; Jin, Z. Evaluation of Antioxidant Activity of Australian Tea Tree (Melaleuca alternifolia) Oil and Its Components. *J. Agric. Food Chem.* **2004,** *52*, 2849–2854.
60. Kirjavainen, M.; Mönkkönen, J.; Saukkosaari, M.; Valjakka-Koskela, R.; Kiesvaara, J.; Urtti, A. Phospholipids Affect Stratum Corneum Lipid Bilayer Fluidity and Drug Partitioning into the Bilayers. *J. Control Rel.* **1999,** *58*, 207–214.
61. Kogan, A.; Garti, N. Microemulsions as Transdermal Drug Delivery Vehicles. *Adv. Colloid Interface Sci.* **2006,** *123–126*, 369–385.
62. Kogan, A.; Rozner, S.; Mehta, S.; Somasundaran, P.; Aserin, A.; Garti, N.; Ottaviani, M. F. Characterization of the Nonionic Microemulsions by EPR. I. Effect of Solubilized Drug on Nanostructure. *J. Phys. Chem. B* **2009,** *113*, 691–699.
63. Kreilgaard, M. Influence of Microemulsions on Cutaneous Drug Delivery. *Adv. Drug Deliv. Rev.* **2002,** *54 Suppl 1*, S77–S98.
64. Lawrence, M. J.; Rees, G. D. Microemulsion-based Media as Novel Drug Delivery Systems. *Adv. Drug Deliv. Rev.* **2000,** *45*, 89–121.
65. Lee, J. W.; Gadiraju, P.; Park, J. H.; Allen, M. G.; Prausnitz, M. R. Microsecond Thermal Ablation of Skin for Transdermal Drug Delivery. *J. Control. Release* **2011,** *154*, 58–68.
66. Lee, S. H.; Jeong, S. K.; Ahn, S. K. An Update of the Defensive Barrier Function of Skin. *Yonsei Med. J.* **2006,** *47*, 293–306.
67. Liu, C. H.; Chang, F. Y.; Hung, D. K. Terpene Microemulsions for Transdermal Curcumin Delivery: Effects of Terpenes and Cosurfactants. *Colloids Surf. B Biointerfaces* **2011,** *82*, 63–70.
68. Lopes, L. B. Overcoming the Cutaneous Barrier with Microemulsions. *Pharmaceutics* **2014,** *6*, 52–77.
69. Marwah, H.; Garg, T.; Goyal, A. K.; Rath, G. Permeation Enhancer Strategies in Transdermal Drug Delivery. *Drug Deliv.* **2016,** *23*, 564–578.

70. Mcclements, D. J. Nanoemulsions Versus Microemulsions: Terminology, Differences and Similarities. *Soft Matter* **2012,** *8*, 1719–1729.
71. Megrab, N. A.; Williams, A. C.; Barry, B. W. Oestradiol Permeation Across Human Skin, Silastic and Snake Skin Membranes: The Effects of Ethanol/water Co-solvent systems. *Int. J. Pharm.* **1995,** *116*, 101–112.
72. Montenegro, L.; Carbone, C.; Condorelli, G.; Drago, R.; Puglisi, G. Effect of Oil Phase Lipophilicity on In vitro Drug Release from o/w Microemulsions with Low Surfactant Content. *Drug Dev. Indus. Pharm.* **2006,** *32*, 539–548.
73. Morais, A. R. V.; Alencar, E. N; Júnior, F. H. X.; Oliveira, C. M.; Marcelino, H. R.; Barratt, G.; Fessi, H.; Egito, E. S. T.; Elaissari, A. Freeze-drying of Emulsified Systems: A Review. *Int. J. Pharm.* **2016,** *503*, 102–114.
74. Mura, P.; Faucci, M. T.; Bramanti, G.; Corti, P. Evaluation of Transcutol as a Clonazepam Transdermal Permeation Enhancer from Hydrophilic Gel Formulations. *Eur. J. Pharm. Sci.* **2000,** *9*, 365–372.
75. Mura, P.; Bragagni, M.; Mennini, N.; Cirri, M.; Maestrelli, F. Development of Liposomal and Microemulsion Formulations for Transdermal Delivery of Clonazepam: Effect of Randomly Methylated Beta-cyclodextrin. *Int. J. Pharm.* **2014,** *475*, 306–314.
76. Nanda, S.; Saroha, K.; Sharma, B. Sonophoresis: An Eminent Advancement for Transdermal Drug Delivery System. *Int. J. Pharm. Technol.* **2011,** *3*, 1285–1307.
77. Nandi, I.; Bari, M.; Joshi, H. Study of Isopropyl Myristate Microemulsion Systems Containing Cyclodextrins to Improve the Solubility of 2 Model Hydrophobic Drugs. *AAPS PharmSciTech* **2003,** *4*, E10.
78. Newby, C. S.; Barr, R. M.; Greaves, M. W.; Mallet, A. I. Cytokine Release and Cytotoxicity in Human Keratinocytes and Fibroblasts Induced by Phenols and Sodium Dodecyl Sulfate. *J. Invest. Dermatol.* **2000,** *115*, 292–298.
79. Pakpayat, N.; Nielloud, F.; Fortune, R.; Tourne-Peteilh, C.; Villarreal, A.; Grillo, I.; Bataille, B. Formulation of Ascorbic Acid Microemulsions with Alkyl Polyglycosides. *Eur. J. Pharm. Biopharm.* **2009,** *72*, 444–452.
80. Paolino, D.; Ventura, C. A.; Nistico, S.; Puglisi, G.; Fresta, M. Lecithin Microemulsions for the Topical Administration of Ketoprofen: Percutaneous Adsorption Through Human Skin and in vivo Human Skin Tolerability. *Int. J. Pharm.* **2002,** *244*, 21–31.
81. Pappas, A. Epidermal Surface Lipids. *Dermato-Endocrinol.* **2009,** *1*, 72–76.
82. Patel, H. K.; Barot, B. S.; Parejiya, P. B.; Shelat, P. K.; Shukla, A. Topical Delivery of Clobetasol Propionate Loaded Microemulsion Based Gel for Effective Treatment of Vitiligo–part II: Rheological Characterization and In vivo Assessment Through Dermatopharmacokinetic and Pilot Clinical Studies. *Colloids Surf. B Biointerfaces* **2014,** *119*, 145–153.
83. Patel, H. K.; Barot, B. S.; Parejiya, P. B.; Shelat, P. K.; Shukla, A. Topical Delivery of Clobetasol Propionate Loaded Microemulsion Based Gel for Effective Treatment of Vitiligo: Ex vivo Permeation and Skin Irritation Studies. *Colloids Surf. B Biointerfaces* **2013,** *102*, 86–94.
84. Patel, Rr; Patel, Kr; Patel, Mr. Formulation and Characterization of Microemulsion Based Gel of Antifungal Drug. *PharmaTutor Mag.* **2014,** *2*, 79–89.
85. Patzelt, A.; Lademann, J.; Richter, H.; Darvin, M. E.; Schanzer, S.; Thiede, G.; Sterry, W.; Vergou, T.; Hauser, M. In vivo Investigations on the Penetration of Various Oils and Their Influence on the Skin Barrier. *Skin Res. Technol.* **2012,** *18*, 364–369.

86. Paul, S.; Panda, A. K. Physico-chemical Studies on Microemulsion: Effect of Cosurfactant Chain Length on the Phase Behavior, Formation Dynamics, Structural Parameters and Viscosity of Water/(Polysorbate-20 + n-alkanol)/n-Heptane Water-in-oil Microemulsion. *J. Surf. Deter.* **2011,** *14*, 473–486.
87. Peltola, S.; Saarinen-Savolainen, P.; Kiesvaara, J.; Suhonen, T. M.; Urtti, A. Microemulsions for Topical Delivery of Estradiol. *Int. J. Pharm.* **2003,** *254*, 99–107.
88. Proksch, E.; Brandner, J. M.; Jensen, J. M. The Skin: An Indispensable Barrier. *Exp. Dermatol.* **2008,** *17*, 1063–1072.
89. Qadir, A.; Faiyazuddin, M. D.; Talib, M. D. H.; Alshammari, T. M.; Shakeel, F. Critical Steps and Energetics Involved in a Successful Development of a Stable Nanoemulsion. *J. Mol. Liquids* **2016,** *214*, 7–18.
90. Rastogi, S. K.; Singh, J. Effect of Chemical Penetration Enhancer and Iontophoresis on the In vitro Percutaneous Absorption Enhancement of Insulin Through Porcine Epidermis. *Pharm. Dev. Technol.* **2005,** *10*, 97–104.
91. Rhee, Y. S.; Choi, J. G.; Park, E. S. ; Chi, S. C. Transdermal Delivery of Ketoprofen Using Microemulsions. *Int. J. Pharm.* **2001,** *421*, 161–170.
92. Sahoo, S.; Pani, N. R.; Sahoo, S. K. Effect of Microemulsion in Topical Sertaconazole Hydrogel: In vitro and In vivo Study. *Drug Deliv.* **2016,** *23*, 338–345.
93. Sahoo, S.; Pani, N. R.; Sahoo, S. K. Microemulsion Based Topical Hydrogel of Sertaconazole: Formulation, Characterization and Evaluation. *Colloids Surf. B Biointerfaces* **2014,** *120*, 193–199.
94. Sangale, P. T.; Manoj, G. V. Organogel: A Novel Approach for Transdermal Drug Delivery System. *World J. Pharm. Res.* **2015,** *4*, 423–442.
95. Savic, S.; Tamburic, S.; Savic, M. M. From Conventional Towards New - Natural Surfactants in Drug Delivery Systems Design: Current Status and Perspectives. *Exp. Opinion Drug Deliv.* **2010,** *7*, 353–369.
96. Shah, D. O., *Micelles, Microemulsions and Monolayers: Science and Technology*; CRC Press: New York, 1998.
97. Shakeel, F.; Ramadan, W.; Faisal, M. S.; Rizwan, M.; Faiyazuddin, M.; Mustafa, G.; Shafiq, S. Transdermal and Topical Delivery of Anti-inflammatory Agents Using Nanoemulsion/Microemulsion: An Updated Review. *Curr. Nanosci.* **2010,** *6*, 184–198.
98. Sintov, A. C. Transdermal Delivery of Curcumin via Microemulsion. *Int. J. Pharm.* **2015,** *481*, 97–103.
99. Sintov, A. C.; Shapiro, L. New Microemulsion Vehicle Facilitates Percutaneous Penetration In vitro and Cutaneous Drug Bioavailability In vivo. *J. Control. Release* **2004,** *95*, 173–183.
100. Tashtoush, B. M.; Bennamani, A. N.; Al-Taani, B. M. Preparation and Characterization of Microemulsion Formulations of Nicotinic Acid and its Prodrugs for Transdermal Delivery. *Pharm. Dev. Technol.* **2013,** *18*, 834–843.
101. Teichmann, A.; Heuschkel, S.; Jacobi, U.; Presse, G.; Neubert, R. H. H.; Sterry, W.; Lademann, J. Comparison of Stratum Corneum Penetration and Localization of a Lipophilic Model Drug Applied in an o/w Microemulsion and an Amphiphilic Cream. *Eur. J. Pharm. Biopharm.* **2007,** *67*, 699–706.
102. Valenta, C.; Schultz, K. Influence of Carrageenan on the Rheology and Skin Permeation of Microemulsion Formulations. *J. Control. Rel.* **2004,** *95*, 257–265.

103. Wang, F.; Fang, B.; Zhang, Z.; Zhang, S.; Chen, Y. The Effect of Alkanol Chain on the Interfacial Composition and Thermodynamic Properties of Diesel Oil Microemulsion. *Fuel* **2008,** *87*, 2517–2522.
104. Williams, A. C.; Barry, B. W. Penetration Enhancers. *Adv. Drug Deliv. Rev.* **2004,** *56*, 603–618.
105. Wu, H.; Ramachandran, C.; Weiner, N. D.; Roessler, B. J. Topical Transport of Hydrophilic Compounds using Water-in-oil Nanoemulsions. *Int. J. Pharm.* **2001,** *220*, 63–75.
106. Wuttikul, K.; Boonme, P. Formation of Microemulsions for Using as Cosmeceutical Delivery Systems: Effects of Various Components and Characteristics of Some Formulations. *Drug Deliv. Trans. Res.* **2016,** *6*, 254–262.
107. Xu, J.; Fan, Q. J.; Yin, Z. Q.; Li, X. T.; Du, Y. H.; Jia, R. Y.; Wang, K. Y.; Lv, C.; Ye, G.; Geng, Y.; Su, G.; Zhao, L.; Hu, T. X.; Shi, F.; Zhang, L.; Wu, C. L.; Tao, C.; Zhang, Y. X.; Shi, D. X. The Preparation of Neem Oil Microemulsion (Azadirachta indica) and the Comparison of Acaricidal Time Between Neem Oil Microemulsion and Other Formulations In vitro. *Vet. Parasitol.* **2010,** *169*, 399–403.
108. Yotsawimonwat, S.; Okonoki, S.; Krauel, K.; Sirithunyalug, J.; Sirithunyalug, B.; Rades, T. Characterisation of Microemulsions Containing Orange Oil With Water and Propylene Glycol as Hydrophilic Components. *Die Pharmazie Int. J. Pharm. Sci.* **2006,** *61*, 920–926.
109. Yu, A.; Guo, C.; Zhou, Y.; Cao, F.; Zhu, W.; Sun, M.; Zhai, G. Skin Irritation and the Inhibition Effect on HSV-1 In vivo of Penciclovir-loaded Microemulsion. *Int. Immunopharmacol.* **2010,** *10*, 1305–1309.
110. Zhang, X.; Wu, Y.; Hong, Y.; Zhu, X.; Lin, L.; Lin, Q. Preparation and Evaluation of dl-praeruptorin A Microemulsion Based Hydrogel for Dermal Delivery. *Drug Deliv.* **2015,** *22*, 757–764.
111. Zhao, X.; Liu, J. P.; Zhang, X.; Li, Y. Enhancement of Transdermal Delivery of Theophylline Using Microemulsion Vehicle. *Int. J. Pharm.* **2006,** *327*, 58–64.
112. Zhu, W.; Yu, A.; Wang, W.; Dong, R.; Wu, J.; Zhai, G. Formulation Design of Microemulsion for Dermal Delivery of Penciclovir. *Int. J. Pharm.* **2008,** *360*, 184–190.
113. Zhu, W.; Guo, C.; Yu, A.; Gao, Y.; Cao, F.; Zhai, G. Microemulsion-based Hydrogel Formulation of Penciclovir for Topical Delivery. *Int. J. Pharm.* **2009,** *378*, 152–158.

PART II

Nanoparticles and Materials for Food, Health, and Pharmaceutical Applications

CHAPTER 3

NANOPARTICLES AND THEIR POTENTIAL APPLICATIONS IN AGRICULTURE, BIOLOGICAL THERAPIES, FOOD, BIOMEDICAL, AND PHARMACEUTICAL INDUSTRY: A REVIEW

KAVITHA PATHAKOTI, LAVANYA GOODLA, MANJUNATH MANUBOLU, and HUEY-MIN HWANG*

*Corresponding author. E-mail: huey-min.hwang@jsums.edu

ABSTRACT

Nanotechnology is a rapidly growing field of research with diverse applications in various sectors of the industry. Nanoparticles (NPs) are extensively used for a broad variety of industrial, biomedical, pharmaceutical, and commercial applications due to their unique physicochemical properties. NPs exist in different forms of metals, metal oxides, quantum dots, polymers, carbon materials, and other NPs. This chapter discusses research efforts and potential application of nanotechnology in agriculture, food, biomedical, biological, and pharmaceutical applications. Nanotechnology applications to the agri-food sector are rather recent compared with their use in drug delivery and pharmaceuticals. Applications of nanotechnology in food industry include: food packaging, food processing, encapsulation of nutrients, and development of new functional foods. Nanosensors and nano-barcodes are used to assure food quality and safety. Besides these topics, this review also addresses biomedical applications such as nano-diagnostics, and developments in the discovery, design, and delivery of drugs, including nano-pharmaceuticals. Furthermore, topics on biological therapies such as vaccination, cell therapy, and gene therapy are also covered. In the end, the implications and perspectives of nanotechnology are discussed.

3.1 INTRODUCTION

Nanotechnology is a fastest developing field of research with diverse applications in various sectors of the industry. It is a multi-interdisciplinary technology based on scientific knowledge from other sister disciplines, including biology, chemistry, and physics.[135] It has been rising extensively globally in the past few years and it has been estimated that the global market for the nanotechnology was $7.6 billion in 2008 and was estimated to be at least $3 trillion by 2020.[200] Nanoparticles (NPs, 1–100 nm in size) have unique physical and chemical properties compared to their bulk counterparts, which endow their beneficial characteristics. Nanomaterials have smaller size, high surface energy, large surface-to-volume ratio and unique mechanical, thermal, electrical, magnetic, and optical behaviors.[253] Nanotechnology is referred to as the production, characterization, and manipulation of such materials.[6] The potential applications of nanomaterials are very attractive in many industrial sectors including biomedical, defense, energy and storage, pharmaceutical, agricultural, food, and environmental remediation.[6] Nanotechnology with these promising new insights and innovations is expected to mass usage by 2020, which can revolutionize many aspects of human life.

Due to the diverse properties and complex structure of nanomaterials, in general engineered NPs can be categorized into different forms of metals (e.g., gold, silver, iron and aluminum), metal oxides (e.g., TiO_2, ZnO, CuO, CoO, Al_2O_3, Fe_3O_4, SnO_2, Fe_2O_3, CeO_2, and others), quantum dots (QDs), polymers, fullerenes (C_{60}) carbon materials (single-walled and multi-walled carbon nanotubes [CNTs]) and other NPs. They exist in different sizes and shapes and also in different chemical compositions such as single or multi-elements. Different types of nanomaterials, including NPs, nanocomposites, nanoclays, and nano-emulsions are used in the agri-food industry.

Nanomaterials are already in commercial use and the range of commercial products existing today are very broad, comprising cosmetics, sunscreens, electronics, paints, varnishes, stain resistant, antibacterial dressings, wrinkle-free textiles, stain resistant fabrics, scratch free paints for cars, and self-cleaning windows. Nano TiO_2 and ZnO are used in sun blocking creams and self-cleaning windows. Silver and copper nanomaterials are used as antimicrobial agents.

In the scope of this chapter, authors have reviewed the research efforts and potential applications of nanotechnology in agriculture, food, biomedical, biological, and pharmaceutical applications. In addition, some of the industrial applications of nanomaterials in these areas and their future perspectives are also discussed.

3.2 NANOTECHNOLOGY IN AGRICULTURE

As the world's population is growing constantly and is expected to reach 9.6 billion by 2050, there is an increasing demand for global agricultural production with sustainable development strategies. Moreover, a continuous innovation is strongly needed in agricultural sector due to the challenges from global climate change. Earlier, many applied technologies have been used to increase the crop production, synthetic chemicals, genetically modified crops, and hybrid varieties. Recently, there has been great attention in the use of nanotechnology in agriculture to enhance agricultural productivity and sustainability.[62,106,182,214,218] There has also been growing trend in both scientific articles and patents in agricultural nanotechnology mostly in disease management and crop protection.[80,86,211] The competent delivery of herbicides, pesticides, fertilizers, and phytohormones can be done by using nanoscale carriers. Detailed overview of the applications in agriculture is provided in Figure 3.1.

3.2.1 NANO-PESTICIDES

Although nanotechnology has several potential applications in agriculture sector, yet its application in pesticide delivery is relatively new and is in infancy stage.[24,88] In addition, nanotechnology applications in agriculture are limited and have not made into the market to a large extent when related to the other industrial sectors. Nonetheless, several possible applications obtainable by nanotechnology are continuously being explored. This could be due to the higher cost incurred for the production of nanotechnology products, which are required in large volumes in agricultural sector.[182] Pesticides of nanoscale can enhance their dispersion, permeability, solubility, durability, and reduce the number of active ingredients used.[28,188] Some of the nano-pesticide delivery techniques, effective for crop protection, include: nano-emulsions, nanocapsules (polymers), and nanoclays.[28,107] Other release

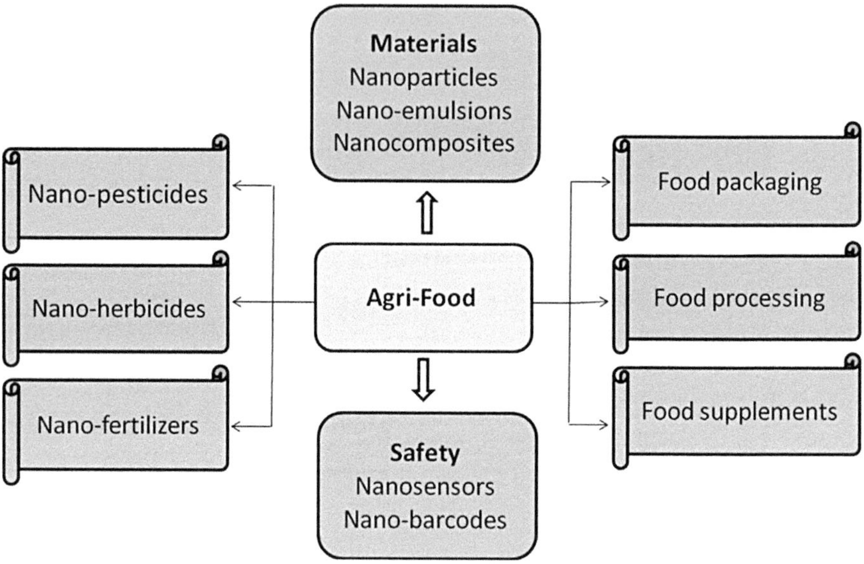

FIGURE 3.1 Nanotechnology applications in agriculture and food.

mechanisms include: biodegradation, osmotic pressure with specific pH, and dissolution. Furthermore, amorphous nanosilica has also been used as a biopesticide.[23]

3.2.2 NANO-HERBICIDES

Sustainable agriculture demands minimum usage of agrochemicals to minimize the extent of environmental pollution. Nano-encapsulation of herbicides in polymeric core–shell NPs can result in suitable management of herbicides, which promise environmental safety. The controlled release of herbicides and fertilizers is done by nano-encapsulation, which promotes the release of pesticide with low dosage over a prolonged period.[5,115] An eco-friendly approach can be used to eradicate weeds by using nano-herbicides, devoid of any toxic residues in soil and other environments.[188]

However, due to the lack of available field application data on nano-pesticides and stringent regulations,[121] it is still controversial whether registered pesticides contain nanomaterials.[75] Nevertheless, nano-pesticides are most likely to emerge into the market in the near future, which consist of formulations of existing active ingredients.[85,106,108]

3.2.3 NANO-FERTILIZERS

Nano-fertilizers are nanomaterials, which can provide nutrients to plants to augment their yield and growth.[143] Higher production costs are incurred due to the low efficiency of conventional fertilizers, which may also cause environmental pollution. Therefore, use of nano-fertilizers may minimize these effects by ensuring the controlled release of fertilizers thereby increasing the fertilizer efficiency. Moreover, the amount of fertilizer applied and the frequency of application are also reduced considerably, thus resulting in reduced labor costs.[67] These nano-fertilizers have shown improved performance when compared to the conventional fertilizers.[65,67,171]

3.2.4 NANOSENSORS

Nanosensors are used in agricultural sector for pesticide residue detection and toxin and pathogen detection. The real-time detection of the pathogen and location can be monitored using these nanosensors, which may be helpful in predicting the environmental and field conditions of the crop.[46] Recently, a chemical based nanosensor was developed using carbon nanomaterials for detection of pesticide residues in plants.[220] These nanosensors show high sensitivity and low-concentration detection limits[144]; and some of the other nanosensors for pesticide residue detection have been reported recently.[71,109,127,228,249,263] A highly sensitive organophosphorus pesticide biosensor was developed by using amino-functionalized CNTs to control the competent immobilization of AChE.[263] A relatively inexpensive nanosensor has been developed for detection of pesticide (organophosphates) residues using acetylcholinesterase.[244] These electrodes have been effectively employed for direct analysis of vegetable samples. Another AChE biosensor was developed by using multiwall CNT onto liposome bioreactors for detection of organophosphate pesticides.[257] Quantum dots (QDs) are used as fluorescent probes for sensitive detection of methylparathion.[49] In addition, highly fluorescent silica molecularly imprinted nanospheres embedded CdTeQDs (water soluble) was developed that can be used as a biosensor for determination of deltamethrin in fruit and vegetable samples.[84]

Mesoporous Silica NPs can deliver DNA and chemicals into plant cells, mediating gene transfer, and thus providing new possibilities in target-specific delivery in plant biotechnology.[236] Furthermore, fluorescent nanoprobes of silica NPs conjugated with goat antibody to detect *Xanthomonas axonopodis*—that causes bacterial spot disease in solanaceous plants—was developed.[259] A nano-gold based immunosensor for detecting Kernel bunt disease in wheat lots by using surface plasmon resonance was developed for establishing seed certification and plant quarantines.[224]

3.2.5 NANO-BARCODES

Nano-barcodes are specialized nanoscale tags and are used for tracking and verification in agricultural food and livestock farming products.[35] Nano-barcodes are also used to monitor the quality control of the agricultural product.[226]

3.2.6 PRECISION FARMING

Precision farming aims at higher crop yields while reducing the usage of pesticides, fertilizers, and herbicides through examining the environmental variables and applying relevant actions. In precision farming, computers, remote sensing devices, geographical information system, and global satellite positioning systems are used to detect specifically localized environmental conditions. This precisely determines whether crops are being grown at high proficiency or recognizing the nature and site of problems. By using integrated data to regulate soil conditions and plant growth, seeding, chemical, fertilizer, and water use, it can be adjusted to reduce manufacture costs and potentially enhance production and others to benefit the farmer.[197] The foremost stage in precision agriculture is the creation of soil maps with its features. Remote sensing coupled with geosynchronous positioning system (GPS) regulates and produces precise maps and models of the agricultural fields. Usage of autonomous sensors associated into a GPS for real-time monitoring is one of the important roles for nanotechnology empowered devices. These nanosensors can be dispersed throughout the field where they can observe plant growth and soil conditions. In certain parts of the Australia and the United States, these wireless sensors are already in use.

3.3 NANOTECHNOLOGY IN FOOD

Several recent reports and reviews have identified prospective applications of nanotechnology for the food sector to improve food safety, to improve packaging, improved processing, and nutrition.[62,42,92,122,174,191,217,218] Increasing the shelf-life of the food (preservation), food safety, coloring, flavoring, nutritional additives, and using the antimicrobial ingredients for food packaging are the important uses of nanotechnology in food industry.[42,72]

Nanotechnology has major advantages in its usage for packaging in comparison to the conventionally ways using polymers, which may include merits such as enhanced barrier, mechanical, and heat resistant properties, along with biodegradability.[133] In addition to enhanced antimicrobial effects, nanomaterials can also be used for detection of food spoilage with nanosensors.[63] Various forms of "nanosystems" such as solid NPs, nanotubes, nanocapsules, nanofibers, nanocomposites, nanosensors, and nano-barcodes are some of the major nanomaterials used in food processing, packaging, and preservation sectors.[72]

3.3.1 FOOD PACKAGING

Currently, there is a growing interest to develop different food packaging materials to meet the increasing demand for foods with longer shelf-life and minimum processing. Apparently, such new materials should have excellent mechanical, thermal, and optical properties. In addition, significant antimicrobial and barrier properties are important to prevent migration of oxygen, carbon dioxide, vapor, and aroma. These properties play a key role in promoting the shelf-life of fresh and processed foods.[14] Among the recent advancements of nanotechnology in food packaging, nanocomposite materials are used in active packaging and smart packaging.

3.3.1.1 NANOCOMPOSITES FOR IMPROVED PACKAGING

Using the polymers in nanocomposite food packaging is one of the good alternatives for conventional packaging materials (glass, paper, and metals) due to their functionality and low cost.[99] Nanocomposites are polymer matrices reinforced in the nanofillers (nanoclays, nano-oxides, CNTs, and cellulose microfibrils), where one of the phases has at least one, two, or three dimensions less than 100 nm.[223] Several synthetic (polyamide, polystyrene, nylon, and polyolefins) and natural polymers (chitosan, cellulose, and carrageenan) have been used in food packaging.[25,223] However, due to environmental concerns, there is a growing demand for biodegradable packaging with the usage of biopolymers that are either natural or synthetic (polyvinyl alcohol, polylactide, and polyglycolic acid)[195].

Graphene nanoplates based nanocomposites are reported for their heat resistance and barrier properties, which has an improved food-packaging application.[193] CNTs and nanofibers are used due to their electrical and mechanical properties, but their application in food packaging is limited because of cost factor and the difficulty in processing dispersions.[14] Nanoclays with montmorillonite NPs in different starch-based materials (biodegradable polymers) are developed to improve barrier and mechanical properties.[18] Most commonly used nanoclay material montmorillonite (also known as bentonite) is used to achieve the gas barrier properties and it can restrict the permeation of gases when incorporated into a polymer. It is also widely available and is relatively inexpensive. Some of the biodegradable nanocomposites in starch–clay have been reported for several applications including food packaging.[44,57,100,101,183]

3.3.1.2 ACTIVE PACKAGING

Traditional food packaging systems are believed to passively protect the food by acting as a barrier among the food and the nearby environment. Active packaging has desirable role in food preservation other than providing a passive barrier to the external conditions. It mainly refers to the packaging systems that respond to changes in the environment. They act by discharging required molecules such as antimicrobial or antioxidant agents or act as gas scavengers. These interactions result in improved food stability and some of these packaging systems include: the antimicrobials, oxygen scavengers, and enzyme immobilization systems. Another application in active packaging is the controlled-release packaging, where nanocomposites can also be used as delivery systems, thereby helping the migration of functional additives, such as minerals, probiotics, and vitamins into food.[202]

3.3.1.3 ANTIMICROBIAL SYSTEMS

Active packaging includes usage of metal and metal oxide NPs as antimicrobial agents in the form of nanocomposites for food packaging. Titanium dioxide, zinc oxide, copper, copper oxide, and silver-based nanofillers are used due to their antimicrobial properties.[72,102] TiO_2 and SiO_2 based nanofillers are used for application in self-cleaning surfaces.[72] Among them silver NPs are the most common NPs, which are effective against a wide variety of microorganisms.[26,63] Adhesion to cell surface, disruption of cell membrane, DNA damage and release of silver ions are the mechanisms of silver antimicrobial agent.[72,58,166] TiO_2 has a broad range of applications, such as UV blocker, pigment, photocatalyst, and antimicrobial agent.[260] Moreover, TiO_2 NPs are effective against food spoilage related bacteria[43,199] and they are also used in food packaging.[72,260] Besides TiO_2, it was reported that a plastic wrap containing ZnO NPs is used to keep the packaging surfaces hygienic under indoor lighting conditions.[42]

3.3.1.4 OXYGEN SCAVENGERS

Oxygen in food causes several degradative oxidation reactions like rancid flavors and browning reactions, and also produces some harmful compounds. Adding oxygen scavengers into food packaging systems can reduce oxygen concentrations and thereby preserving the food. Oxygen scavenger films were developed by adding TiO_2 NPs into different polymers[255] under UV illumination; however, one major drawback of TiO_2 NPs is due to its photocatalytic mechanisms; that is, becoming active only under UV light because of the large bandgap.[158]

3.3.1.5 SMART PACKAGING

Smart packaging includes nano-devices or polymer associated nanosensors to monitor the quality of the food while storage and processing.[19,256] Moreover, smart packaging ensures true nutritional value and authenticity of the packed food product. Furthermore, these devices may also track the history of time, temperature, and expiration date. Some recent reports indicated that nanosensors are able to detect toxins and food pathogens in the packaging.[136,223,258] Beside antimicrobial properties, NPs are also applied as vehicles to enhance flavor, enzymes carriers, antioxidants, anti-browning agents, and additional materials to increase shelf-life, even when the packaging is unwrapped.[39,129,253] Nanocoatings of TiO_2 are used as nanosensors and nano-barcodes for food safety and authenticity, respectively.[41] Another application is shown with the plastic films containing aluminum NPs. In their usage as the covering, there was no evidence of NPs migration from plastic polymers into food.[40] Nanosensors have also been advanced for analysis of food, drinking water, flavors, and clinical diagnostics[141] and NPs can also be integrated as nanostructured transducers of the electrochemical sensors.[145]

Nanobarcodes are used as a biological fingerprint created by NPs, which generate unique reading stripes for every food item.[176] Moreover, these nano-barcodes can also detect pathogens in food products.[91,140]

3.3.2 FOOD PROCESSING

Food processing includes toxin removal, nutrition preservation, and pathogen prevention for maintaining food nutritional quality or to alter the food matrix, as per the consumer needs. Development of new functional materials for product development with improved food safety and biosecurity is essential in food processing industries. Nanocapsule delivery systems play a major role in processing sector and the functional properties are conserved by encapsulating colloids, emulsions and biopolymers into foods.[1] These collective food processing constituents like nanoemulsions and nanocapsules possibly are employed so that functional food materials are integrated, absorbed, or dispersed in nanostructures.

3.3.2.1 NANO-ENCAPSULATION

Nano-encapsulation is the method of packing the things at nanoscale with the help of nanocapsules and it affords end product performance that comprises controlled release of the core. Thus encapsulated forms of constituents have extended shelf-life; enhanced stability with consecutive delivery of multiple active ingredients including pH-triggered controlled release. Functional ingredients (such as vitamins, antioxidants, probiotics, carotenoids, preservatives, omega fatty acids, proteins, peptides, and lipids as well as carbohydrates) are incorporated into a nano-delivery system.[74] This increases the functionality and stability of these foods, as they are not used in their pure form. Lipid-based nano-encapsulation can intensely increase the strength, solubility, and bioavailability of foods, thus avoiding undesirable contacts with added food components. Nanoliposomes and nanocochleates are a few of the most encouraging lipid-based carriers for antioxidants. Nanoliposomes also help in controlled and specific delivery of nutraceuticals, nutrients, enzymes, vitamins, antimicrobials, and additives.[231] Nanocochleates have the ability to stabilize the micronutrients and the processed foods nutritional value can be enhanced. Probiotic encapsulation with nanoparticles can be delivered to the target area of the intestinal tract, where they have the capability to modulate immune responses.[242] Tip-Top Up bread in Western Australia is fortified with omega-3 fatty acids, which make it one of the best examples in the present market. HydraCel®, a regular mineral product (5 nm size), can be used to improve the water and nutrients absorption in the body by decreasing the surface tension of drinking water.[155] Casein micelle is a carrier of sensitive food products,[219] dextrins for bioactive compounds and hydrophobically altered starch for encapsulation of curcumin.[264]

3.3.2.2 NANO-EMULSIONS

Nano-emulsions are colloidal dispersions with droplet sizes ranging from 50 to 1000 nm. They are used to produce food products for flavored oils, salad dressing, personalized beverages, sweeteners, and other processed foods.[125] Nano-emulsions compromise several benefits such as cleansing of equipment and high precision without compromising product appearance and flavor. Nano-sized functional compounds—that are encapsulated by the self-assembled nano-emulsions are used for targeted delivery of lutein, β-carotene, lycopene, vitamins A, D, E_3, co-enzymeQ10, and omega-3-fatty acids[83] Stable double-layered capsaicin-loaded nano-emulsions were stabilized with natural polymers such as alginate and chitosan for use as a functional ingredient delivery system.[226]

3.3.2.3 NANO-FILTRATION

Nano-filtration is a method for selective passage of materials on the basis of shape and sizes, thus separating the materials from a watery medium. Nano-filtration membranes require a great porousness for monovalent salts (sodium and potassium salts) and organic compounds (e.g., proteins, lactose, and urea) with very low permeability. Nano-filtration has remained effectively practical in treatment plants for drinking water.[33,56,95] Moreover, nano-filtration combined with powdered activated carbon has been directly used to eliminate effluent organic material from municipal wastewater[52,113] Overall a purity of over 91% can be achieved by using nano-filtration, which is much greater than the traditional methods.[241]

3.3.3 FOOD SUPPLEMENTS

The potential nanotechnology applications in functional food design of dietary supplements and nutraceuticals comprising nanosized materials and additives like vitamins, antimicrobials, antioxidants, and preservatives presently exist for enriched taste, absorption, and bioavailability.[164] To avoid accumulation of cholesterol, some of the nutraceuticals integrated in the carriers consist of beta-carotenes, lycopene, and phytosterols.[169] A green tea product containing nanoselenium has many health benefits arising after better uptake of selenium. Nanocalcium salts are used in chewing gums and have a patent application.[212] Nano-salts of iron, calcium, and magnesium are accessible as health supplements. Nano-iron is also used to disinfect water by degradation of organic pollutants and killing pathogenic microorganisms. Recently, bulk amounts of SiO_2 and TiO_2 oxides have been permitted to be used as food additives (E551 and E171, respectively).[76] Furthermore, SiO_2 is used as an anti-caking agent[168] and these materials, including TiO_2, are often not deliberately created in the nano-sized range, but may include a fraction in that size range.[64,189] These NPs are weakly soluble and their components may be gradually dissolved through digestion.[40,54] Consequently, these nanomaterials may be absorbed, retained, and accumulated within the body.[168]

3.4 BIOLOGICAL THERAPIES INVOLVING NANOTECHNOLOGY

The integration of nanomaterials with molecular biology has led to therapeutic applications, and biological therapies are intended to precisely target a biological phenomenon where a gene or a protein (a group of genes or proteins) is believed to be involved in the disease. Recent developments in nanotechnology are focused on using molecular techniques to generate nanoscale products that might find usage in biological therapies. It is a rapidly developing field, which is principally beneficial for delivery of biological therapies. These alternative approaches involving nanotechnology can overcome several limitations of conventional therapies. The objective of this part of the chapter is to present the potential applications of nanotechnology in biological therapies mainly on vaccines, cell therapy, and gene therapy (Fig. 3.2) that have been discussed under this section.

3.4.1 VACCINES

Traditional vaccines are prepared to employ live microorganisms (cellular vaccines), deactivated proteins, or toxins (acellular vaccines), which provide active acquired immunity to a particular disease. The major drawback of live vaccines comprises a high risk of reverting back to their virulent form and intrinsic instability, creating them tough to deliver, while inactivated vaccines produce a weaker immune response and multiple doses are required.[51] Synthetic peptide-based vaccines have shown several benefits like enhanced safety and preservation over the traditional vaccines. Nonetheless, a weaker immune response is produced by using peptide-based vaccines; and to boost the immunogenicity addition of adjuvants and novel delivery systems are highly needed.[209] The majority of vaccines represent "minimalist" composition, which typically exhibits less immunogenicity.[149] The role of nanotechnology in vaccine development has been growing greatly which is leading to nano-vaccinology.[149]

The development of virus-like particles (VLPs) and the resurgence of NPs (QDs and magnetic NPs) mark a convergence of protein biotechnology with inorganic nanotechnology that promises an era of significant progress for nanomedicine.[149,235] In therapeutic approaches, NPs can be used as a delivery system to increase antigen presentation or as an immune-stimulant adjuvant to activate or augment immunity (Fig. 3.2). Additionally, hypersensitivity caused by the NPs can be improved by controlling the shape and size or by

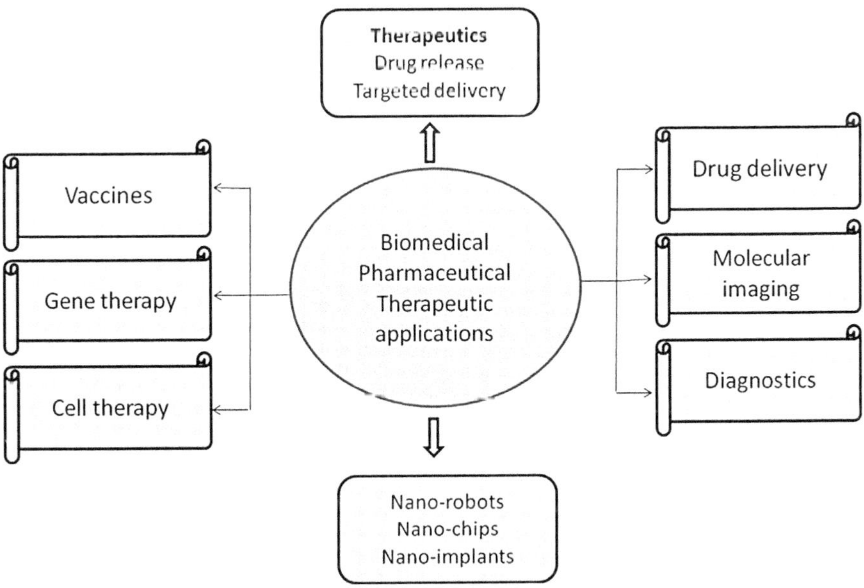

FIGURE 3.2 Diagrammatic representation of nanotechnology involved in biomedical, pharmaceutical, and therapeutic applications.

reducing the infusion rate of the delivery nanovaccine system.[161] Therapeutic nanovaccinology is typically used for cancer treatment.[32,90,124] Furthermore, there is ongoing number of accepted nano-vaccine and drug delivery systems that emphasize on the revolution in prevention and treatment.[51,69,154,203]

According to the nanomaterial composition,[150] the vaccine-related NPs[268] can be classified as polymers, VLPs, liposomes, nano-emulsions, and non-degradable nanospheres. Over the past decade, nanocarriers (such as VLPs,[201] liposomes,[265] polymeric NPs,[229] and non-degradable nanospheres[68]) were deemed as prospective delivery carriers for vaccine antigens. These carriers stabilize vaccine antigens and can act alone as adjuvants. These NP systems also enable entry into antigen-presenting cells (APCs) by diverse pathways and modulate immune responses.[89,98]

3.4.1.1 POLYMERS

A large variety of synthetic polymers like poly(D,L-lactic-co-glycolic acid) (PLGA),[116] polyethylene glycol (PEG),[243] polyester bio-beads,[185] and polystyrene[159]) are used to prepare NPs. PLG and PLGA are extensively investigated due to their exceptional biocompatibility and biodegradability[59] Some of the natural polymers based on polysaccharide are used to prepare NP adjuvants, such as inulin,[205] chitosan,[77] alginate,[137] and pullulan[213]. Hydrogel NPs are synthesized from synthetic polymers, Poly L-lactic acid (PLA), PLGA, PEG, and natural polymers, which have amazing properties such as broad surface area aimed at multivalent conjugation, flexible mesh size, high moisture content, and antigen loading capacity.[230]

3.4.1.2 INORGANIC NPs

Inorganic NPs are non-biodegradable, but they offer advantages such as rigid structure and controllable synthesis.[268] Gold NPs and carbon NPs are used in vaccine delivery[89,186] and mesoporous carbon NPs have been considered for application as an oral vaccine adjuvant.[250] Silica-based NPs (SiNPs) are biocompatible and they are used as nanocarriers for various applications, such as selective tumor targeting,[181] real-time multimodal imaging[250] and vaccine delivery. In the near future, vaccine formulations may possibly be developed from mesoporous silica NPs (MSNPs) into highly efficient and controlled-release nanocarriers.[237]

3.4.1.3 LIPOSOMES

Liposomes are biodegradable phospholipids consisting of phospholipid bilayer with an aqueous core. They can include antigens within the core for delivery and integrate viral envelope glycoproteins to form virosomes. One of the most used adjuvant delivery systems in DNA vaccine studies is the combination of a cationic polymer (protamine) with a cationic liposome condensed DNA (liposome-polycation-DNA NPs [LPD]).[138]

3.4.1.4 NANO-EMULSIONS

Nano-emulsions can bear antigens within their core to enhance the competence of vaccine passage or simply they are combined with an antigen.[268] For example, malaria vaccines,[126,179] Montanide ISA 201 and 206 have been used in foot-and-mouth disease vaccines.[60]

3.4.1.5 VIRUS-LIKE PARTICLES

VLPs are self-assembling NPs to exclude infectious components, which are naturally improved for contact with the immune system. Even in the absence of adjuvant, they also can stimulate effective immune responses.[267] The first VLP based vaccine (hepatitis B) was the first NP class to reach the market,[10] which was commercialized in 1986. Other VLP vaccines for human papillomavirus[55] and hepatitis E[184] were approved recently for use in humans in 2006 and 2011, respectively. Another example is Engerix-B® (Hep-B[Eng]), which is a noninfectious recombinant DNA vaccine containing hepatitis B surface antigen (HBsAg).[146]

3.4.1.6 IMMUNOSTIMULATING COMPLEX

Immunostimulating complex (ISCOMs) are cage-like particles (40 nm) comprising the saponin adjuvant and immunizing peptides, which permit selective inclusion of viral envelope proteins by hydrophobic interactions.[165] Diverse antigens have been used to form ISCOMs, including antigens derived from influenza,[50] herpes simplex virus,[162] HIV,[4] and Newcastle disease.[97]

3.4.1.7 SELF-ASSEMBLED PROTEINS

Self-assembled proteins are used in the preparation of NP-based vaccines, in an attempt to provide enhanced immunological properties. Self-assembled protein NPs confer highly stable and symmetric structure, which can be synthesized from a range of natural proteins.[184]

3.4.2 STEM CELL THERAPY

Recently, nanotechnology applications in stem cell investigations have developed immense advancement. For instance, magnetic NPs (MNPs) have been efficiently used for isolation and sorting of stem cells.[105] Tracking of stem cells and molecular imaging can be done by using QDs.[178] In addition, nanomaterials such as CNTs,[53] fluorescent CNTs,[221] and fluorescent MNPs[262] are used in drug delivery or genes into stem cells. Tysseling et al.[239] reported that self-assembling nanofibers (peptide amphiphile molecules) can promote axon elongation after spinal cord injury. Perivascular application of nitric oxide (NO) releasing self-assembling nanofiber gels is an efficient and simple therapy to avert neointimal hyperplasia after arterial injury.[111] This NO-based therapy has immense clinical potential to avoid neointimal hyperplasia after open vascular interventions. Another study was reported by Silva et al.[222] featuring the selective differentiation of neural progenitor cells by high-epitope density nanofibers.[222]

3.4.3 GENE THERAPY

Initially, gene therapy was restricted to augmentation of a specific missing/malfunctioning gene (gene augmentation therapy). Nowadays, gene therapy broadly includes several types of genetically induced modalities aimed at both treatment (therapeutics) and prevention (prophylaxis) of diseases. Another strategy of gene therapy involving RNA interference is intended at targeting and impairing the posttranscriptional expression of undesirable genes and disease-related signaling pathways.[73] Traditional gene therapy directly affects gene-directed processes of transcription or translation. Nanoparticulate nonviral vectors are considered to be safe and efficient delivery agents. Newer NPs have been developed for gene therapy, which are used for nonviral carrier systems including liposomes, polymers, and inorganic NPs. These NPs are assembled for the combination of gene delivery with other imaging or therapeutic capability.[45]

3.4.3.1 LIPOSOMES

Liposomes for gene delivery usually comprise a cationic lipid and a "helper" lipid. The cationic lipids (e.g., 1,2-dioleoyl-3-trimethylammoniumpropane [DOTAP]) carry quaternary ammonium groups that facilitate them to merge with the anionic genetic components.[190]

3.4.3.2 POLYMERS

Polymeric gene delivery is typically carried out by using the cationic polymer polyethylenimine in both the branched (b-PEI) and the linear (l-PEI) forms.[147]

3.4.3.3 INORGANIC NANOPARTICLES

Some of the inorganic-based NPs, with suitable surface modification, have been investigated for their applications in gene delivery. Major examples include silica-based NPs,[204] gold nanorods,[34] metal oxide NPs,[15] QDs,[261] and calcium phosphate NPs.[31] Covalent conjugation of gold nanorod (GNR) with a thiolated siRNA was used for siRNA delivery, utilizing the strong binding affinity of thiol groups on the gold surface.[34] Iron oxide NPs coated with a lipid-like material for delivery of both DNA and siRNA offer a platform for magnetically based gene delivery.[104]

3.5 BIOMEDICAL APPLICATIONS OF NANOTECHNOLOGY

The nanomaterials employed as constituents in designing nanomedical equipment have given rise to enhanced performance and biocompatibility, crafting them remarkable for regenerating and substituting the lost tissues through the process of tissue engineering, cell repair, and prostheses.[82,153]

3.5.1 NANO-ROBOTS

The wide range of biomedical applications involving nano-robots include diagnosis of diseases at primitive stage and drug delivery to the targeted sites, nanosized biomedical devices for surgical procedures, cancer treatment, monitoring of diabetes, gene therapy, and drug delivery.[51,132] The processes via engineering of the biological systems like adhering NPs to microorganisms by employing the platforms inside genetic machinery of bacteria[12,151] and computer designing of deoxyribonucleic acid (DNA) that react to coherent mixtures of constituents are able to create the nanorobots.[27] One of the major beneficial effects of nanorobots is that they are able to diffuse through the smallest blood vessels and can, thereby, access each single cell of tissues to extend the preferred regions and adhere on to the targeted cells through the process of engineering.[7,96] Nanorobots may be installed into the circulatory system or body cavities, and then they can be computed and composed distantly through a specialist doctor to execute different diagnostic and therapeutic methods accurately with minimal irruption. Arrangement like this will lead to speedy retrieval while permitting contacts with the operating expert via indications. During the surgery, nanorobots are employed for nano manipulations at the level of nanoscale.[114, 207]

Unlike regular clinical research systems, medical robots like, medical micro-devices that are implanted in tissues could deliberately congregate the diagnostic data and adjust the mechanisms of usages progressively over a lengthy duration of time.[96] Additionally, small mini cameras for viewing the digestive tract, together with embedded glucose and bone growth monitors can be used in the management of diabetes and joint substitutes. Nanotechnology has been extended to manufacturing the nano-size in vivo devices, which has the perspective to modernize health care[167] with the device tiny adequate to approach and interrelate with each and every single cell of the body.[120,246] Moreover, a wide range of studies has been concentrated on the cell imaging, which is specific to target cells and drug delivery attributing to the nanomaterials applications and progress[112,157] For instance, in a magnetic resonance imaging (MRI) system, the outer magnetic fields can transfer ferromagnetic particles comprising microrobots over circulatory system.[79,152]

Intensive Investigations on additional complicated equipment that comprise multicomponent nano-devices known as astecto dendrimers that process a dendrimer having single-core to auxiliary dendrimer modules of various kinds are attached to achieve precise functioning of smart therapeutic nanomachines.[29] Likewise, these nano-devices are created to offer the advantages such as external control of metabolism within the cells by the practice of nano-machines acting as minute nano radiofrequency antennas adhered to DNA to control hybridization.

3.5.2 NANO-CHIPS AND NANO-IMPLANTS

To design nanomaterials for preparing nano-chips and nano-implants, understanding of the structures of biological tissues is important.[142] Nanomaterial characterization based on structure and size is extremely essential in considering the association among the structures of nanomaterials and normal tissues at different levels. Currently, different types of nanomaterials are being generated for manufacturing various kinds of nano-chips, nano-implants, and even for regenerating the stem cells.[225] Nano-electrodes implanted in brains are extensively employed to manage neurological disorders. Because of their physical and chemical properties, NPs can robustly influence different inflammatory processes.[131] As the progression of angiogenesis is related to inflammation,[247] therefore, it is mandatory for having sufficient vascularization of scaffolds after inserting into the host tissue.[131]

The bone is the strongest cellular matrix which contributes mechanical strength to our body, shielding other internal organs, synthesizing, and storage of hemopoietic cells in the bone marrow.[238] Bone tissue engineering scaffolds need supply an osteoconductive surface to boost new bone ingrowth after embedding into the defective bone that could be reached by hydroxyapatite (HA) filling of specific scaffold biomaterials. The in vivo and in vitro effects of a unique hydroxyapatite NPs/polyester-urethane composite scaffold which was synthesized with a salt leaching–phase inverse technique was reported by Lashke et al.[131] Nano hydroxyapatite strengthened polymer or ceramic nanocomposites have acquired accreditation as bone scaffolds due to their configuration and structural resemblance to natural bones[13]

In 1999, an increase in the adhesion of osteoblasts on nanostructured materials was reported employing alumina and titania to design robust nano-implants.[252] Studies on investigation and comparison of the osteointegration of zirconia and titanium dental implants in mini pigs have been reported by Stadlinger et al.[227] Another alluring material that can aid in the preparation of implants is ceramics, due to their magnificent biocompatibility and strong resistance to wear. Ceramics can be used in an extensive array of clinical applications similar in orthopedics for femoral head and hip substitutions.[66] The mechanisms of enhanced cellular activities of bone forming cells such as osteoblast, and chondrocyte on nanophase materials are well documented.[251] Studies on the mechanism of action of vitronectin (a protein in human serum) demonstrated that its concentration, conformation, and bioactivity were responsible for the increasing adhering ability of osteoblasts on nanophase ceramic formulations.

Another broadly studied nanostructures as strengthening agents for bone scaffolds and in composite materials are CNTs and carbon nanofibers (CNFs).[20,38] Unlike ceramic and other metallic-based bone scaffolds employed in orthopaedics, single-walled CNTs are minimal thickness, which enable creating lighter scaffolds with greater strength. Recently, zirconia implants have been introduced into dental implantology, which might offer beneficial alternative to titanium.[227] Reiotelli et al.[11] reported potential of zirconia as an outstanding implant material but recommended for further clinical studies to provide additional evidence to support its clinical application.

In biomedical treatments, the cartilage regeneration is one of the requirements. One of the most studied nanomaterial as scaffold for cartilage tissue engineering is electrospun polymeric nanofibers. As these nanofibers greatly mimic the natural extracellular matrix (ECM), they have emerged as potential candidates for cartilage tissue engineering[139,234] Poly-ε-caprolactone (PCL) nanofiber sheets could serve as excellent candidates in covering the surface of the joints as they are malleable and can be rolled or folded and molded. Liu et al.[145] reported that it is viable to implant PCL nanofiber scaffolds with periosteal cells in vivo and consequently synthesize engineered cartilage in vitro.

Polymeric nanomaterials bolster response of vascular cells including endothelial and smooth muscle cells (SMCs).[142] The first tissue-based engineered blood vessel substitute was generated in 1986. Ravi et al.[194] indicate that the blood vessel was generated by culturing the bovine endothelial cells, SMCs, and fibroblasts in layers of collagen gel entrenched on a Dacron mesh. At present, synthetic vascular grafts are fabricated by employing expanded polytetrafluoroethylene (ePTFE), polyethylene terephthalate (Dacron), and polyurethane.[110] Synthetic resources are being used in designing the vascular grafts for diverse causes, largely due to the simplicity and malleability of altering their mechanical characteristics. At present, use for the replacement of aorta Dacron is most frequently employed and to a minimum extent it is used as a canal/duct for femoropopliteal bypass surgery. For the purpose of hemodialysis, Nakagawa et al.[172] created a poly(ether-urethane) graft strengthened with joined polyester fibers, which was recognized to be extra robust than ePTFE. Additional advancement in the research has resulted in a poly(carbonate-urea) urethane vascular graft that illustrates a confirmed structure identical to human arteries.[215]

Nanomaterials are also being investigated for treating the bladder diseases.[232] Greater than 90% of bladder cancers are initiated in the urothelium layer. These classified seeming bladder cancers require the replacement of bladder tissue because they generally need the evacuation of enormous sections of the bladder or the whole bladder wall.[233] Micro dimensional polymer structures have been used in designing the substitutes of synthetic bladder wall. Nanostructured polymers employed in creating synthetic bladder imitate the topography of normal bladder tissue and have been examined in vitro. Thapa et al.[233] first reported that adhesion and proliferation activities of bladder smooth muscle cell were enhanced on nano-dimensional polymeric surfaces, compared to polymeric surfaces with micro dimensional features.

Neural prostheses have been employed for applying and tracking the electrical signals in the neural tissues.[142] Nanomaterials with remarkable electrical properties can serve as excellent contender to receive and transfer electrical signals. Furthermore, nanomaterials—that assist and augment the extension of nerve cell neurite or axon due to their exemplary biocompatibility properties and conductivity can also serve as promising neural prostheses. Investigations on the biomaterial usage to boost the nervous tissue injury function of have furnished encouraging results but not without noteworthy obstacles, like incomplete or delayed regeneration of tissue.[216] Recent advances in scaffold tissue engineering have integrated nanoscale characteristic surface magnitudes that mimic normal neural tissue to meet the requirement for improved nervous system biomaterials. Nanoscale composite material synthesis with piezoelectric zinc oxide particles are encapsulated into a polymer matrix from electrically active nanomaterials. Zinc oxide can ideally administer an electrical stimulus that is a recognized stimulatory indication for regeneration of neural tissue.

3.6 PHARMACEUTICAL APPLICATIONS OF NANOPARTICLES

NPs are engineered to identify the disease circumstances and identify pathogens; recognize quintessential pharmaceutical agents to cure the disease ailment or pathogens; fuel enormous fabrication of compared pharmaceuticals (potentially in vivo); locate, enter, or attach to the target tissue, pathogens, or structures; and disburse the quintessential mass of accorded biological compound to the target areas.

3.6.1 NANOPARTICLES AS DIAGNOSTIC AGENTS

Detection of disease condition at a very early stage by identifying biomarkers/indicators in a lab sample can be achieved by attaching NPs to proteins or other molecules. Several efforts are underway to develop NP disease detection systems. MRI is one of the most mature applications of nanotechnology to diagnose the cancer. A wide range of NPs have been developed as MRI contrast agents; and these NP-based contrast agents are produced with an iron oxide crystals core, with or without an encapsulation of organic material, including polyethylene glycol (PEG).[160,187]

Researchers have established the method for acquisition of each and every single cancerous cell circulating in the bloodstream by employing nanofibers coated with antibodies that adhere to the cancer cells, trapping them for analysis. Compared to the conventional diagnostic tools which need the presence of hundreds of cancer cells to detect, this method allows the acquisition of individual cancer cells at an earlier stage. The targeted molecular markers present inside the cells can be recognized by injecting the biomolecule coated ultra-small superparamagnetic iron oxide (USPIO) particles into the bloodstream, which can generate an explicit signal for detection by MRI. For example, in human epidermoid carcinoma cells flow cytometric detection of epidermal growth factor receptor (EGFR) can be performed by employing in vitro streptavidin-coated fluorescent polystyrene NPs and also well documented was that these NPs enhance the sensitivity to identify EGFR related to the conjugate streptavidin—fluorescein.[30]

In extension, minute quantities of serum cancer biomarkers (prostate-specific antigen) were detected by unique technique known as NP oligonucleotide bio-barcode assay. This unique technique offers high ratio of PCR-amplifiable DNA to labeling antibodies that can substantially improve the sensitivity of the assay.[173] Therefore, a minute quantity of serum-free prostate-specific antigen can be identified in the victims of prostate cancer and even in women patients with breast cancer with a huge development in tumor diagnosis and screening.[130] At present, NPs are being used in molecular imaging analyses to obtain a most rigorous detection with greater image quality. Indeed, for tumor and atherosclerosis diagnosis, contrast agents have been loaded onto NPs. Wide range of NPs are being employed for molecular imaging with computed tomography, fluorescence, MRI, nuclear, and ultrasound imaging.[130,254]

During the last 15 years, nanotechnology has contributed more to the oncology. Nano-carriers occupied a significant area in curing cancer with the first nanocarrier drug for injectable therapeutics using liposomes is available in the market.[93,198] FDA approval of liposomal doxorubicin for use against Kaposi's sarcoma was granted in 1990[208] and later this drug was also approved for the therapeutic application of breast and recurrent ovarian cancers. To date, different kinds of nano-carrier centered drug-delivery systems with diversified structures, geometry, physicochemical characteristics, and surface functionalization have been created and are at the various phases of reinforcement.[78,248] In the near future, QDs may become the most employed robust NPs for screening and diagnosis of cancer tumors in patients.

Raman Spectroscopy, method validated to permit identification of viruses and bacteria in less than an hour, employs silver nanorods in the diagnostic system.[177] Detection of virus left after conventional drug therapy may be possible with gold-coated ferromagnetic NPs tagged with HIV antibodies. Metal NPs have also been functionalized with diverse biomolecules to identify specific proteins, antibodies and other disease indicators in the form of dendrimers. These dendrimers can also be attached to the fluorescent markers. Diagnosis of flu virus can be provided by gold NPs that have antibodies attached. The extent of light reflected back rises when it is focused on a sample comprising virus particles and the NPs because of the clustering around virus particles.[2]

Stem cell imaging with MRI using superparamagnetic NPs, is an emerging area of great interest in the last few years, which is used in treating stem cells in vitro. On injecting the stem cells to a definite location in the body, these cells ingest the NPs by endocytosis resulting in their accumulation at intracellular level that can use a local effect for in vivo detection.[81,123]

3.6.2 NANOPARTICLES AS THERAPEUTIC AGENTS

Nanomaterials are beneficial as efficient vaccine adjuvants for the treatment of cancer and other infectious diseases by modulating cellular and humoral immune responses.[21] Various approaches have been offered to utilize nanomaterials as efficient vehicles, such as conjugation to antigens accepted by specific receptors; incorporating antigens within nanomaterials may eventually guard the antigen degradation; and labeling nanomaterials to be accepted by specific receptors and permit their passage tracing off in the host system.[117] Applications in medicine such as immunotherapy for cancer, vaccine delivery, and allergen immunotherapy can be leveraged to these diverse approaches to control the immune system for improved therapeutic benefits with minimum side effects.

Chitosan is a biocompatible polysaccharide derived from the crustacean shell chitin, which is used in the form of NPs (100–200 nm) to deliver plasmids.[148,196] In humans, this Chitosan is nontoxic, non-haemolytic, weakly immunogenic, and slowly biodegradable, which justifies its application in controlled drug delivery.[17,134] The other attributed properties of Chitosan include anticoagulation, immune system stimulation, wound healing, and antimicrobial activities.[119,170] It is an ideal gene delivery agent for effective gene expression therapy because of its muco-adhesive property which increases the paracellular and transcellular transport through the mucosal epithelium, positively facilitating gene delivery to mucosa and bronchus-related lymphoid tissue.[16,119]

The search for new generation antibiotics, antifungal, antiprotozoal, and antiviral agents to treat resistant strains, in the challenge to the difficulty offered by the global emergence of multidrug-resistant microorganisms, is paramount.[94,118,128,175] A variety of multidrug-resistant bacteria that have resisted maximum antibiotic treatment can be inhibited by NPs, the potential broad-spectrum antibiotics. It has been demonstrated by Adeli et al.[3] that *E. coli*, *K. pneumoniae*, *P. aeruginosa,* and *S. aureus* being resistant to all the antibiotic drugs were inhibited by AgNPs.[253] The AgNPs-coated condom has antiviral, antibacterial, and anti-fungi properties as shown by the findings of Fayaz et al.,[22] which was also demonstrated the same results with CNTs.[163] A high antioxidant activity was also exhibited by the nanomaterial hybrids such as cholesterol comprising liposomes of phyto nanosilver and CNTs.

Parasitic infections (such as dengue fevers, filariasis, giardiasis, Japanese encephalitis, leishmaniasis, malaria, schistosomiasis, and trypanosomiasis) remain to raise in tropical and low income countries, even though great efforts made in their eradication.[210,206] Significant improvements in the treatment of parasitic infections have been shown by the newer approaches such as nano-biotechnology which is centered on the exclusive characteristics of NPs that demonstrate excellent inhibitory effect against parasitic infections.[9,61,206,210] The NPs microbicides as well as those of nanoscale dendrimer microbicides hold prospective safety efficiency against viruses.[70,87,156] For example, the VivaGel™ (SPL7013Gel) is a dendrimer specifically prepared against HIV and HSV, which is physiologically compatible with the acidic nature of vagina.[192] HIV blocking potential in an in vitro model, epithelial monolayer cells was exhibited by Carbosilane dendrimer microbicide.[48] Heparan sulfate-binding peptide, another dendrimer, has been found to be potent antiviral microbicide by inhibiting human papillomavirus.[70] Dendrimers, solid lipid NPs, Polymeric NPs, and polymer micelles are potent antimicrobials that have ligand conjugate properties which will advance the drug's therapeutical efficiency to incorporate or cross the cellular membranes, and provide competent antibacterial activity.[8,103,240,266]

3.7 IMPLICATIONS AND PERSPECTIVES OF NANOTECHNOLOGY

Investment in nanotechnology has prospective breakthroughs in areas such as medicine and healthcare, environment and energy, materials and manufacturing, biotechnology, food and agriculture, information technology, and national security. Without a doubt, many more nanotechnology products are being researched and developed and they are expected to be released into the market in the near future. While taking the advantage of this new technology, the possible environmental, social and ethical implications need to be considered. Such implications can be revealed by scrutinizing several issues such as: safety design of nanomaterials; toxicity testing of manufactured NPs on fish, animals, plant and also potential ecological impacts; and complete life-cycle assessments. Therefore, risk assessment and the impact of risk factors associated with their usage in the environment and possible adverse effects on non-target organisms and mammals, especially humans are needed. Furthermore, it is essential to have regulatory frameworks that appropriately address and explicitly manage the forthcoming risks of nanotechnology. Pertinent education on nanotechnology and nanoscience, along with novel and pioneering ideas and accurate risk-assessment policies, will definitely help in fulfilling the prospects of nanotechnology and nanomaterials to develop the quality of human life, while alleviating negative environmental impacts.

3.8 SUMMARY

Nanotechnology applications to the agri-food sector are rather recent compared with their use in drug delivery and pharmaceuticals. Agri-food subject centers on sustainability and fortification of agriculturally produced foods, as well as yields for human consumption and animal nourishing. It also covers novel agrochemical agents and new transport mechanisms to enhance crop output, and it assures to reduce pesticide usage. Precision farming aims at higher crop yields while reducing the use of fertilizers, pesticides, and herbicides through monitoring of environmental variables and implementing target specific action. Applications of nanotechnology in food industry include food packaging, food processing, encapsulation of nutrients, and development of new functional foods. Nanosensors and nanobarcodes are used to assure food quality and safety. Besides these topics, this review also addresses biomedical applications such as nanodiagnostics, and developments in the discovery, design, and delivery of drugs, including nanopharmaceuticals. Furthermore, topics on biological therapies such as vaccination, cell therapy, and gene therapy are also covered. In the end, the implications and perspectives of nanotechnology are discussed.

ACKNOWLEDGEMENTS

This study was supported by NSF-CREST program (the National Science Foundation-Centers of Research Excellence in Science and Technology [NSF-CREST]) with grant #HRD-1547754 to Jackson State University.

KEYWORDS

- **nanotechnology**
- **nano-food**
- **nano-sensors**
- **nano-fertilizers**
- **nano-pesticides**
- **nano-diagnostics**

REFERENCES

1. Abbas, K. A.; Saleh, A. M.; Mohamed, A.; MohdAzhan, N. The Recent Advances in the Nanotechnology and its Applications in Food Processing: A Review. *J. Food Agric. Environ.* **2009,** *7* (3–4), 14–17.
2. Abraham, A.; Kannangai, R.; Sridharan, G. Nanotechnology: A New Frontier in Virus Detection in Clinical Practice. *Ind. J. Med. Microbiol.* **2008,** *26* (4), 297–301.
3. Adeli, M.; Hosainzadegan, H.; Pakzad, I.; Zabihi, F.; Alizadeh, M.; Karimi, F. Preparation of the Silver Nanoparticle Containing Starch Foods and Evaluation of Antimicrobial Activity. *Jundishapur J. Microbiol.* **2013,** *6* (4), e5075.
4. Agrawal, L.; Haq, W.; Hanson, C. V.; Rao, D. N. Generating Neutralizing Antibodies, Th1 Response and MHC Non Restricted Immunogenicity of HIV-I Env and Gag Peptides in Liposomes and ISCOMs with in-built Adjuvant City. *J. Immune Based Ther. Vacc.* **2003,** *1* (1), 1.
5. Agrawal, S.; Rathore, P. Nanotechnology Pros and Cons to Agriculture: A Review. *Int. J. Curr. Microbiol. Appl. Sci.* **2014,** *3* (3), 43–55.
6. Aitken, R. J.; Chaudhry, M. Q.; Boxall, A. B.; Hull, M. Manufacture and Use of Nanomaterials: Current Status in the UK and Global Trends. *Occup. Med.* **2006,** *56* (5), 300–306.
7. Akerman, M. E.; Chan, W. C.; Laakkonen, P.; Bhatia, S. N.; Ruoslahti, E. Nanocrystal Targeting in vivo. *Proc. Natl. Acad. Sci. USA* **2002,** *99* (20), 12617–12621.

8. Alizadeh, H.; Salouti, M.; Shapouri, R. Bactericidal Effect of Silver Nanoparticles on Intramacrophage Brucella abortus 544. *Jundishapur J. Microbiol.* **2014,** *7* (3), 1.
9. Allahverdiyev, A. M.; Abamor, E. S.; Bagirova, M.; Ustundag, C. B.; Kaya, C.; Kaya, F.; Rafailovich, M. Antileishmanial Effect of Silver Nanoparticles and their Enhanced Antiparasitic Activity Under Ultraviolet Light. *Int. J. Nanomed.* **2011,** *6,* 2705–2714.
10. Andre, F. Overview of a 5-year Clinical Experience with a Yeast-derived Hepatitis B Vaccine. *Vaccine* **1990,** *8,* S74–S78.
11. Andreiotelli, M.; Wenz, H. J.; Kohal, R. Are Ceramic Implants a Viable Alternative to Titanium Implants? A Systematic Literature Review. *Clin. Oral Implants Res.* **2009,** *20,* 32–47.
12. Andrianantoandro, E.; Basu, S.; Karig, D. K.; Weiss, R. Synthetic Biology: New Engineering Rules for an Emerging Discipline. *Mol. Syst. Biol.* **2006,** *2,* e2006.0028.
13. Armentano, I.; Dottori, M.; Fortunati, E.; Mattioli, S.; Kenny, J. M. Biodegradable Polymer Matrix Nanocomposites for Tissue Engineering: A Review. *Polym. Degrad. Stab.* **2010,** *95* (11), 2126–2146.
14. Arora, A.; Padua, G. Review: Nanocomposites in Food Packaging. *J. Food Sci.* **2010,** *75* (1), R43–R49.
15. Arsianti, M.; Lim, M.; Marquis, C. P.; Amal, R. Assembly of Polyethylenimine-based Magnetic Iron Oxide Vectors: Insights Into Gene Delivery. *Langmuir* **2010,** *26* (10), 7314–7326.
16. Artursson, P.; Lindmark, T.; Davis, S. S.; Illum, L. Effect of Chitosan on the Permeability of Monolayers of Intestinal Epithelial Cells (Caco-2). *Pharm. Res.* **1994,** *11* (9), 1358–1361.
17. Aspden, T. J.; Mason, J. D.; Jones, N. S.; Lowe, J.; Skaugrud, O.; Illum, L. Chitosan as a Nasal Delivery System: The Effect of Chitosan Solutions on in vitro and in vivo Mucociliary Transport Rates in Human Turbinates and Volunteers. *J. Pharm. Sci.* **1997,** *86* (4), 509–513.
18. Avella, M.; De Vlieger, J. J.; Errico, M. E.; Fischer, S.; Vacca, P.; Volpe, M. G. Biodegradable Starch/Clay Nanocomposite Films for Food Packaging Applications. *Food Chem.* **2005,** *93* (3), 467–474.
19. Baeumner, A. Nanosensors Identify Pathogens in Food. *Food Technol.* **2004,** *58*, 51–52.
20. Balani, K.; Anderson, R.; Laha, T.; Andara, M.; Tercero, J. E.; Crumpler, E. T.; Agarwal, A. Plasma-sprayed Carbon Nanotube Reinforced Hydroxyapatite Coatings and their Interaction with Human Osteoblasts *in vitro*. *Biomaterials* **2007,** *28* (4), 618–624.
21. Banchereau, J.; Palucka, A. K. Dendritic Cells as Therapeutic Vaccines Against Cancer. *Nature Rev. Immunol.* **2005,** *5* (4), 296–306.
22. Barbinta-Patrascu, M. E.; Ungureanu, C.; Iordache, S. M.; Iordache, A. M.; Bunghez, I. R.; Ghiurea, M.; Badea, N.; Fierascu, R. C.; Stamatin, I. Eco-designed Biohybrids Based on Liposomes, Mint-nanosilver and Carbon Nanotubes for Antioxidant and Antimicrobial Coating. *Mater. Sci. Eng. C* **2014,** *39*, 177–185.
23. Barik, T. K.; Sahu, B.; Swain, V. Nanosilica-from medicine to Pest Control. *Parasitol. Res.* **2008,** *103* (2), 253–258.
24. Baruah, S.; Dutta, J. Nanotechnology Applications in Pollution Sensing and Degradation in Agriculture: A Review. *Environ. Chem. Lett.* **2009,** *7* (3), 191–204.
25. Bastarrachea, L.; Dhawan, S.; Sablani, S. S. Engineering Properties of Polymeric-based Antimicrobial Films for Food Packaging: A Review. *Food Eng. Rev.* **2011,** *3* (2), 79–93.

26. Becaro, A. A.; Puti, F. C.; Correa, D. S.; Paris, E. C.; Marconcini, J. M.; Ferreira, M. D. Polyethylene Films Containing Silver Nanoparticles for Applications in Food Packaging: Characterization of Physico-chemical and Anti-microbial Properties. *J. Nanosci. Nanotechnol.* **2015,** *15* (3), 2148–2156.
27. Benenson, Y.; Gil, B.; Ben-Dor, U.; Adar, R.; Shapiro, E. An Autonomous Molecular Computer for Logical Control of Gene Expression. *Nature* **2004,** *429* (6990), 423–429.
28. Bergeson, L. L. Nanosilver Pesticide Products: What does the Future Hold? *Environ. Qual. Manag.* **2010,** *19* (4), 73–82.
29. Betley, T. A.; Hessler, J. A.; Mecke, A.; Holl, M. M. B.; Orr, B .G.; Uppuluri, S.; Tomalia, D. A.; Baker, J. R. Tapping Mode Atomic Force Microscopy Investigation of Poly(amidoamine) Core–Shell Tecto(dendrimers) Using Carbon Nanoprobes. *Langmuir* **2002,** *18* (8), 3127–3133.
30. Bhalgat, M. K.; Haugland, R. P.; Pollack, J. S.; Swan, S.; Haugland, R. P. Green- and Red-fluorescent Nanospheres for the Detection of Cell Surface Receptors by Flow Cytometry. *J. Immunol. Methods* **1998,** *219* (1), 57–68.
31. Bisht, S.; Bhakta, G.; Mitra, S.; Maitra, A. pDNA Loaded Calcium Phosphate Nanoparticles: Highly Efficient Non-viral Vector for Gene Delivery. *Int. J. Pharm.* **2005,** *288* (1), 157–168.
32. Bolhassani, A.; Safaiyan, S.; Rafati, S. Improvement of Different Vaccine Delivery Systems for Cancer Therapy. *Mol. Cancer* **2011,** *10*, 3.
33. Bonne, P.; Beerendonk, E.; Van der Hoek, J.; Hofman, J. Retention of Herbicides and Pesticides in Relation to Aging of RO Membranes. *Desalination* **2000,** *132* (1), 189–193.
34. Bonoiu, A. C.; Mahajan, S. D.; Ding, H.; Roy, I.; Yong, K. -T.; Kumar, R.; Hu, R.; Bergey, E. J.; Schwartz, S. A.; Prasad, P. N. Nanotechnology Approach for Drug Addiction Therapy: Gene Silencing Using Delivery of Gold Nanorod-siRNA Nanoplex in Dopaminergic Neurons. *Proc. Natl. Acad. Sci.* **2009,** *106* (14), 5546–5550.
35. Branton, D.; Deamer, D. W.; Marziali, A.; Bayley, H.; Benner, S. A.; Butler, T.; Di Ventra, M.; Garaj, S.; Hibbs, A.; Huang, X.; Jovanovich, S. B.; Krstic, P. S.; Lindsay, S.; Ling, X. S.; Mastrangelo, C. H.; Meller, A.; Oliver, J. S.; Pershin, Y. V.; Ramsey, J. M.; Riehn, R.; Soni, G. V.; Tabard-Cossa, V.; Wanunu, M.; Wiggin, M.; Schloss, J. A. The Potential and Challenges of Nanopore Sequencing. *Nature Biotechnol.* **2008,** *26* (10), 1146–1153.
36. Bugusu, B.; Mejia, C.; Magnuson, B.; Tafazoli, S. Global Regulatory Policies on Food Nanotechnology. *Food Technol.* **2009,** *63* (5), 24–28.
37. Burda, C.; Chen, X.; Narayanan, R.; El-Sayed, M. A. Chemistry and Properties of Nanocrystals of Different Shapes. *Chem. Rev.* **2005,** *105* (4), 1025–1102.
38. Cadek, M.; Coleman, J. N.; Barron, V.; Hedicke, K.; Blau, W. J. Morphological and Mechanical Properties of Carbon-nanotube-reinforced Semicrystalline and Amorphous Polymer Composites. *Appl. Phys. Lett.* **2002,** *81* (27), 5123–5125.
39. Cha, D. S.; Chinnan, M. S. Biopolymer-based Antimicrobial Packaging: A Review. *Crit. Rev. Food Sci. Nutr.* **2004,** *44* (4), 223–237.
40. Chaudhry, Q.; Castle, L. Food Applications of Nanotechnologies: An Overview of Opportunities and Challenges for Developing Countries. *Trends Food Sci. Technol.* **2011,** *22* (11), 595–603.
41. Chaudhry, Q.; Castle, L.; Watkins, R. Nanotechnologies in Food Arena: New Opportunities, New Question, and New Concerns. Chapter 1, In *Nanotechnologies in*

Food; Chaudhry, Q.; Castle, L.; Watkins, R. Eds.; Royal Society of Chemistry: Great Britain, 2010,pp 1–17.

42. Chaudhry, Q.; Scotter, M.; Blackburn, J.; Ross, B.; Boxall, A.; Castle, L.; Aitken, R.; Watkins, R. Applications and Implications of Nanotechnologies for the Food Sector. *Food Add. Contam.* **2008,** *25* (3), 241–258.
43. Chawengkijwanich, C.; Hayata, Y. Development of TiO_2 Powder-coated Food Packaging Film and its Ability to Inactivate *Escherichia coli in vitro* and in Actual Tests. *Int. J. Food Microbiol.* **2008,** *123* (3), 288–292.
44. Chen, B.; Evans, J. R. Thermoplastic Starch–clay Nanocomposites and Their Characteristics. *Carbohydr. Polym.* **2005,** *61* (4), 455–463.
45. Chen, G.; Roy, I.; Yang, C.; Prasad, P. N. Nanochemistry and Nanomedicine for Nanoparticle-based Diagnostics and Therapy. *Chem. Rev.* **2016,** *116* (5), 2826–2885.
46. Chen, H.; Yada, R. Y. Nanotechnologies in Agriculture: New Tools for Sustainable Development. *Trends Food Sci. Technol.* **2011,** *22* (11), 585–594.
47. Choi, A. -J.; Kim, C. -J.; Cho, Y. -J.; Hwang, J. -K.; Kim, C. -T. Characterization of Capsaicin-loaded Nanoemulsions Stabilized with Alginate and Chitosan by Self-assembly. *Food Bioproc. Technol.* **2011,** *4* (6), 1119–1126.
48. Chonco, L.; Pion, M.; Vacas, E.; Rasines, B.; Maly, M.; Serramia, M. J.; Lopez-Fernandez, L.; De la Mata, J.; Alvarez, S.; Gomez, R.; Munoz-Fernandez, M. A. Carbosilane Dendrimer Nanotechnology Outlines of the Broad HIV Blocker Profile. *J. Contr. Release* **2012,** *161* (3), 949–958.
49. Chouhan, R. S.; Vinayaka, A. C.; Thakur, M. S. Thiol-stabilized Luminescent CdTe Quantum Dot as Biological Fluorescent Probe for Sensitive Detection of Methyl Parathion by a Fluoroimmunochromatographic Technique. *Anal. Bioanal. Chem.* **2010,** *397* (4), 1467–1475.
50. Coulter, A.; Harris, R.; Davis, R.; Drane, D.; Cox, J.; Ryan, D.; Sutton, P.; Rockman, S.; Pearse, M. Intranasal Vaccination with ISCOMATRIXÂ® Adjuvanted Influenza Vaccine. *Vaccine* **2003,** *21* (9), 946–949.
51. Couvreur, P.; Vauthier, C. Nanotechnology: Intelligent Design to Treat Complex Disease. *Pharm. Res.* **2006,** *23* (7), 1417–1450.
52. Cuartas-Uribe, B.; Alcaina-Miranda, M.; Soriano-Costa, E.; Bes-Pia, A. Comparison of the Behavior of Two Nanofiltration Membranes for Sweet Whey Demineralization. *J. Dairy Sci.* **2007,** *90* (3), 1094–1101.
53. Cui, D.; Zhang, H.; Wang, Z.; Asahi, T.; Osaka, T. Effects of Dendrimer-Functionalized Multi-walled Carbon Nanotubes on Murine Embryonic Stem Cells. *ECS Trans.* **2008,** *13* (14), 111–116.
54. Cushen, M.; Kerry, J.; Morris, M.; Cruz-Romero, M.; Cummins, E. Nanotechnologies in the Food Industry: Recent Developments, Risks and Regulation. *Trends Food Sci. Technol.* **2012,** *24* (1), 30–46.
55. Cutts, F.; Franceschi, S.; Goldie, S.; Castellsague, X.; De Sanjose, S.; Garnett, G.; Edmunds, W.; Claeys, P.; Goldenthal, K.; Harper, D. Human Papillomavirus and HPV Vaccines: A Review. *Bull. World Health Org.* **2007,** *85* (9), 719–726.
56. Cyna, B.; Chagneau, G.; Bablon, G.; Tanghe, N. Two Years of Nanofiltration at the Mery-sur-Oise Plant, France. *Desalination* **2002,** *147* (1), 69–75.
57. Cyras, V. P.; Manfredi, L. B.; Ton-That, M. T.; Vazquez, A. Physical and Mechanical Properties of Thermoplastic Starch/Montmorillonite Nanocomposite Films. *Carbohydr. Polym.* **2008,** *73* (1), 55–63.

58. Dallas, P.; Sharma, V. K.; Zboril, R. Silver Polymeric Nanocomposites as Advanced Antimicrobial Agents: Classification, Synthetic Paths, Applications, and Perspectives. *Adv. Colloid Interface Sci.* **2011,** *166* (1), 119–135.
59. Danhier, F.; Ansorena, E.; Silva, J. M.; Coco, R.; Le Breton, A.; Preat, V. PLGA-based Nanoparticles: An Overview of Biomedical Applications. *J. Control. Release* **2012,** *161* (2), 505–522.
60. Dar, P.; Kalaivanan, R.; Sied, N.; Mamo, B.; Kishore, S.; Suryanarayana, V.; Kondabattula, G. Montanide ISA™ 201 Adjuvanted FMD Vaccine Induces Improved Immune Responses and Protection in Cattle. *Vaccine* **2013,** *31* (33), 3327–3332.
61. Das, S.; Bhattacharya, A.; Debnath, N.; Datta, A.; Goswami, A. Nanoparticle-induced Morphological Transition of Bombyx mori Nucleopolyhedro Virus: A Novel Method to Treat Silkworm Grasserie Disease. *Appl. Microbiol. Biotechnol.* **2013,** *97* (13), 6019–6030.
62. Dasgupta, N.; Ranjan, S.; Mundekkad, D.; Ramalingam, C.; Shanker, R.; Kumar, A. Nanotechnology in Agro-food: From Field to Plate. *Food Res. Int.* **2015,** *69,* 381–400.
63. De Azeredo, H. M. Nanocomposites for Food Packaging Applications. *Food Res. Int.* **2009,** *42* (9), 1240–1253.
64. Dekkers, S.; Krystek, P.; Peters, R. J.; Lankveld, D. I. P.; Bokkers, B. G.; van Hoeven-Arentzen, P. H.; Bouwmeester, H.; Oomen, A. G. Presence and Risks of Nanosilica in Food Products. *Nanotoxicology* **2011,** *5* (3), 393–405.
65. Derosa, M. C.; Monreal, C. M.; Schnitzer, M.; Walsh, R.; Sultan, Y. Nanotechnology in Fertilizers. *Nature Nanotechnol.* **2010,** *5* (2), 91–91.
66. Dion, I.; Rouais, F.; Baquey, C.; Lahaye, M.; Salmon, R.; Trut, L.; Cazorla, J.; Huong, P. V.; Monties, J. R.; Havlik, P. Physico-chemistry and Cytotoxicity of Ceramics: Part I: Characterization of Ceramic Powders. *J. Mater. Sci. Mater. Med.* **1997,** *8* (5), 325–332.
67. Ditta, A.; Arshad, M.; Ibrahim, M., Nanoparticles in Sustainable Agricultural Crop Production: Applications and Perspectives. In *Nanotechnology and Plant Sciences: Nanoparticles and Their Impact on Plants*; *Siddiqui M. H., Al-Whaibi M. H., Mohammad F., Eds.*; Springer International Publishing: Cham, 2015,pp 55–75.
68. Ditto, A. J.; Shah, P. N.; Yun, Y. H. Non-viral Gene Delivery Using Nanoparticles. *Exp. Opinion Drug Del.* **2009,** *6* (11), 1149–1160.
69. Dobrovolskaia, M. A.; McNeil, S. E. Immunological Properties of Engineered Nanomaterials. *Nature Nanotechnol.* **2007,** *2* (8), 469–478.
70. Donalisio, M.; Rusnati, M.; Cagno, V.; Civra, A.; Bugatti, A.; Giuliani, A.; Pirri, G.; Volante, M.; Papotti, M.; Landolfo, S.; Lembo, D. Inhibition of Human Respiratory Syncytial Virus Infectivity by a Dendrimeric Heparan Sulfate-binding Peptide. *Antimicrob. Agents Chemother.* **2012,** *56* (10), 5278–5288.
71. Du, D.; Chen, S.; Cai, J.; Zhang, A. Electrochemical Pesticide Sensitivity Test Using Acetylcholinesterase Biosensor Based on Colloidal Gold Nanoparticle Modified Sol-Gel Interface. *Talanta* **2008,** *74* (4), 766–772.
72. Duncan, T. V. Applications of Nanotechnology in Food Packaging and Food Safety: Barrier Materials, Antimicrobials and Sensors. *J. Colloid Interface Sci.* **2011,** *363* (1), 1–24.
73. Elbashir, S. M.; Harborth, J.; Lendeckel, W.; Yalcin, A.; Weber, K.; Tuschl, T. Duplexes of 21-nucleotide RNAs Mediate RNA Interference in Cultured Mammalian Cells. *Nature* **2001,** *411* (6836), 494–498.
74. Elliott, R.; Ong, T. J. Nutritional Genomics. *Br. Med. J.* **2002,** *324* (7351), 1438–1442.

75. European Food Safety Authority. Updating the Opinion Related to the Revision of Annexes II and III to Council Directive 91/414/EEC Concerning the Placing of Plant Protection Products on the Market—Analytical Methods—Prepared by the Panel on Plant Protection Products and their Residues. *Eur. Food Safety Author. J.* **2009,** 1174, 1–6.
76. European Food Safety Authority. The Potential Risks Arising from Nanoscience and Nanotechnologies on Food and Feed Safety. *Eur. Food Safety Author. J.* **2009,** 958, 1–39.
77. Feng, G.; Jiang, Q.; Xia, M.; Lu, Y.; Qiu, W.; Zhao, D.; Lu, L.; Peng, G.; Wang, Y. Enhanced Immune Response and Protective Effects of Nano-chitosan-based DNA Vaccine Encoding T Cell Epitopes of Esat-6 and FL Against Mycobacterium Tuberculosis Infection. *PLoS One* **2013,** *8* (4), e61135.
78. Ferrari, M. Cancer Nanotechnology: Opportunities and Challenges. *Nature Rev. Cancer* **2005,** *5* (3), 161–171.
79. Firtel, M.; Henderson, G. S.; Sokolov, I. Nanosurgery: Observation of Peptidoglycan Strands in Lactobacillus Helveticus Cell Walls. *Ultramicroscopy* **2004,** *101* (2), 105–109.
80. Fraceto, L. F.; Grillo, R.; de Medeiros, G. A.; Scognamiglio, V.; Rea, G.; Bartolucci, C. Nanotechnology in Agriculture: Which Innovation Potential does it have? *Front. Environ. Sci.* **2016,** *4* (20), 110–115.
81. Frank, J. A.; Miller, B. R.; Arbab, A. S.; Zywicke, H. A.; Jordan, E. K.; Lewis, B. K.; Bryant, L. H., Jr.; Bulte, J. W. Clinically Applicable Labeling of Mammalian and Stem Cells by Combining Superparamagnetic Iron Oxides and Transfection Agents. *Radiology* **2003,** *228* (2), 480–487.
82. Fritz, J.; Baller, M.; Lang, H.; Rothuizen, H.; Vettiger, P.; Meyer, E.; GÃ¼ntherodt, H.-J.; Gerber, C.; Gimzewski, J. Translating Biomolecular Recognition into Nanomechanics. *Science* **2000,** *288* (5464), 316–318.
83. Garti, N., 2008. *Delivery and Controlled Release of Bioactives in Foods and Nutraceuticals*; CRC Press: New York, 2008, 496.
84. Ge, S.; Lu, J.; Ge, L.; Yan, M.; Yu, J. Development of a Novel Deltamethrin Sensor Based on Molecularly Imprinted Silica Nanospheres Embedded CdTe Quantum Dots. *Spectrochimica Acta Part A Mol. Biomol. Spectrosc.* **2011,** *79* (5), 1704–1709.
85. Gewin, V. Everything You Need to Know About Nanopesticides. http://modernfarmer.com/2015/01/everything-need-know-nanopesticides/Accessed on July 23, 2017.
86. Gogos, A.; Knauer, K.; Bucheli, T. D. Nanomaterials in Plant Protection and Fertilization: Current State, Foreseen Applications, and Research Priorities. *J. Agric. Food Chem.* **2012,** *60* (39), 9781–9792.
87. Gong, E.; Matthews, B.; McCarthy, T.; Chu, J.; Holan, G.; Raff, J.; Sacks, S. Evaluation of Dendrimer SPL7013, A Lead Microbicide Candidate Against Herpes Simplex Viruses. *Antiviral Res.* **2005,** *68* (3), 139–146.
88. Gonzalez-Melendi, P.; Fernandez-Pacheco, R.; Coronado, M. J.; Corredor, E.; Testillano, P. S.; Risueno, M. C.; Marquina, C.; Ibarra, M. R.; Rubiales, D.; Perez-de-Luque, A. Nanoparticles as Smart Treatment-delivery Systems in Plants: Assessment of Different Techniques of Microscopy for their Visualization in Plant Tissues. *Annals Botany* **2008,** *101* (1), 187–195.
89. Gregory, A. E.; Williamson, D.; Titball, R. Vaccine Delivery Using Nanoparticles. *Front. Cell. Infect. Microbiol.* **2013,** *3,* 13.

90. Hamdy, S.; Haddadi, A.; Hung, R. W.; Lavasanifar, A. Targeting Dendritic Cells with Nano-particulate PLGA Cancer Vaccine Formulations. *Adv. Drug Deliv. Rev.* **2011,** *63* (10), 943–955.
91. Han, D.; Hong, J.; Kim, H. C.; Sung, J. H.; Lee, J. B. Multiplexing Enhancement for the Detection of Multiple Pathogen DNA. *J. Nanosci. Nanotechnol.* **2013,** *13* (11), 7295–7299.
92. He, X.; Hwang, H. M. Nanotechnology in Food Science: Functionality, Applicability, and Safety Assessment. *J. Food Drug Anal.* **2016,** *24* (4), 671–681.
93. Heath, J. R.; Davis, M. E. Nanotechnology and Cancer. *Ann. Rev. Med.* **2008,** *59* (1), 251–265.
94. Hoffmann, H. H.; Kunz, A.; Simon, V. A.; Palese, P.; Shaw, M. L. Broad-spectrum Antiviral that Interferes with de novo Pyrimidine Biosynthesis. *Proc. Natl. Acad. Sci. USA* **2011,** *108* (14), 5777–5782.
95. Hofman, J.; Beerendonk, E.; Folmer, H.; Kruithof, J. Removal of Pesticides and Other Micropollutants with Cellulose-acetate, Polyamide and Ultra-low Pressure Reverse Osmosis Membranes. *Desalination* **1997,** *113* (2), 209–214.
96. Hogg, T.; Freitas, R. A. Chemical Power for Microscopic Robots in Capillaries. *Nanomedicine* **2010,** *6* (2), 298–317.
97. Homhuan, A.; Prakongpan, S.; Poomvises, P.; Maas, R. A.; Crommelin, D. J.; Kersten, G. F.; Jiskoot, W. Virosome and ISCOM Vaccines Against Newcastle Disease: Preparation, Characterization and Immunogenicity. *Eur. J. Pharm. Sci.* **2004,** *22* (5), 459–468.
98. Hua, S.; Cabot, P. J. Targeted Nanoparticles that Mimic Immune Cells in Pain Control Inducing Analgesic and Anti-inflammatory Actions: A Potential Novel Treatment of Acute and Chronic Pain Conditions. *Pain Phys.* **2013,** *16* (3), E199–E216.
99. Huang, J. -Y.; Li, X.; Zhou, W. Safety Assessment of Nanocomposite for Food Packaging Application. *Trends Food Sci. Technol.* **2015,** *45* (2), 187–199.
100. Huang, J.; He, C.; Liu, X.; Xu, J.; Tay, C. S.; Chow, S. Y. Organic-inorganic Nanocomposites from Cubic Silsesquioxane Epoxides: Direct Characterization of Interphase, and Thermomechanical Properties. *Polymer* **2005,** *46* (18), 7018–7027.
101. Huang, J.; Xiao, Y.; Mya, K. Y.; Liu, X.; He, C.; Dai, J.; Siow, Y. P. Thermomechanical Properties of Polyimide-epoxy Nanocomposites from Cubic Silsesquioxane Epoxides. *J. Mater. Chem.* **2004,** *14* (19), 2858–2863.
102. Huang, Y.; Chen, S.; Bing, X.; Gao, C.; Wang, T.; Yuan, B. Nanosilver Migrated into Food-simulating Solutions from Commercially Available Food Fresh Containers. *Packag. Technol. Sci.* **2011,** *24* (5), 291–297.
103. Imbuluzqueta, E.; Gamazo, C.; Ariza, J.; Blanco-Prieto, M. J. Drug Delivery Systems for Potential Treatment of Intracellular Bacterial Infections. *Front. Biosci.* **2010,** *15,* 397–417.
104. Jiang, S.; Eltoukhy, A. A.; Love, K. T.; Langer, R.; Anderson, D. G. Lipidoid-coated Iron Oxide Nanoparticles for Efficient DNA and siRNA Delivery. *Nano Lett.* **2013,** *13* (3), 1059–1064.
105. Jing, Y.; Moore, L. R.; Williams, P. S.; Chalmers, J. J.; Farag, S. S.; Bolwell, B.; Zborowski, M. Blood Progenitor Cell Separation from Clinical Leukapheresis Product by Magnetic Nanoparticle Binding and Magnetophoresis. *Biotechnol. Bioeng.* **2007,** *96* (6), 1139–1154.
106. Kah, M. Nanopesticides and Nanofertilizers: Emerging Contaminants or Opportunities for Risk Mitigation? *Front. Chem.* **2015,** *3,* 64.

107. Kah, M.; Beulke, S.; Tiede, K.; Hofmann, T. Nanopesticides: State of Knowledge, Environmental Fate, and Exposure Modeling. *Crit. Rev. Environ. Sci. Technol.* **2013,** *43* (16), 1823–1867.
108. Kah, M.; Weniger, A.; Hofmann, T. Impacts of (Nano) Formulations on the Fate of an Insecticide in Soil and Consequences for Environmental Exposure Assessment. *Environ. Sci. Technol.* **2016.**
109. Kang, T. F.; Wang, F.; Lu, L. P.; Zhang, Y.; Liu, T. S. Methyl Parathion Sensors Based on Gold Nanoparticles and Nafion Film Modified Glassy Carbon Electrodes. *Sensors Actuat. B Chem.* **2010,** *145* (1), 104–109.
110. Kannan, R. Y.; Salacinski, H. J.; Butler, P. E.; Hamilton, G.; Seifalian, A. M. Current Status of Prosthetic Bypass Grafts: A Review. *J. Biomed. Mater. Res. Part B Appl. Biomater.* **2005,** *74* (1), 570–581.
111. Kapadia, M. R.; Chow, L. W.; Tsihlis, N. D.; Ahanchi, S. S.; Eng, J. W.; Murar, J.; Martinez, J.; Popowich, D. A.; Jiang, Q.; Hrabie, J. A. Nitric Oxide and Nanotechnology: A Novel Approach to Inhibit Neointimal Hyperplasia. *J. Vasc. Sur.* **2008,** *47* (1), 173–182.
112. Kateb, B.; Chiu, K.; Black, K. L.; Yamamoto, V.; Khalsa, B.; Ljubimova, J. Y.; Ding, H.; Patil, R.; Portillaarias, J.; Modo, M. Nanoplatforms for Constructing New Approaches to Cancer Treatment, Imaging, and Drug Delivery: What Should be the Policy? *Neuroimage* **2011,** *54* (1), S106–S124.
113. Kazner, C.; Wintgens, T.; Melin, T.; Baghoth, S.; Sharma, S.; Amy, G. Comparing the Effluent Organic Matter Removal of Direct NF and Powdered Activated Carbon/NF as High Quality Pretreatment Options for Artificial Groundwater Recharge. *Water Sci. Technol.* **2008,** *57* (6), 821–827.
114. Khawaja, A. M. The Legacy of Nanotechnology: Revolution and Prospects in Neurosurgery. *Int. J. Surg.* **2011,** *9* (8), 608–614.
115. Khot, L. R.; Sankaran, S.; Maja, J. M.; Ehsani, R.; Schuster, E. W. Applications of Nanomaterials in Agricultural Production and Crop Protection: A Review. *Crop Prot.* **2012,** *35,* 64–70.
116. Kim, S. -Y.; Doh, H. -J.; Jang, M. -H.; Ha, Y. -J.; Chung, S. -I.; Park, H. -J. Oral Immunization with Helicobacter Pylori-loaded Poly (D, L-lactide-co-glycolide) Nanoparticles. *Helicobacter* **1999,** *4* (1), 33–39.
117. Klippstein, R.; Pozo, D. Nanotechnology-based Manipulation of Dendritic Cells for Enhanced Immunotherapy Strategies. *Nanomedicine* **2010,** *6* (4), 523–529.
118. Kollef, M. H. Broad-spectrum Antimicrobials and the Treatment of Serious Bacterial Infections: Getting it Right Up Front. *Clin. Infect. Dis.* **2008,** *47* (1), S3–S13.
119. Kong, X.; Hellermann, G. R.; Zhang, W.; Jena, P.; Kumar, M.; Behera, A.; Behera, S.; Lockey, R.; Mohapatra, S. S. Chitosan Interferon Nanogene Therapy for Lung Disease: Modulation of t-cell and Dendritic Cell Immune Responses. *Aller. Asthma Clin. Immun Off. J. Can. Soc. Aller. Clin. Immun.* **2008,** *4* (3), 95–105.
120. Koo, O. M. Y.; Rubinstein, I.; Onyuksel, H. Role of Nanotechnology in Targeted Drug Delivery and Imaging: A Concise Review. *Nanomed. Nanotechnol. Biol. Med.* **2005,** *1* (3), 193–212.
121. Kookana, R. S.; Boxall, A. B.; Reeves, P. T.; Ashauer, R.; Beulke, S.; Chaudhry, Q.; Cornelis, G.; Fernandes, T. F.; Gan, J.; Kah, M.; Lynch, I.; Ranville, J.; Sinclair, C.; Spurgeon, D.; Tiede, K.; Van den Brink, P. J. Nanopesticides: Guiding Principles for

Regulatory Evaluation of Environmental Risks. *J. Agric. Food Chem.* **2014,** *62* (19), 4227–4240.

122. Kour, H.; Malik, A.; Ahmad, N.; Wani, T. A.; Kaul, R. K.; Bhat, A. Nanotechnology—New Lifeline For Food Industry. *Crit. Rev. Food Sci. Nutr.* **2015,** *15*, 1–10.
123. Kraitchman, D. L.; Heldman, A. W.; Atalar, E.; Amado, L. C.; Martin, B. J.; Pittenger, M. F.; Hare, J. M.; Bulte, J. W. *In vivo* Magnetic Resonance Imaging of Mesenchymal Stem Cells in Myocardial Infarction. *Circulation* **2003,** *107* (18), 2290–2293.
124. Krishnamachari, Y.; Geary, S. M.; Lemke, C. D.; Salem, A. K. Nanoparticle Delivery Systems in Cancer Vaccines. *Pharm. Res.* **2011,** *28* (2), 215–236.
125. Kumar, M. N. R. A Review of Chitin and Chitosan Applications. *React. Funct. Polym.* **2000,** *46* (1), 1–27.
126. Kumar, S.; Jones, T. R.; Oakley, M. S.; Zheng, H.; Kuppusamy, S. P.; Taye, A.; Krieg, A. M.; Stowers, A. W.; Kaslow, D. C.; Hoffman, S. L. CpG Oligodeoxynucleotide and Montanide ISA 51 Adjuvant Combination Enhanced the Protective Efficacy of a Subunit Malaria Vaccine. *Infect. Immun.* **2004,** *72* (2), 949–957.
127. Kumaravel, A.; Chandrasekaran, M. A Biocompatible Nano TiO_2/nafion Composite Modified Glassy Carbon Electrode for the Detection of Fenitrothion. *J. Electroanal. Chem.* **2011,** *650* (2), 163–170.
128. Kwon, D. S.; Mylonakis, E. Posaconazole: A New Broad-spectrum Antifungal Agent. *Exp. Opin. Pharmacother.* **2007,** *8* (8), 1167–1178.
129. LaCoste, A.; Schaich, K. M.; Zumbrunnen, D.; Yam, K. L. Advancing Controlled Release Packaging Through Smart Blending. *Packag. Technol. Sci.* **2005,** *18* (2), 77–87.
130. Lanza, G. M.; Wickline, S. A. Targeted Ultrasonic Contrast Agents for Molecular Imaging and Therapy. *Curr. Probl. Cardiol.* **2003,** *28* (12), 625–653.
131. Laschke, M. W.; Strohe, A.; Menger, M. D.; Alini, M.; Eglin, D. In Vitro and in Vivo Evaluation of a Novel Nanosize Hydroxyapatite Particles/Poly(ester-urethane) Composite Scaffold for Bone Tissue Engineering. *Acta Biomater.* **2010,** *6* (6), 2020–2027.
132. LaVan, D. A.; McGuire, T.; Langer, R. Small-scale Systems for In vivo Drug Delivery. *Nature Biotechnol.* **2003,** *21* (10), 1184–1191.
133. Lee, K. T. Quality and Safety Aspects of Meat Products as Affected by Various Physical Manipulations of Packaging Materials. *Meat Sci.* **2010,** *86* (1), 138–150.
134. Lee, K. Y.; Kwon, I. C.; Kim, Y. H.; Jo, W. H.; Jeong, S. Y. Preparation of Chitosan Self-aggregates as a Gene Delivery System. *J. Contr. Release* **1998,** *51* (2–3), 213–220.
135. Lehn, J. M. Toward Self-organization and Complex Matter. *Science* **2002,** *295* (5564), 2400–2403.
136. Lerner, M. B.; Goldsmith, B. R.; McMillon, R.; Dailey, J.; Pillai, S.; Singh, S. R.; Johnson, A. C. A Carbon Nanotube Immunosensor for Salmonella. *AIP Adv.* **2011,** *1* (4), 042127.
137. Li, P.; Luo, Z.; Liu, P.; Gao, N.; Zhang, Y.; Pan, H.; Liu, L.; Wang, C.; Cai, L.; Ma, Y. Bioreducible Alginate-poly (ethylenimine) Nanogels as an Antigen-delivery System Robustly Enhance Vaccine-elicited Humoral and Cellular Immune Responses. *J. Control. Release* **2013,** *168* (3), 271–279.
138. Li, S.; Rizzo, M.; Bhattacharya, S.; Huang, L. Characterization of Cationic Lipid-protamine-DNA (LPD) Complexes for Intravenous Gene Delivery. *Gene Ther.* **1998,** *5* (7), 930–937.

139. Li, W. J.; Cooper, J. A., Jr.; Mauck, R. L.; Tuan, R. S. Fabrication and Characterization of Six Electrospun Poly(alpha-hydroxy ester)-based Fibrous Scaffolds for Tissue Engineering Applications. *Acta Biomater.* **2006,** *2* (4), 377–385.
140. Li, Y.; Cu, Y. T. H.; Luo, D. Multiplexed Detection of Pathogen DNA with DNA-based Fluorescence Nanobarcodes. *Nature Biotechnol.* **2005,** *23* (7), 885–889.
141. Li, Z.; Sheng, C. Nanosensors for Food Safety. *J. Nanosci. Nanotechnol.* **2014,** *14* (1), 905–912.
142. Liu, H.; Webster, T. Nanomedicine for Implants: A Review of Studies and Necessary Experimental Tools. *Cell. Mol. Biol. Tech. Biomater. Eval.* **2007,** *28* (2), 354–369.
143. Liu, R.; Lal, R. Potentials of Engineered Nanoparticles as Fertilizers For Increasing Agronomic Productions. *Sci. Total Environ.* **2015,** *514*, 131–139.
144. Liu, S.; Yuan, L.; Yue, X.; Zheng, Z.; Tang, Z. Recent Advances in Nanosensors for Organophosphate Pesticide Detection. *Adv. Powder Technol.* **2008,** *19* (5), 419–441.
145. Liu, X.; Jin, X.; Ma, P. X. Nanofibrous Hollow Microspheres Self-assembled from Star-shaped Polymers as Injectable Cell Carriers for Knee Repair. *Nature Mater.* **2011,** *10* (5), 398–406.
146. Lopez-Sagaseta, J.; Malito, E.; Rappuoli, R.; Bottomley, M. J. Self-assembling Protein Nanoparticles in the Design of Vaccines. *Comput. Struct. Biotechnol. J.* **2016,** *14,* 58–68.
147. Lungwitz, U.; Breunig, M.; Blunk, T.; Gapferich, A. Polyethylenimine Based Non-viral Gene Delivery Systems. *Eur. J. Pharm. Biopharm.* **2005,** *60* (2), 247–266.
148. Ma, P. X. Biomimetic Materials for Tissue Engineering. *Adv. Drug Del. Rev.* **2008,** *60* (2), 184–198.
149. Mamo, T.; Poland, G. A. Nanovaccinology: The Next Generation of Vaccines Meets 21st Century Materials Science and Engineering. *Vaccine* **2012,** *30* (47), 6609–6611.
150. Mateescu, A. L.; Dimov, T. V.; Grumezescu, A. M.; Gestal, M. C.; Chifiriuc, M. C. Nanostructured Bioactive Polymers Used in Food-packaging. *Curr. Pharm. Biotechnol.* **2015,** *16* (2), 121–127
151. Martel, S.; Felfoul, O.; Mohammadi, M. In *Flagellated Bacterial Nanorobots for Medical Interventions in the Human Body*, 2008 2nd IEEE RAS & EMBS International Conference on Biomedical Robotics and Biomechatronics, IEEE, 2008; pp 264–269.
152. Martel, S.; Mathieu, J.; Felfoul, O.; Chanu, A.; Aboussouan, E.; Tamaz, S.; Pouponneau, P.; Yahia, L. H.; Beaudoin, G.; Soulez, G. Automatic Navigation of an Untethered Device in the Artery of a Living Animal Using a Conventional Clinical Magnetic Resonance Imaging System. *Appl. Phys. Lett.* **2007,** *90* (11), 114105.
153. Martel, S.; Mohammadi, M.; Felfoul, O.; Lu, Z.; Pouponneau, P. Flagellated Magnetotactic Bacteria as Controlled MRI-trackable Propulsion and Steering Systems for Medical Nanorobots Operating in the Human Microvasculature. *Int. J. Robot. Res.* **2009,** *28* (4), 571–582.
154. Maurer, P.; Jennings, G. T.; Willers, J. R.; Rohner, F.; Lindman, Y.; Roubicek, K.; Renner, W. A.; Muller, P.; Bachmann, M. F. A Therapeutic Vaccine for Nicotine Dependence: Preclinical Efficacy, and Phase I Safety and Immunogenicity. Eur. J. Immun. 2005, 35 (7), 2031–2040.
155. Mazzocchi, S. Five Things You Need to Know About Nanofoods, 2011. http://www.pbs.org/wnet/need-to-know/five-things/nanofoods/6682/ Accessed July 23, 2017.
156. McCarthy, T. D.; Karellas, P.; Henderson, S. A.; Giannis, M.; O'Keefe, D. F.; Heery, G.; Paull, J. R.; Matthews, B. R.; Holan, G. Dendrimers as Drugs: Discovery and

Preclinical and Clinical Development of Dendrimer-based Microbicides for HIV and STI Prevention. *Mol. Pharmacol.* **2005,** *2* (4), 312–318.
157. Medina, O. P.; Pillarsetty, N.; Glekas, A.; Punzalan, B.; Longo, V. A.; Gonen, M.; Zanzonico, P.; Smithjones, P.; Larson, S. M. Optimizing Tumor Targeting of the Lipophilic EGFR-binding Radiotracer SKI 243 Using a Liposomal Nanoparticle Delivery System. *J. Control. Release* **2011,** *149* (3), 292–298.
158. Mills, A.; Doyle, G.; Peiro, A. M.; Durrant, J. Demonstration of a Novel, Flexible, Photocatalytic Oxygen-scavenging Polymer Film. *J. Photochem. Photobiol. Chem.* **2006,** *177* (2), 328–331.
159. Minigo, G.; Scholzen, A.; Tang, C. K.; Hanley, J. C.; Kalkanidis, M.; Pietersz, G. A.; Apostolopoulos, V.; Plebanski, M. Poly-L-lysine-coated Nanoparticles: A Potent Delivery System to Enhance DNA Vaccine Efficacy. *Vaccine* **2007,** *25* (7), 1316–1327.
160. Moffat, B. A.; Reddy, G. R.; McConville, P.; Hall, D. E.; Chenevert, T. L.; Kopelman, R. R.; Philbert, M.; Weissleder, R.; Rehemtulla, A.; Ross, B. D. A Novel Polyacrylamide Magnetic Nanoparticle Contrast Agent for Molecular Imaging Using MRI. *Mol. Imag.* **2003,** *2* (4), 324–332.
161. Moghimi, S. M.; Hunter, A. C.; Murray, J. C. Nanomedicine: Current Status and Future Prospects. *FASEB J.* **2005,** *19* (3), 311–330.
162. Mohamedi, S.; Brewer, J.; Alexander, J.; Heath, A.; Jennings, R. Antibody Responses, Cytokine Levels and Protection of Mice Immunized with HSV-2 Antigens Formulated into NISV or ISCOM Delivery Systems. *Vaccine* **2000,** *18* (20), 2083–2094.
163. Mohammed Fayaz, A.; Ao, Z.; Girilal, M.; Chen, L.; Xiao, X.; Kalaichelvan, P.; Yao, X. Inactivation of Microbial Infectiousness by Silver Nanoparticles-coated Condom: A New Approach to Inhibit HIV- and HSV-transmitted Infection. *Int. J. Nanomed.* **2012,** *7,* 5007–5018.
164. Momin, J. K.; Jayakumar, C.; Prajapati, J. B. Potential of Nanotechnology in Functional Foods. *Emir. J. Food Agric.* **2013,** *25* (1), 10.
165. Morein, B.; Lovgren, K.; Hoglund, S.; Sundquist, B. The ISCOM: An Immunostimulating Complex. *Immunol. Today* **1987,** *8* (11), 333–338.
166. Morones, J. R.; Elechiguerra, J. L.; Camacho, A.; Holt, K.; Kouri, J. B.; Ramirez, J. T.; Yacaman, M. J. The Bactericidal Effect of Silver Nanoparticles. *Nanotechnology* **2005,** *16* (10), 2346.
167. Morris, K. Macrodoctor, Come Meet the Nanodoctors. *Lancet* **2001,** *357* (9258), 778.
168. Morris, V. Emerging Roles of Engineered Nanomaterials in the Food Industry. *Trends Biotechnol.* **2011,** *29* (10), 509–516.
169. Mozafari, M. R.; Flanagan, J.; Merino, L.; Awati, A.; Omri, A.; Suntres, Z. E.; Singh, H. Recent Trends in the Lipid-based Nanoencapsulation of Antioxidants and Their Role in Foods. *J. Sci. Food Agric.* **2006,** *86* (13), 2038–2045.
170. Muzzarelli, R.; Baldassarre, V.; Conti, F.; Ferrara, P.; Biagini, G.; Gazzanelli, G.; Vasi, V. Biological Activity of Chitosan: Ultrastructural Study. *Biomaterials* **1988,** *9* (3), 247–252.
171. Naderi, M.; Danesh-Shahraki, A. Nanofertilizers and Their Roles in Sustainable Agriculture. *Int. J. Agric. Crop Sci.* **2013,** *5* (19), 2229.
172. Nakagawa, Y.; Ota, K.; Sato, Y.; Teraoka, S.; Agishi, T. Clinical Trial of New Polyurethane Vascular Grafts for Hemodialysis: Compared with Expanded Polytetrafluoroethylene Grafts. *Artif. Organs* **1995,** *19* (12), 1227–1232.

173. Nam, J.; Thaxton, C. S.; Mirkin, C. A. Nanoparticle-based Bio-bar Codes for the Ultrasensitive Detection of Proteins. *Science* **2003,** *301* (5641), 1884–1886.
174. Narayanan, A.; Sharma, P.; Moudgil, B. M. Applications of Engineered Particulate Systems in Agriculture and Food Industry. *Kona Powder Particle J.* **2012,** *30*, 221–235.
175. Navarrete-Vazquez, G.; Chavez-Silva, F.; Argotte-Ramos, R.; Rodriguez-Gutierrez Mdel, C.; Chan-Bacab, M. J.; Cedillo-Rivera, R.; Moo-Puc, R.; Hernandez-Nunez, E. Synthesis of Benzologues of Nitazoxanide and Tizoxanide: A Comparative Study of their *In vitro* Broad-spectrum Antiprotozoal Activity. *Bioorg. Med. Chem. Lett.* **2011,** *21* (10), 3168–3171.
176. Neethirajan, S.; Jayas, D. S. Nanotechnology for the Food and Bioprocessing Industries. *Food Bioproc. Technol.* **2011,** *4* (1), 39–47.
177. Negri, P.; Dluhy, R. A. Ag Nanorod Based Surface-enhanced Raman Spectroscopy Applied to Bioanalytical Sensing. *J. Biophoton.* **2013,** *6* (1), 20–35.
178. Ohyabu, Y.; Kaul, Z.; Yoshioka, T.; Inoue, K.; Sakai, S.; Mishima, H.; Uemura, T.; Kaul, S. C.; Wadhwa, R.. Stable and Nondisruptive In vitro/In vivo Labeling of Mesenchymal Stem Cells by Internalizing Quantum Dots. *Human Gene Ther.* **2009,** *20* (3), 217–224.
179. Oliveira, G. A.; Wetzel, K.; Calvo-Calle, J. M.; Nussenzweig, R.; Schmidt, A.; Birkett, A.; Dubovsky, F.; Tierney, E.; Gleiter, C. H.; Boehmer, G. Safety and Enhanced Immunogenicity of a Hepatitis B Core Particle Plasmodium Falciparum Malaria Vaccine Formulated in Adjuvant Montanide ISA 720 in a Phase I Trial. *Infect. Immun.* **2005,** *73* (6), 3587–3597.
180. Pankhurst, Q. A.; Connolly, J.; Jones, S. K.; Dobson, J. Applications of Magnetic Nanoparticles in Biomedicine. *J. Phys. D Appl. Phys.* **2003,** *36* (13), R167.
181. Pantarotto, D.; Partidos, C. D.; Hoebeke, J.; Brown, F.; Kramer, E.; Briand, J. -P.; Muller, S.; Prato, M.; Bianco, A. Immunization with Peptide-functionalized Carbon Nanotubes Enhances Virus-specific Neutralizing Antibody Responses. *Chem. Biol.* **2003,** *10* (10), 961–966.
182. Parisi, C.; Vigani, M.; Rodriguez-Cerezo, E. Agricultural Nanotechnologies: What are the Current Possibilities? *Nano Today* **2015,** *10* (2), 124–127.
183. Park, H. M.; Li, X.; Jin, C. Z.; Park, C. Y.; Cho, W. J.; Ha, C. S. Preparation and Properties of Biodegradable Thermoplastic Starch/Clay Hybrids. *Macromol. Mater. Eng.* **2002,** *287* (8), 553–558.
184. Park, S. B. Hepatitis E Vaccine Debuts. *Nature* **2012,** *491*, 21–22.
185. Parlane, N. A.; Rehm, B. H.; Wedlock, D. N.; Buddle, B. M. Novel Particulate Vaccines Utilizing Polyester Nanoparticles (Bio-beads) for Protection Against Mycobacterium Bovis Infection: A Review. *Vet. Immunol. Immunopathol.* **2014,** *158* (1), 8–13.
186. Peek, L. J.; Middaugh, C. R.; Berkland, C. Nanotechnology in Vaccine Delivery. *Adv. Drug Deliv. Rev.* **2008,** *60* (8), 915–928.
187. Peira, E.; Marzola, P.; Podio, V.; Aime, S.; Sbarbati, A.; Gasco, M. R. *In vitro* and *In vivo* Study of Solid Lipid Nanoparticles Loaded with Superparamagnetic Iron Oxide. *J. Drug Target* **2003,** *11* (1), 19–24.
188. Perez-de-Luque, A.; Rubiales, D. Nanotechnology for Parasitic Plant Control. *Pest Manag. Sci.* **2009,** *65* (5), 540–545.
189. Peters, R. J.; van Bemmel, G.; Herrera-Rivera, Z.; Helsper, H. P.; Marvin, H. J.; Weigel, S.; Tromp, P. C.; Oomen, A. G.; Rietveld, A. G.; Bouwmeester, H. Characterization of Titanium Dioxide Nanoparticles in Food Products: Analytical Methods to Define Nanoparticles. *J. Agric. Food Chem.* **2014,** *62* (27), 6285–6293.

190. Porteous, D. J.; Dorin, J. R.; McLachlan, G.; Davidson-Smith, H.; Davidson, H.; Stevenson, B.; Carothers, A.; Wallace, W.; Moralee, S.; Hoenes, C. Evidence for Safety and Efficacy of DOTAP Cationic Liposome Mediated CFTR Gene Transfer to the Nasal Epithelium of Patients with Cystic Fibrosis. *Gene Ther.* **1997,** *4* (3), 210–218.
191. Pradhan, N.; Singh, S.; Ojha, N.; Shrivastava, A.; Barla, A.; Rai, V.; Bose, S. Facets of Nanotechnology as Seen in Food Processing, Packaging, and Preservation Industry. *BioMed Res. Int.* **2015,** *2015*, 365672.
192. Price, C. F.; Tyssen, D.; Sonza, S.; Davie, A.; Evans, S.; Lewis, G. R.; Xia, S.; Spelman, T.; Hodsman, P.; Moench, T. R.; Humberstone, A.; Paull, J. R.; Tachedjian, G. SPL7013 Gel (VivaGel(R)) Retains Potent HIV-1 and HSV-2 Inhibitory Activity Following Vaginal Administration in Humans. *PLoS One*, **2011,** *6* (9), 15.
193. Ramanathan, T.; Abdala, A.; Stankovich, S.; Dikin, D.; Herrera-Alonso, M.; Piner, R.; Adamson, D.; Schniepp, H.; Chen, X.; Ruoff, R. Functionalized Graphene Sheets for Polymer Nanocomposites. *Nature Nanotechnol.* **2008,** *3* (6), 327–331.
194. Ravi, S.; Chaikof, E. L. Biomaterials for Vascular Tissue Engineering. *Regener. Med.* **2010,** *5* (1), 107.
195. Rhim, J. W.; Park, H. M.; Ha, C. S. Bio-nanocomposites for Food Packaging Applications. *Prog. Polym. Sci.* **2013,** *38* (10), 1629–1652.
196. Richardson, S. W.; Kolbe, H. J.; Duncan, R. Potential of Low Molecular Mass Chitosan as a DNA Delivery System: Biocompatibility, Body Distribution and Ability to Complex and Protect DNA. *Int. J. Pharm.* **1999,** *178* (2), 231–243.
197. Rickman, D.; Luvall, J.; Shaw, J.; Mask, P.; Kissel, D.; Sullivan, D. Precision Agriculture: Changing the Face of Farming. *Geotimes* **2003,** *48* (11), 28–33.
198. Riehemann, K.; Schneider, S. W.; Luger, T. A.; Godin, B.; Ferrari, M.; Fuchs, H. Nanomedicine - Challenge and Perspectives. *Angewandte Chemie* **2009,** *48* (5), 872–897.
199. Robertson, J. M.; Robertson, P. K.; Lawton, L. A. A Comparison of the Effectiveness of TiO_2 Photocatalysis and UVA Photolysis for the Destruction of Three Pathogenic Micro-organisms. *J. Photochem. Photobiol. A: Chem.* **2005,** *175* (1), 51–56.
200. Roco, M. C.; Mirkin, C. A.; Hersam, M. C. Nanotechnology Research Directions for Societal Needs in 2020: Summary of International Study. *J. Nanopart. Res.* **2011,** *13* (3), 897–919.
201. Rodriguez-Limas, W. A.; Sekar, K.; Tyo, K. E. Virus-like Particles: The Future of Microbial Factories and Cell-free Systems as Platforms for Vaccine Development. *Curr. Opin. Biotechnol.* **2013,** *24* (6), 1089–1093.
202. Rodriguez, F.; Sepulveda, H. M.; Bruna, J.; Guarda, A.; Galotto, M. J. Development of Cellulose Eco-nanocomposites with Antimicrobial Properties Oriented for Food Packaging. *Pack. Technol. Sci.* **2013,** *26* (3), 149–160.
203. Roldao, A.; Mellado, M. C. M.; Castilho, L. R.; Carrondo, M. J.; Alves, P. M. Virus-like Particles in Vaccine Development. *Exp. Rev. Vacc.* **2010,** *9* (10), 1149–1176.
204. Roy, I.; Ohulchanskyy, T. Y.; Bharali, D. J.; Pudavar, H. E.; Mistretta, R. A.; Kaur, N.; Prasad, P. N. Optical Tracking of Organically Modified Silica Nanoparticles as DNA Carriers: A Nonviral, Nanomedicine Approach for Gene Delivery. *Proc. Natl. Acad. Sci. USA* **2005,** *102* (2), 279–284.
205. Saade, F.; Honda-Okubo, Y.; Trec, S.; Petrovsky, N. A Novel Hepatitis B Vaccine Containing Advax™, A Polysaccharide Adjuvant Derived from Delta Inulin, Induces

Robust Humoral and Cellular Immunity with Minimal Reactogenicity in Preclinical Testing. *Vaccine* **2013,** *31* (15), 1999–2007.
206. Said, D. E.; Elsamad, L. M.; Gohar, Y. M. Validity of Silver, Chitosan, and Curcumin Nanoparticles as Anti-Giardia Agents. *Parasitol. Res.* **2012,** *111* (2), 545–554.
207. Saini, R.; Saini, S. Nanotechnology and Surgical Neurology. *Surgical Neurol. Inter.* **2010,** *1* (1), 57–57.
208. Sakamoto, J. H.; van de Ven, A. L.; Godin, B.; Blanco, E.; Serda, R. E.; Grattoni, A.; Ziemys, A.; Bouamrani, A.; Hu, T.; Ranganathan, S. I.; De Rosa, E.; Martinez, J. O.; Smid, C. A.; Buchanan, R. M.; Lee, S. Y.; Srinivasan, S.; Landry, M.; Meyn, A.; Tasciotti, E.; Liu, X.; Decuzzi, P.; Ferrari, M. Enabling Individualized Therapy Through Nanotechnology. *Pharmacol. Res.* **2010,** *62* (2), 57–89.
209. Salvador, A.; Igartua, M.; Hernandez, R. M.; Pedraz, J. L. An Overview on the Field of Micro-and Nanotechnologies for Synthetic Peptide-based Vaccines. *J. Drug Deliv.* **2011,** Article ID-181646, 18.
210. Santos-Magalhaes, N. S.; Mosqueira, V. C. Nanotechnology Applied to the Treatment of Malaria. *Adv. Drug Del. Rev.* **2010,** *62* (4–5), 560–575.
211. Sastry, K.; Rashmi, H.; Rao, N. Nanotechnology Patents as R&D Indicators for Disease Management Strategies in Agriculture. *J. Intel. Prop. Rights* **2010,** *15*, 197–205.
212. Schechner, G.; Braunbarth, C.; Poth, T.; Franke, H., Sweet Containing Calcium-Coated Chewing Gum. Patent Appl. EP20030748025. Germany; 2003, pp 21.
213. Scheerlinck, J. -P. Y.; Gloster, S.; Gamvrellis, A.; Mottram, P. L.; Plebanski, M. Systemic Immune Responses in Sheep, Induced by a Novel Nano-bead Adjuvant. *Vaccine* **2006,** *24* (8), 1124–1131.
214. Scott, N.; Chen, H. Nanoscale Science and Engineering for Agriculture and Food Systems. *Indus. Biotechnol.* **2013,** *9* (1), 17–18.
215. Seifalian, A. M.; Salacinski, H. J.; Tiwari, A.; Edwards, A.; Bowald, S.; Hamilton, G. In vivo Biostability of a Poly(carbonate-urea)urethane Graft. *Biomaterials* **2003,** *24* (14), 2549–2557.
216. Seil, J. T.; Webster, T. J. Electrically Active Nanomaterials as Improved Neural Tissue Regeneration Scaffolds. *Rev. Nanomed. Nanobiotechnol.* **2010,** *2* (6), 635–647.
217. Sekhon, B. S. Food Nanotechnology - An Overview. *Nanotechnol. Sci. Appl.* **2010,** *3* (1), 1–15.
218. Sekhon, B. S. Nanotechnology in Agri-food Production: An Overview. *Nanotechnol. Sci. Appl.* **2014,** *7*, 31–53.
219. Semo, E.; Kesselman, E.; Danino, D.; Livney, Y. D. Casein Micelle as a Natural Nano-capsular Vehicle for Nutraceuticals. *Food Hydrocoll.* **2007,** *21* (5), 936–942.
220. Sharon, M.; Sharon, M. Carbon Nanomaterials: Applications in Physico-chemical Systems and Biosystems. *Def. Sci. J.* **2008,** *58* (4), 460.
221. Shi, D.; Lian, J.; Wang, W.; Liu, G.; He, P.; Dong, Z.; Wang, L. M.; Ewing, R. C. Luminescent Carbon Nanotubes by Surface Functionalization. *Adv. Mater.* **2006,** *18* (2), 189–193.
222. Silva, G. A.; Czeisler, C.; Niece, K. L.; Beniash, E.; Harrington, D. A.; Kessler, J. A.; Stupp, S. I. Selective Differentiation of Neural Progenitor Cells by High-epitope Density Nanofibers. *Science* **2004,** *303* (5662), 1352–1355.
223. Silvestre, C.; Duraccio, D.; Cimmino, S. Food Packaging Based on Polymer Nanomaterials. *Prog. Polym. Sci.* **2011,** *36* (12), 1766–1782.

224. Singh, S.; Singh, M.; Agrawal, V. V.; Kumar, A. An Attempt to Develop Surface Plasmon Resonance Based Immunosensor for Karnal bunt (*Tilletia indica*) Diagnosis Based on the Experience of Nano-gold Based Lateral Flow Immuno-dipstick Test. *Thin Solid Films* **2010,** *519* (3), 1156–1159.
225. Slotkin, J. R.; Chakrabarti, L.; Dai, H. N.; Carney, R. S.; Hirata, T.; Bregman, B. S.; Gallicano, G. I.; Corbin, J. G.; Haydar, T. F. *In vivo* Quantum Dot Labeling of Mammalian Stem and Progenitor Cells. *Dev. Dyn.* **2007,** *236* (12), 3393–3401.
226. Sonkaria, S.; Ahn, S. -H.; Khare, V. Nanotechnology and its Impact on Food and Nutrition: A Review. *Recent Patents Food Nutr. Agric.* **2012,** *4* (1), 8–18.
227. Stadlinger, B.; Hennig, M. H.; Eckelt, U.; Kuhlisch, E.; Mai, R. Comparison of Zirconia and Titanium Implants After a Short Healing Period. A Pilot Study in Minipigs. *Int. J. Oral Maxillofac. Surg.* **2010,** *39* (6), 585–592.
228. Sun, X.; Liu, B.; Xia, K. A Sensitive and Regenerable Biosensor For Organophosphate Pesticide Based on Self-assembled Multilayer Film with CdTe as Fluorescence Probe. *Luminescence* **2011,** *26* (6), 616–621.
229. Syed, S.; Zubair, A.; Frieri, M. Immune Response to Nanomaterials: Implications for Medicine and Literature Review. *Curr. Aller. Asthma Rep.* **2013,** *13* (1), 50–57.
230. Tahara, Y.; Mukai, S. -A.; Sawada, S. -I.; Sasaki, Y.; Akiyoshi, K. Nanocarrier-integrated Microspheres: Nanogel Tectonic Engineering for Advanced Drug-delivery systems. *Adv. Mater.* **2015,** *27* (34), 5080–5088.
231. Taylor, T. M.; Weiss, J.; Davidson, P. M.; Bruce, B. D. Liposomal Nanocapsules in Food Science and Agriculture. *Crit. Rev. Food Sci. Nutr.* **2005,** *45* (7–8), 587–605.
232. Thapa, A.; Miller, D. C.; Webster, T. J.; Haberstroh, K. M. Nano-structured Polymers Enhance Bladder Smooth Muscle Cell Function. *Biomaterials* **2003,** *24* (17), 2915–2926.
233. Thapa, A.; Webster, T. J.; Haberstroh, K. M. Polymers with Nano-dimensional Surface Features Enhance Bladder Smooth Muscle Cell Adhesion. J. Biomed. Mater. Res. A **2003,** *67* (4), 1374–1383.
234. Thorvaldsson, A.; Stenhamre, H.; Gatenholm, P.; Walkenstrom, P. Electrospinning of Highly Porous Scaffolds for Cartilage Regeneration. *Biomacromolecules* **2008,** *9* (3), 1044–1049.
235. Tissot, A. C.; Maurer, P.; Nussberger, J.; Sabat, R.; Pfister, T.; Ignatenko, S.; Volk, H. - D.; Stocker, H.; MÃ¼ller, P.; Jennings, G. T. Effect of Immunization Against Angiotensin II with CYT006-AngQb on Ambulatory Blood Pressure: A Double-blind, Randomized, Placebo-controlled Phase IIa Study. *Lancet* **2008,** *371* (9615), 821–827.
236. Torney, F. O.; Trewyn, B. G.; Lin, V. S. -Y.; Wang, K. Mesoporous Silica Nanoparticles Deliver DNA and Chemicals into Plants. *Nature Nanotechnol.* **2007,** *2* (5), 295–300.
237. Torres-Sangiao, E.; Holban, A. M.; Gestal, M. C. Advanced Nanobiomaterials: Vaccines, Diagnosis and Treatment of Infectious Diseases. *Molecules* **2016,** *21* (7), 867.
238. Tran, P. A.; Zhang, L.; Webster, T. J. Carbon Nanofibers and Carbon Nanotubes in Regenerative Medicine. *Adv. Drug Deliv. Rev.* **2009,** *61* (12), 1097–1114.
239. Tysseling-Mattiace, V. M.; Sahni, V.; Niece, K. L.; Birch, D.; Czeisler, C.; Fehlings, M. G.; Stupp, S. I.; Kessler, J. A. Self-assembling Nanofibers Inhibit Glial Scar Formation and Promote Axon Elongation After Spinal Cord Injury. *J. Neurosci.* **2008,** *28* (14), 3814–3823.
240. Upadhyay, R. K. Drug Delivery Systems, CNS Protection, and the Blood Brain Barrier. *Biomed. Res. Int.* **2014,** *2014,* 869269.

241. Vegas, R.; Moure, A. S.; DomÃnguez, H.; ParajÃ³, J. C.; Alvarez, J. R. N.; Luque, S. Evaluation of Ultra-and Nanofiltration for Refining Soluble Products from Rice Husk Xylan. *Biores. Technol.* **2008,** *99* (13), 5341–5351.
242. Vidhyalakshmi, R.; Bhakyaraj, R.; Subhasree, R. Encapsulation "The Future of Probiotics" - A Review. *Adv. Biolog. Res.* **2009,** *3* (3–4), 96–103.
243. Vila, A.; Sanchez, A.; Evora, C.; Soriano, I.; Vila Jato, J.; Alonso, M. PEG-PLA Nanoparticles as Carriers for Nasal Vaccine Delivery. *J. Aerosol Med.* **2004,** *17* (2), 174–185.
244. Vimala, V.; Clarke, S.; Urvinder Kaur, S. Pesticides Detection Using Acetylcholinesterase Nanobiosensor. *Biosens. J.* **2015,** *5* (133), 2.
245. Vo-Dinh, T.; Cullum, B. M.; Stokes, D. L. Nanosensors and Biochips: Frontiers in Biomolecular Diagnostics. *Sens. Actuat. B Chem.* **2001,** *74* (1), 2–11.
246. Vodinh, T.; Kasili, P. M.; Wabuyele, M. B. Nanoprobes and Nanobiosensors for Monitoring and Imaging Individual Living Cells. *Nanomed. Nanotechnol. Biol. Med.* **2006,** *2* (1), 22–30.
247. Wagner, E. M.; Sanchez, J. M. S.; Mcclintock, J. Y.; Jenkins, J.; Moldobaeva, A. Inflammation and Ischemia-induced Lung Angiogenesis. *Am. J. Physiol. Lung Cell. Mol. Physiol.* **2008,** *294* (2), 112–118.
248. Wagner, V.; Dullaart, A.; Bock, A.; Zweck, A. The Emerging Nanomedicine Landscape. *Nature Biotechnol.* **2006,** *24* (10), 1211–1217.
249. Wang, M.; Li, Z. Nano-composite ZrO 2/Au Film Electrode for Voltametric Detection of Parathion. *Sens. Actuat. B Chem.* **2008,** *133* (2), 607–612.
250. Wang, T.; Zou, M.; Jiang, H.; Ji, Z.; Gao, P.; Cheng, G. Synthesis of a Novel Kind of Carbon Nanoparticle with Large Mesopores and Macropores and its Application as an Oral Vaccine Adjuvant. *Eur. J. Pharm. Sci.* **2011,** *44* (5), 653–659.
251. Webster, T. J.; Schadler, L. S.; Siegel, R. W.; Bizios, R. Mechanisms of Enhanced Osteoblast Adhesion on Nanophase Alumina Involve Vitronectin. *Tissue Eng.* **2004,** *7* (3), 291–301.
252. Webster, T. J.; Siegel, R. W.; Bizios, R. Osteoblast Adhesion on Nanophase Ceramics. *Biomaterials*, **1999,** *20* (13), 1221–1227.
253. Weiss, J.; Takhistov, P.; McClements, D. J. Functional Materials in Food Nanotechnology. *J. Food Sci.* **2006,** *71* (9), R107–R116.
254. Wickline, S. A.; Lanza, G. M. Nanotechnology for Molecular Imaging and Targeted Therapy. *Circulation* **2003,** *107* (8), 1092–1095.
255. Xiao-e, L.; Green, A. N.; Haque, S. A.; Mills, A.; Durrant, J. R. Light-driven Oxygen Scavenging by Titania/Polymer Nanocomposite Films. *J. Photochem. Photobiol. A Chem.* **2004,** *162* (2), 253–259.
256. Yam, K. L.; Takhistov, P. T.; Miltz, J. Intelligent Packaging: Concepts and Applications. *J. Food Sci.* **2005,** *70* (1), R1–R10.
257. Yan, J.; Guan, H.; Yu, J.; Chi, D. Acetylcholinesterase Biosensor Based on Assembly of Multiwall Carbon Nanotubes onto Liposome Bioreactors for Detection of Organophosphates Pesticides. *Pest. Biochem. Physiol.* **2013,** *105* (3), 197–202.
258. Yang, J. -Y.; Li, Y.; Chen, S. -M.; Lin, K. -C. Fabrication of a Cholesterol Biosensor Based on Cholesterol Oxidase and Multiwall Carbon Nanotube Hybrid Composites. *Int. J. Electrochem. Sci.* **2011,** *6* (6), 2223–2234.

259. Yao, K. S.; Li, S.; Tzeng, K.; Cheng, T. C.; Chang, C. Y.; Chiu, C.; Liao, C.; Hsu, J.; Lin, Z. Fluorescence Silica Nanoprobe as a Biomarker for Rapid Detection of Plant Pathogens. *Adv. Mater. Res.* **2009,** *79*, 513–516.

260. Yemmireddy, V. K.; Hung, Y. -C. Effect of Binder on the Physical Stability and Bactericidal Property of Titanium Dioxide (TiO_2) Nanocoatings on Food Contact Surfaces. *Food Control* **2015,** *57,* 82–88.

261. Yezhelyev, M. V.; Qi, L.; O'Regan, R. M.; Nie, S.; Gao, X. Proton-sponge Coated Quantum Dots for siRNA Delivery and Intracellular Imaging. *J. Am. Chem. Soc.* **2008,** *130* (28), 9006–9012.

262. You, X.; He, R.; Gao, F.; Shao, J.; Pan, B.; Cui, D. Hydrophilic High-luminescent Magnetic Nanocomposites. *Nanotechnology* **2007,** *18* (3), 035701.

263. Yu, G.; Wu, W.; Zhao, Q.; Wei, X.; Lu, Q. Efficient Immobilization of Acetylcholinesterase onto Amino Functionalized Carbon Nanotubes for the Fabrication of High Sensitive Organophosphorus Pesticides Biosensors. *Biosens. Bioelectron.* **2015,** *68*, 288–294.

264. Yu, H.; Huang, Q. Enhanced In vitro Anti-cancer Activity of Curcumin Encapsulated in Hydrophobically Modified Starch. *Food Chem.* **2010,** *119* (2), 669–674.

265. Yuba, E.; Harada, A.; Sakanishi, Y.; Watarai, S.; Kono, K. A Liposome-based Antigen Delivery System Using pH-sensitive Fusogenic Polymers for Cancer Immunotherapy. *Biomaterials* **2013,** *34* (12), 3042–3052.

266. Zhang, L.; Pornpattananangku, D.; Hu, C. M.; Huang, C. M. Development of Nanoparticles for Antimicrobial Drug Delivery. *Curr. Med. Chem.* **2010,** *17* (6), 585–594.

267. Zhang, L. F.; Zhou, J.; Chen, S.; Cai, L. L.; Bao, Q. Y.; Zheng, F. Y.; Lu, J. Q.; Padmanabha, J.; Hengst, K.; Malcolm, K. HPV6b Virus Like Particles are Potent Immunogens Without Adjuvant in Man. *Vaccine* **2000,** *18* (11), 1051–1058.

268. Zhao, L.; Seth, A.; Wibowo, N.; Zhao, C. -X.; Mitter, N.; Yu, C.; Middelberg, A. P. Nanoparticle Vaccines. *Vaccine* **2014,** *32* (3), 327–337.

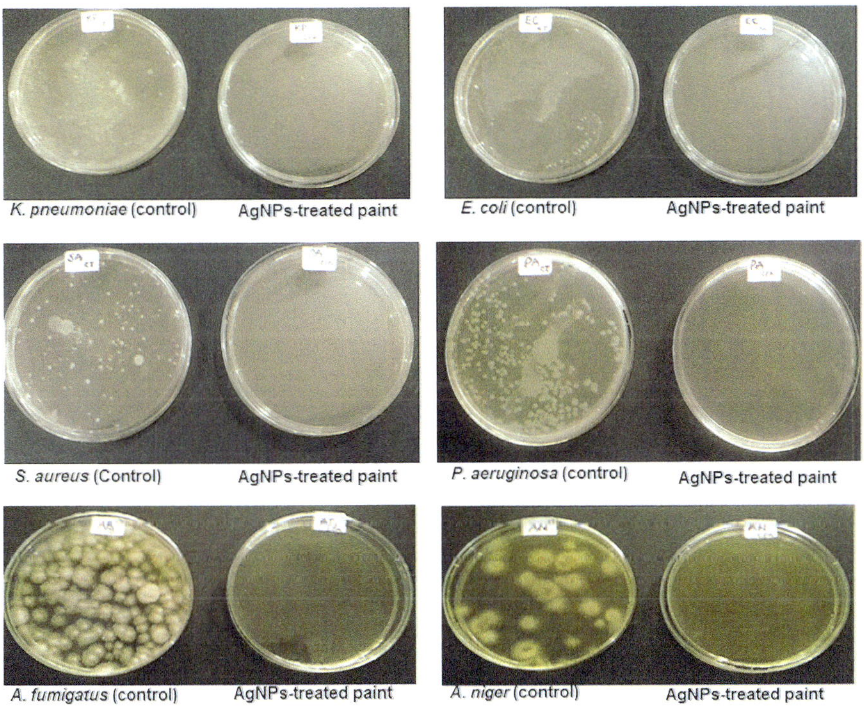

FIGURE 1.3 Antimicrobial activities of biosynthesized CHPE-AgNPs in emulsion paint.

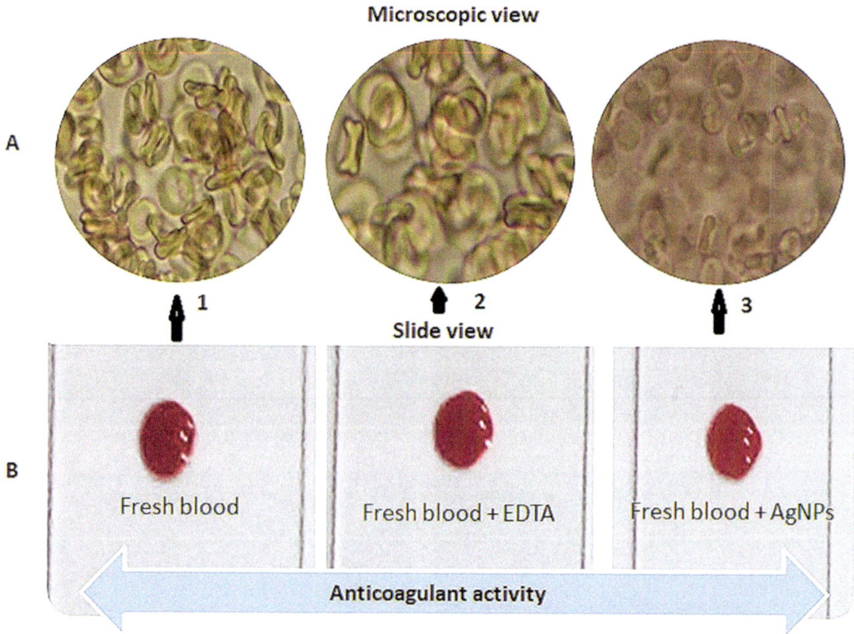

FIGURE 1.4 Anticoagulant activity of AgNPs biosynthesized using the extract of cocoa beans.

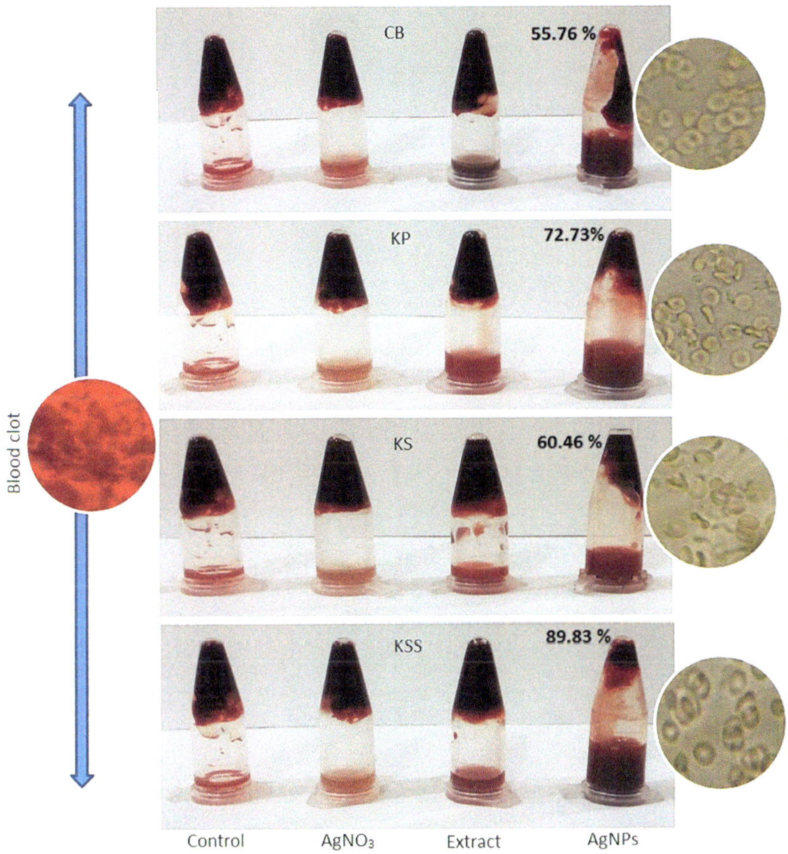

FIGURE 1.5 Thrombolytic activities of CB-, KP-, KS-, and KSS-AgNPs.

Source: Reprinted with permission from Ref [88].

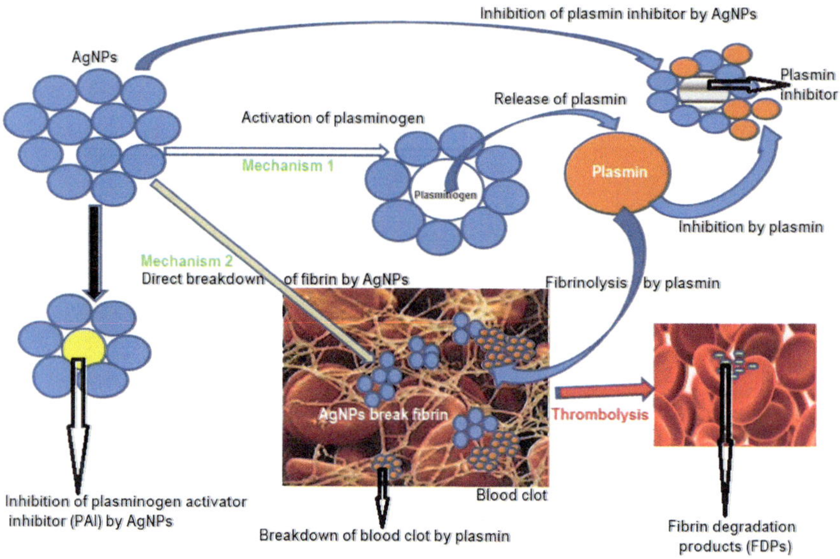

FIGURE 1.6 The possible mechanisms of thrombolytic activity of AgNPs.

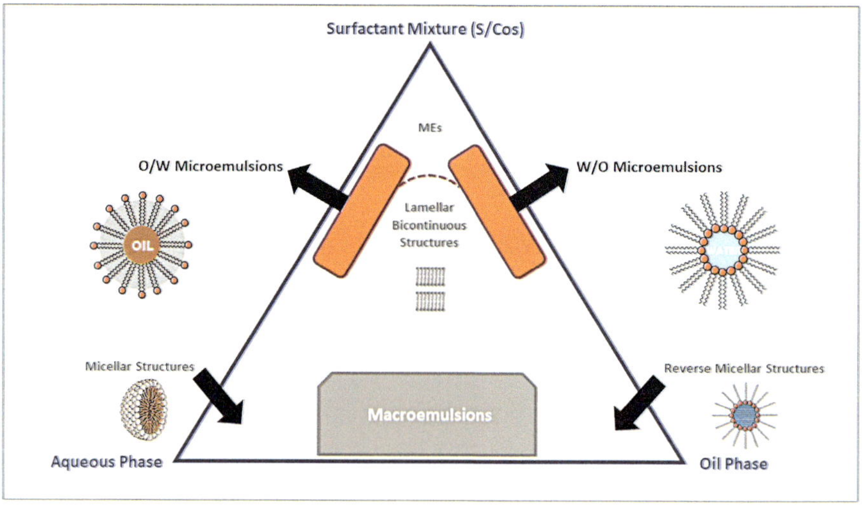

FIGURE 2.1 Diagram showing different drug delivery systems regions in a "*pseudo-ternary phase diagram*". The vertices of the triangle represent: oil phase, aqueous phase, and surfactant/cosurfactant mixture. All area above the curved dotted line corresponds to the possible formation of microemulsions.[42,84]

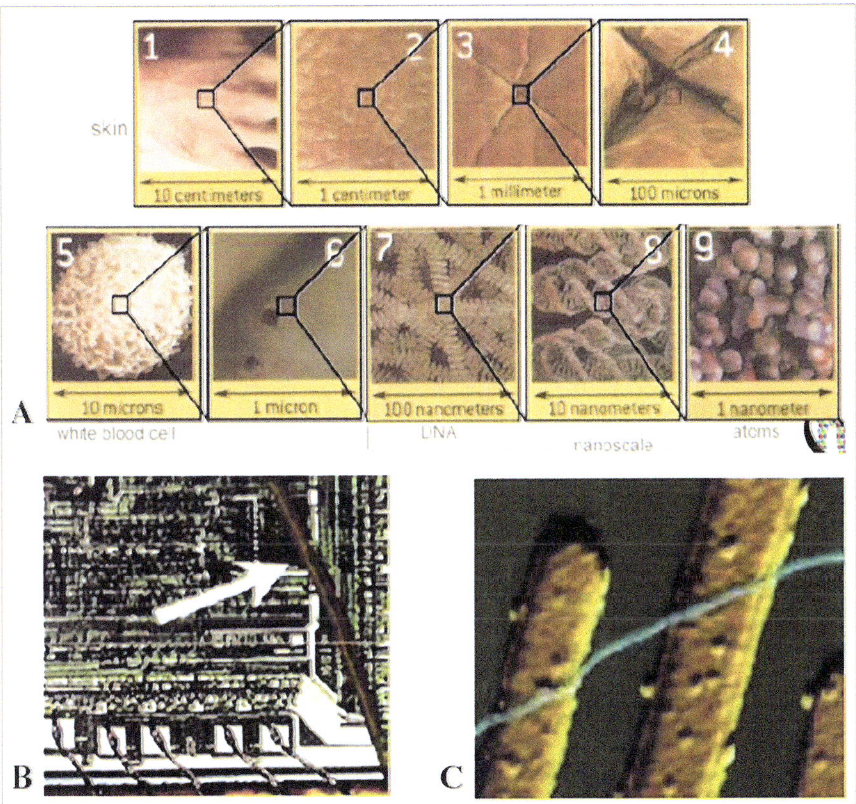

FIGURE 4.2 (A) Magnification of human skin to find out nanoscale-materials, (B) the width of a human hair (placed on a microchip and pointed by a white arrow) is roughly 104 times, and (C) the diameter of carbon nanotubes (placed on the top of some metal electrodes) is yet again 104 times smaller.

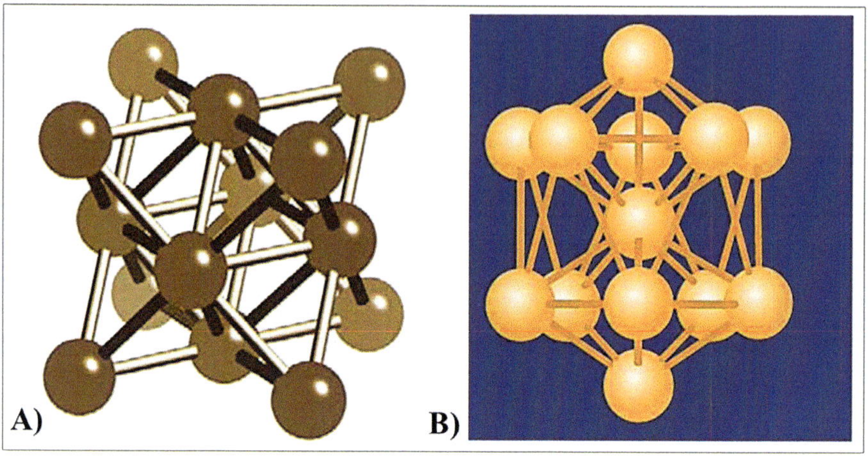

FIGURE 4.5 Crystal structure. (A) bulk (FCC) and (B) nano (icosahedral) gold.

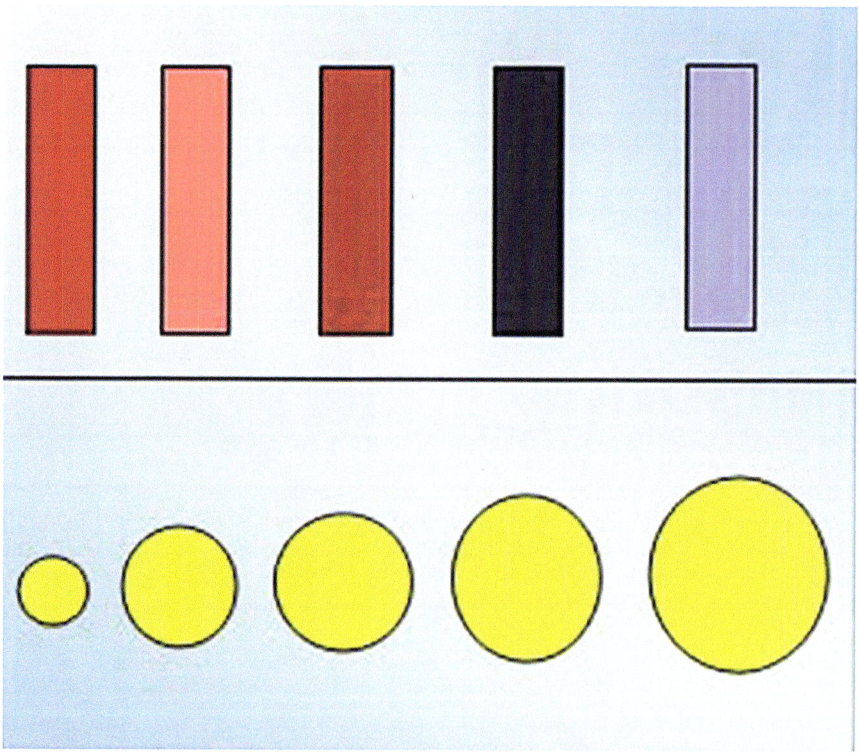

FIGURE 4.8 The diameter of gold nanoparticle with the respective color.

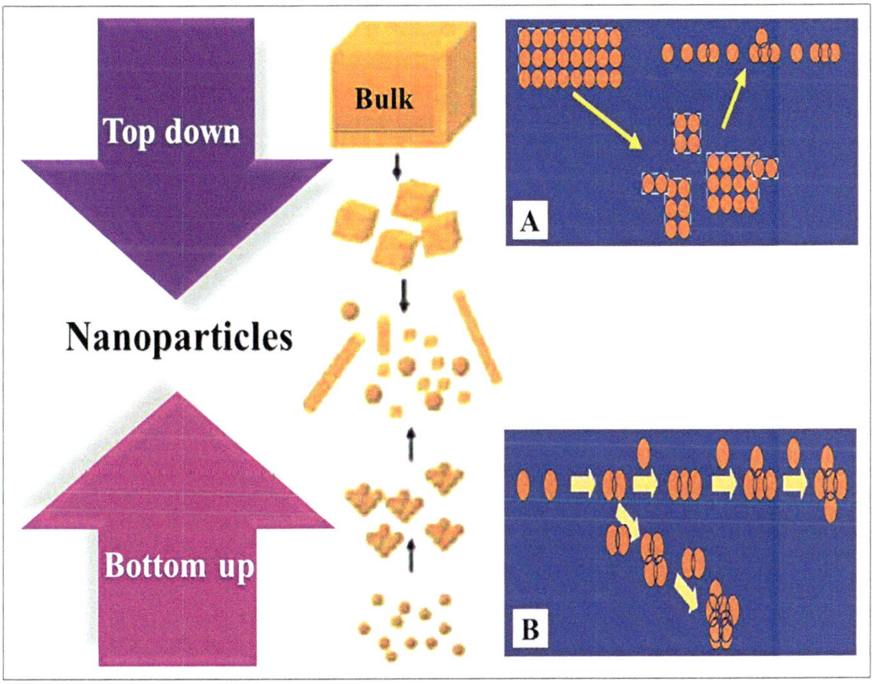

FIGURE 4.11 Representation of nanoparticle fabrication (A) top–down and (B) bottom–up methods.

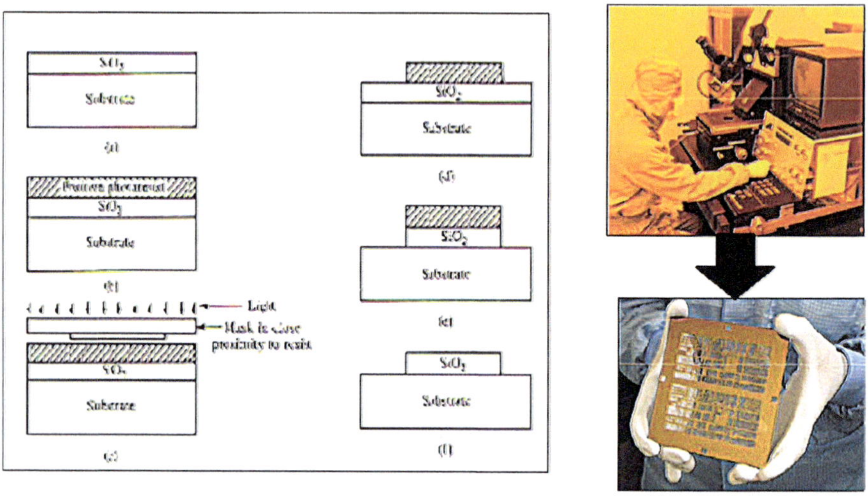

FIGURE 4.13 Photolithography technique. Processes are coat, protect, expose, etch, repeat…, and the result is multiple patterned layers of different materials.

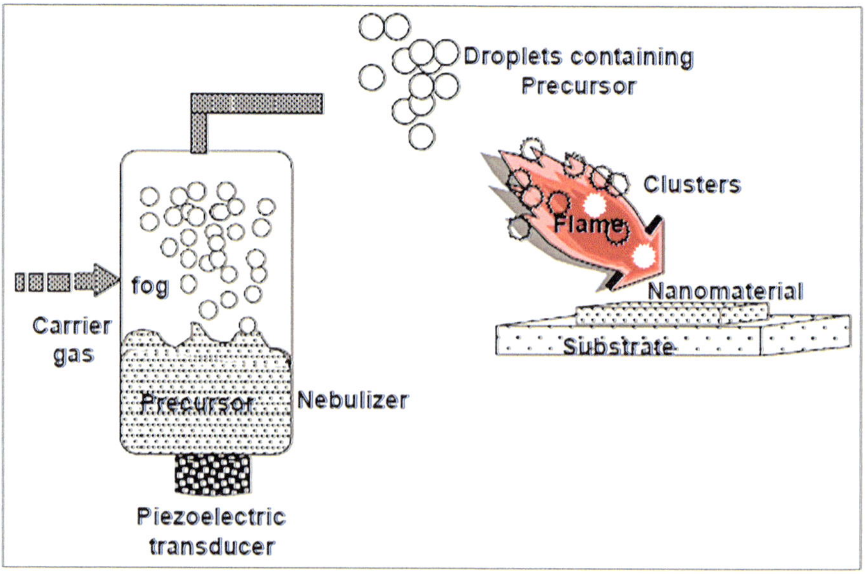

FIGURE 4.16 Flame-based synthesis of nanoparticles.

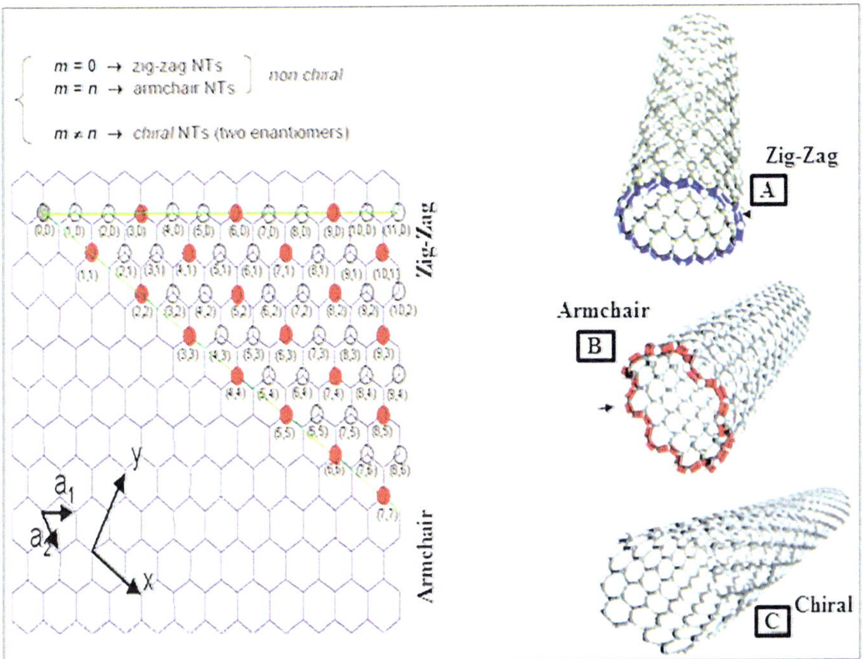

FIGURE 4.20 Representation of graphene sheet and types of graphene sheet folding (A) zigzag, (B) armchair, and (C) chiral nanotubes patterns.

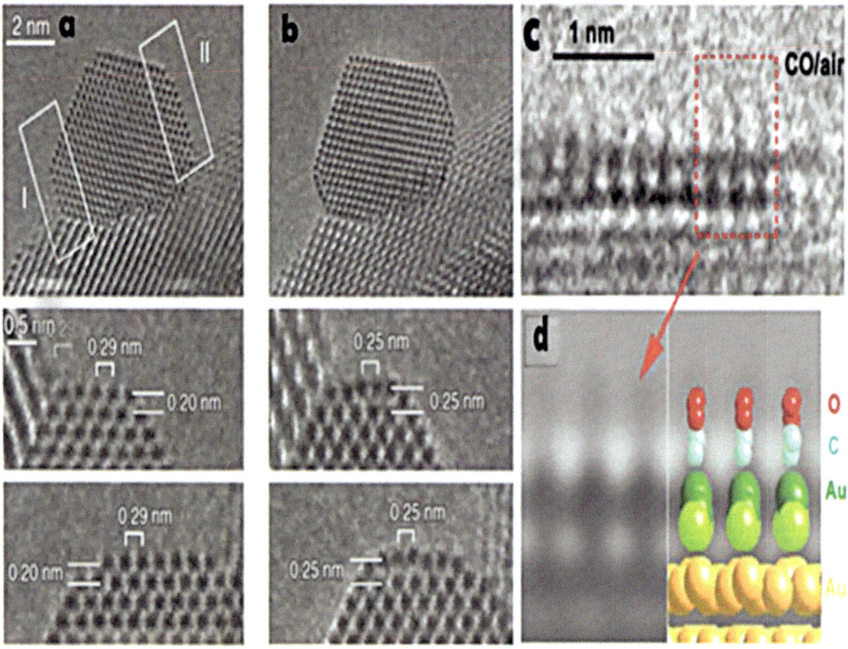

FIGURE 5.5 Schematic diagram of scanning probe microscopy.[72]

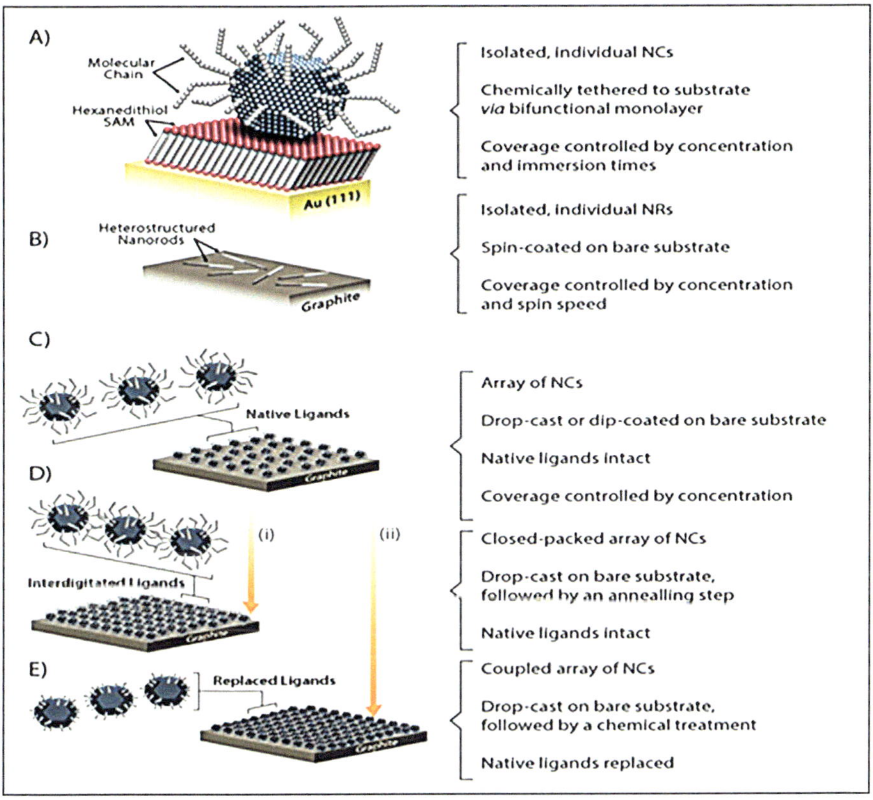

FIGURE 5.6 Sample preparation methods for AFM analysis: (A) Isolated and tethered nanocrystals (NCs) were assembled on Au; (B) Spin coating of nanorods; (C) Easiest way for array construction is by drop casting system; (D) Drop casting, and (E) NCs are prepared via chemical treatment. Reprinted with permission from Ref [38]. © 2015 American Chemical Society.

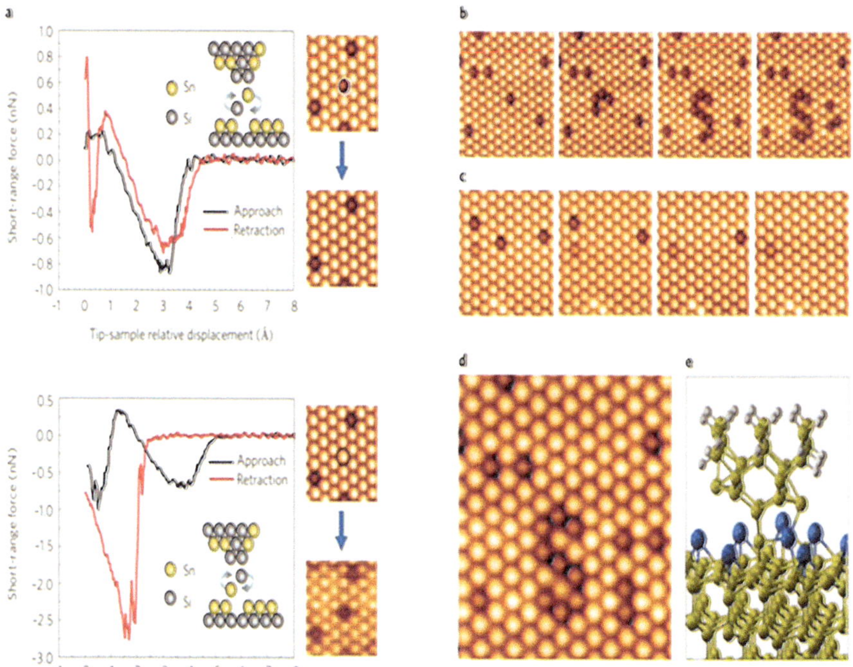

FIGURE 5.7 Vertical interchange exploring in atomic manipulation. Reprinted with permission from Ref [10]. © 2009 Nature Publishing Group.

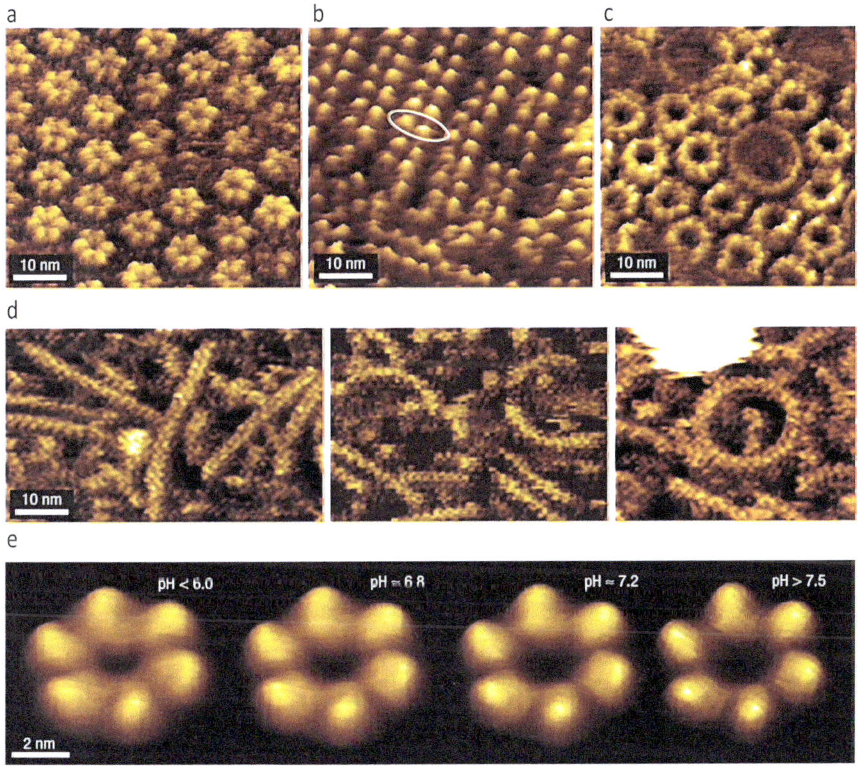

FIGURE 5.8 Meticulous observations of individual cellular activities at ultrahigh surface resolution. Reprinted with permission from Ref [37]. © 2008 Nature Publishing Group.

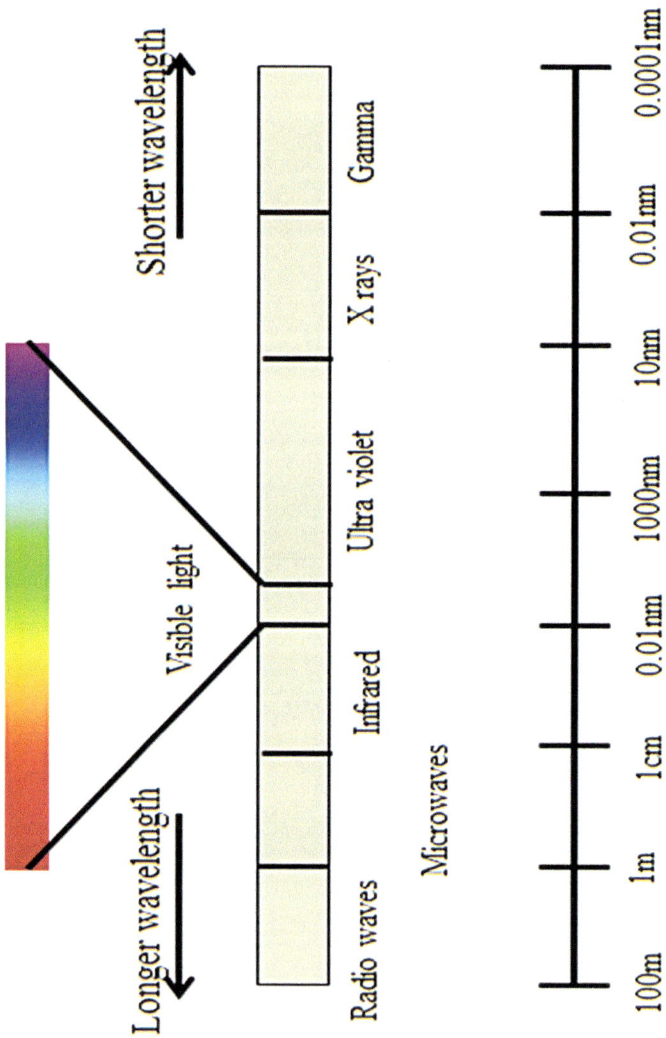

FIGURE 5.9 Electromagnetic spectrum for different types of waves

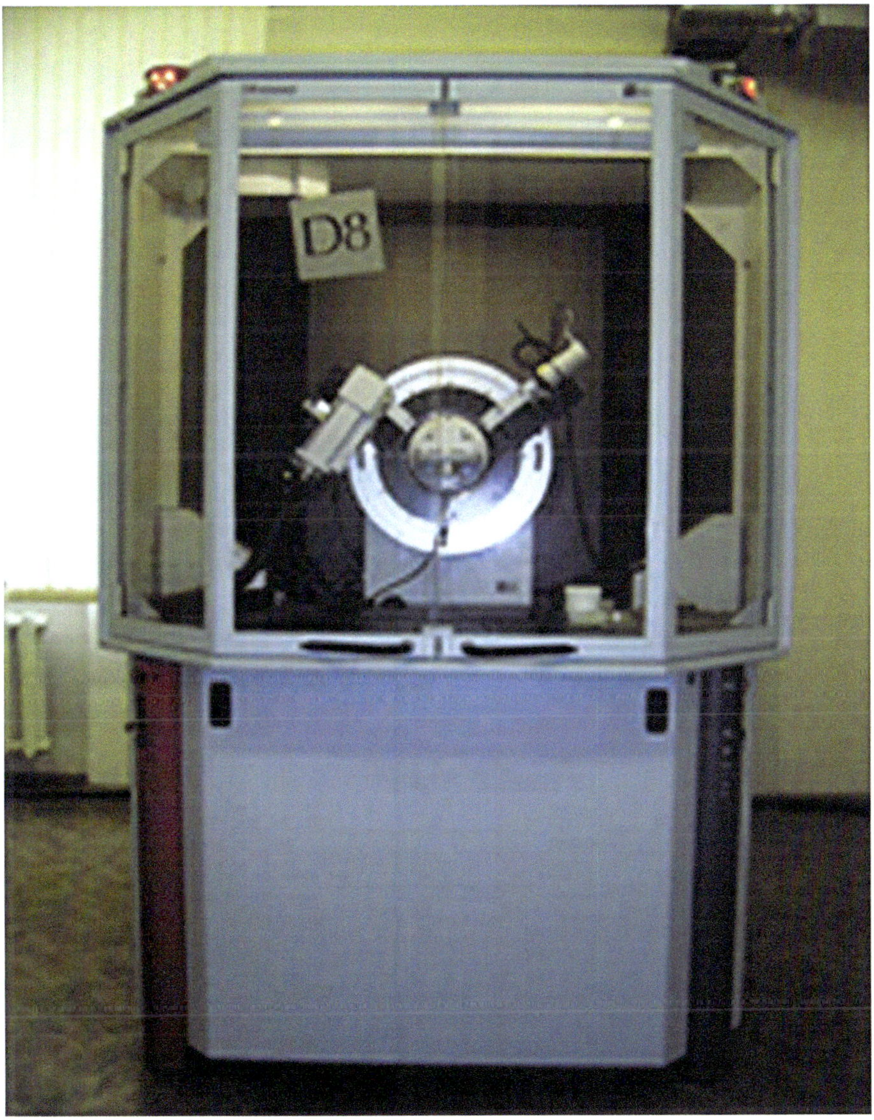

FIGURE 5.12 XRD instrument. Courtesy: Chemistry, IITM.

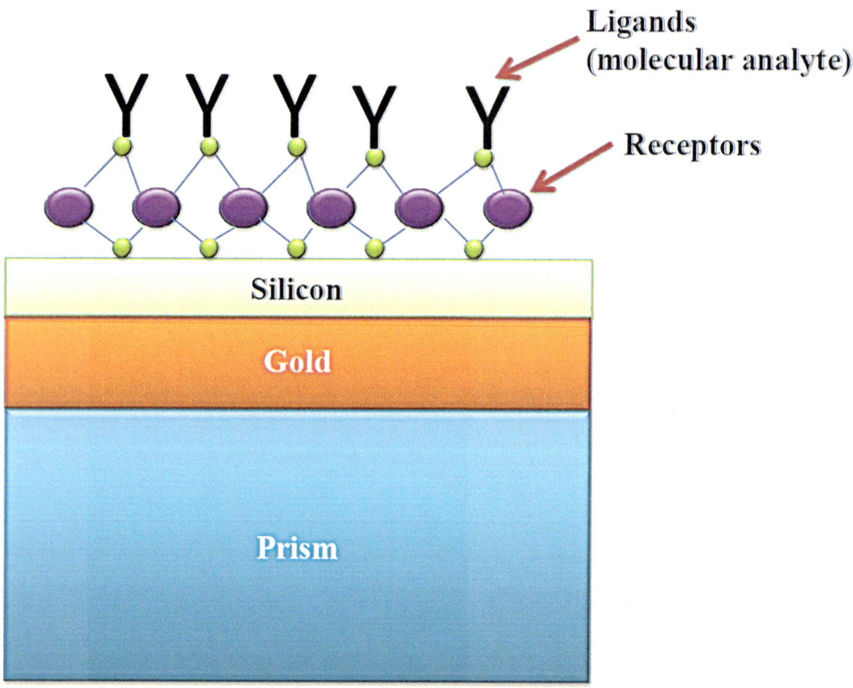

FIGURE 7.1 Design of surface plasmon resonance (SPR) based sensor.

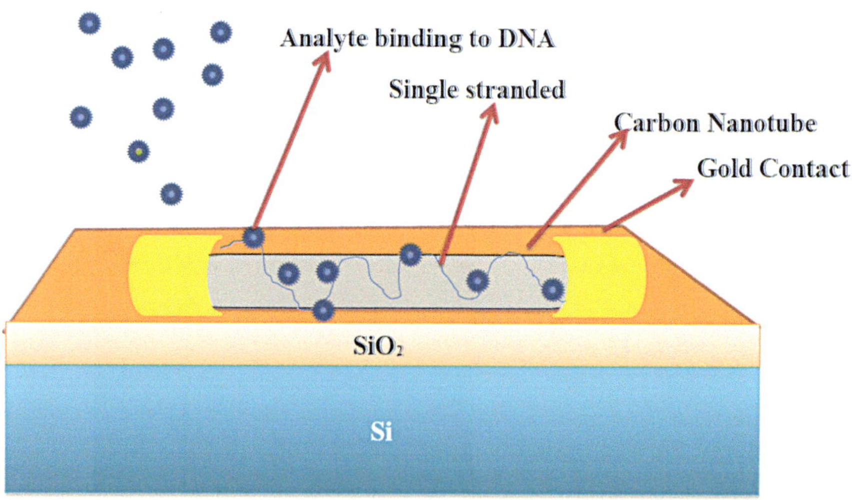

FIGURE 7.2 Basic layout of CNTs based senor.

CHAPTER 4

NANOMATERIALS: FUNDAMENTAL PRINCIPLE AND APPLICATIONS

AJIT BEHERA*, SOUMYA SANJEEB MOHAPATRA, and DEEPAK KUMAR VERMA

*Corresponding author. E-mail: ajit.behera88@gmail.com

ABSTRACT

Nanomaterial and nanotechnologies are the center of interest in various fields, including physics, chemistry, materials science, and engineering and are a real interdisciplinary area covering all engineering and science fields. To achieve miniaturization of devices, reduction of scale from micro- to nano-scale materials or nanoparticles has to be focused on. This describes how nanomaterial can have distinctly different properties than bulk materials. The size-dependence of the physical, electrical, magnetic, and optical properties of nanoparticles and nanostructures materials has been better understood in this chapter. How a relative increase in surface area and quantum confinement plays a vital role in small-scale materials is described here. Nanomaterial are classified according to their dimension—zero-dimensional, one-dimensional, two-dimensional, and three-dimensional nanomaterial—and are briefly described with their structural examples. Again, the current nanomaterials are categorized into four types: carbon-based materials, metal-based nanomaterials, polymer-based nanomaterials, and composite nanomaterials. Top–down approach and Bottom–up approach are the two synthesis procedures for nanomaterials. Nanotechnology has applications in all conceivable areas and has few limitations, which are discussed in this chapter. Currently, nanotechnology is mostly used in cosmetics, medicine/drugs, fabrics, defence, energy, and water purification, but within a few years all other domains will also be using nanotechnology as a potential tool. In the future, the question may not be how much smaller can the size of the computer on our lap be, but how many computers can we fit on our fingers.

This chapter is of great importance in the area of nanotechnology. Nanotechnology is a real interdisciplinary area covering all engineering and science fields. Interactions among researchers about their interdisciplinary subjects will develop new materials and structures with novel technological possibilities. The future application of the nanotechnologies looks bright and intense because of their efficiency and potential to give rise to new efficient products.

4.1 INTRODUCTION

Nanomaterials and nanotechnologies have been the center of interest of recent research. Various research fields, including physics, chemists, materials science, and engineering are involved in this advanced research. A lotus plant has nanometric-sized hairs on its leaf surface that can run off water and act as self-cleaning agents. This is a hint of nanosize given by nature. The American physicist Richard Feynman in 1959 (Nobel awarded in Physics in 1965), inspired in the field of nanotechnology at an American Physical Society's lecture titled "There's Plenty of Room at the Bottom." He told about the arrangement of atoms which can be put atom-by-atom as per the chemist, that help to generate a new substance, whether it is atomic scale or macro scale. Then the definition of nanotechnology has given by Norio Taniguchi, Tokyo Science University in 1970. He stated that "Nanotechnology is defined as the technology associated with different design and production of structures by controlling the size and the shape at nanometer scale." In nanotechnology, objects and phenomena are considered only at a very small scale, around 1–100 nm. Figure 4.1 shows the scale of measurements used for the system in our universe. Here the considered range of nanotechnology has given.[3] To understand how small one nm is let us see a few comparisons:

1. Figure 4.2A showing the magnification of human skin which contains approximately 10 nm DNA,
2. Figure 4.2B showing human hair which is approximately 80,000 nm wide placed on a microelectromechanical system having many metal circuit, and
3. Nanotubes have shown in Figure 4.2C. It is found that the number of macro devices squeezed into nano-devices in each year, which is the revolution of micro-system to nano-system.

4.2 CAN WE MAKE SMALL STRUCTURED DEVICES?

On September 14, 1956, IBM first used only five megabytes hard drive, a secondary magnetic storage which is installed at the US Navy and private corporations. Figure 4.3 shows that hard drive during the shipment. Now you can compare this 5-MB hard drive with our 2-TB portable hard disk. This is the actual revolution till now due to nanotechnology. More than 10 million transistors can be accommodated inside a computer chip and chip size is reduced up to 100 nm range. In 2003, the first high volume 90 nm line width production by Intel was shipped to the market. Now a day, commercial exploration tends to produce below 50 nm nanochips. Also, all the potential consumer products are gifts of nanotechnology.

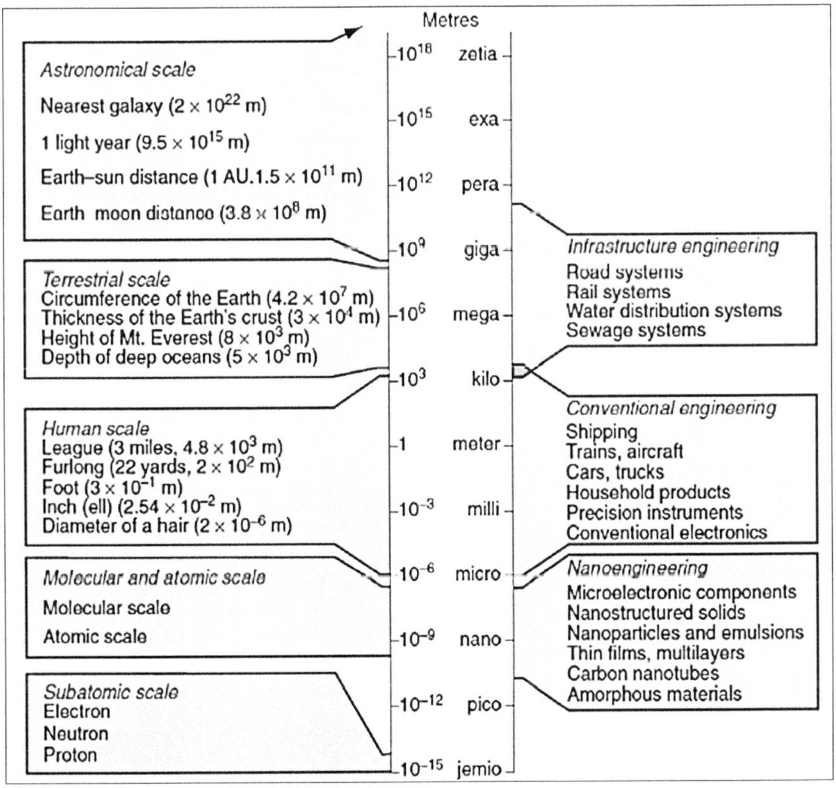

FIGURE 4.1 Representation of comparison scale.

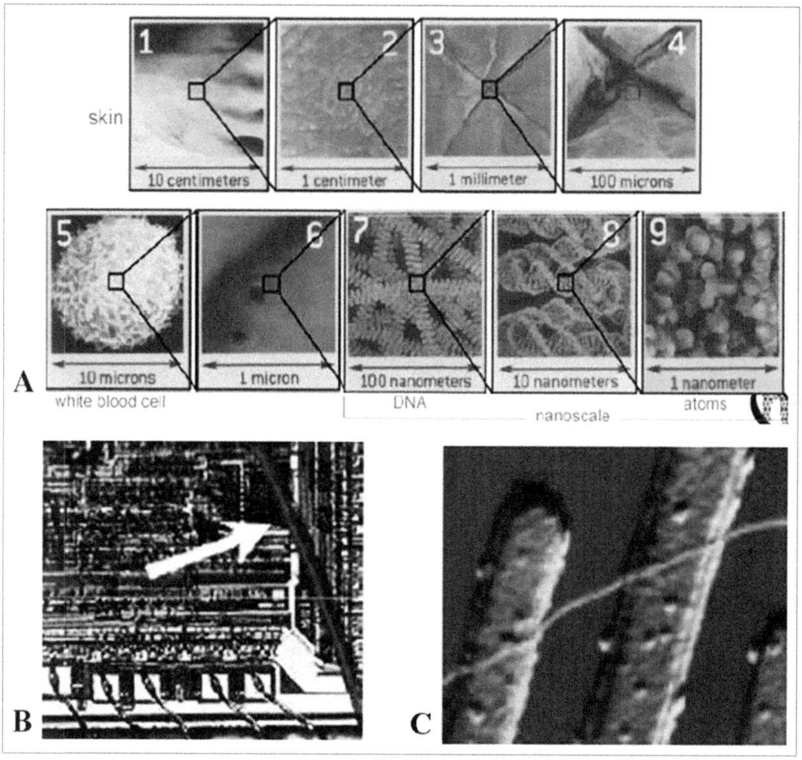

FIGURE 4.2 (See color insert.) (A) Magnification of human skin to find out nanoscale-materials, (B) the width of a human hair (placed on a microchip and pointed by a white arrow) is roughly 104 times, and (C) the diameter of carbon nanotubes (placed on the top of some metal electrodes) is yet again 104 times smaller.

4.3 SIZE EFFECTS

To achieve the nanostructured device, we have to focus on reducing the scale up to nanoscale-materials or nanoparticles. Some significant issues in nanoscale-materials correspond to size effects and shape phenomena which affect significantly on the development of structural, thermodynamic, electronic, spectroscopic, and chemical features.[18] There are two types of effect: internal size effects and external size effects. Internal or intrinsic size effects are related to the particles, which can be determined by the change in particle properties. Internal size effects associated with lattice parameters, chemical activity, absorption, melting temperature, luminescence, hardness, diffusion coefficients, bandgap, and so on. External size effects related with the processes of interaction among different physical parameters with respect to decreasing their building units from (particles, grains, and domains) to a length of physical phenomena (phonons, free length of electrons, screening length, coherent length, etc.).

4.4 WHY ARE THE PROPERTIES OF NANOMATERIALS DIFFERENT?

The properties of nanomaterials are very much different from their larger scale material. There are two aspects that play a major role for the variation of the properties of nanomaterials significantly from their bulk form: (1) the relative surface area and (2) the quantum confinement effect. These are the main reasons, which can alter the properties.

4.4.1 INCREASED RELATIVE SURFACE AREA

Nanomaterials possess relatively higher surface area when compare with the same material in the bulk one having the same volume or mass. Consider a bulk cube having volume of 1 m^3 and surface area of 6 m^2 as shown in Figure 4.4. Let the bulk cube to be divided into 27 small cubes, which results in increment of total surface area to 18 m^2.

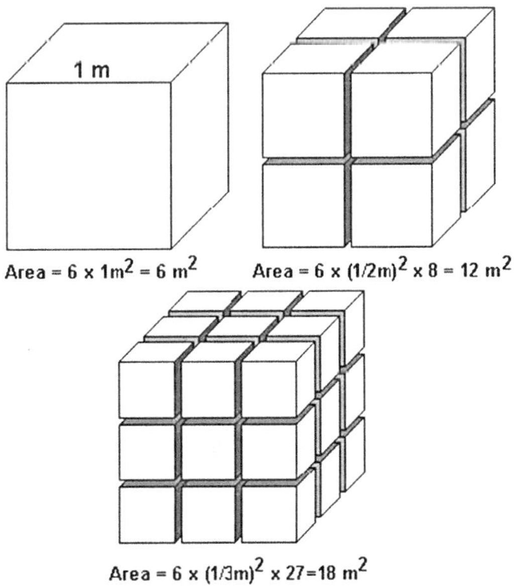

FIGURE 4.4 Calculation of surface area by dividing the cube.

4.4.2 QUANTUM CONFINEMENT

To confine or limit the randomness of electron and restrict its motion in specific energy levels is called quantum confinement. Quantum confinement effect (or discretization of e^-) is the most popular term in the nanoworld. Quantum confinement arises with change in atomic structure (ultra-small length scale) which results in a direct influence on the energy band structure. When the particle dimension of a semiconductor reduced to Bohr excitation radius, then the materials properties became size dependent. As the particle dimension reduces and it gets a discrete energy levels results widens up of their bandgap, and the energy of bandgap also increases. On the basis of electronic wave function on different physical dimensions of the materials, nanomaterials can be categorized as:

1. One-dimensional confinement: thin films, quantum well
2. Two-dimensional confinement: nanotubes, nanorods, quantum wires
3. Three-dimensional confinement: quantum dots, precipitates, colloids.

4.5 PROPERTIES OF NANOMATERIALS

Nanomaterials have distinct physical, mechanical, thermal, chemical, electronic, and magnetic properties different from those of the same chemically composed bulk materials. Some of the important properties of materials in nanoscale are discussed below:

4.5.1 STRUCTURAL PROPERTIES

The crystal structure of nanomaterials may not or may differ as that of bulk one with different lattice parameters. Generally, gold (Au) and aluminum (Al) have face-centered cubic (FCC) crystal structure in bulk form, but in case of nanoform, the crystal structure will be icosahedral (Fig. 4.5). Indium is face-centered tetrahedral and it will be FCC if the size is less than 6.5 nm. The interatomic spacing are lesser in nanomaterials compared to their bulk form because of the presence of electrostatic forces in long range and the core to core repulsion in short range. In case of Al, if the interatomic spacing decreases from 2.86 Å to 2.81 Å, the binding energy subsequently decreases from 3.39 to 2.77 eV. Nanostructure materials have unique characteristics that showing crystalline nature.

4.5.2 THERMAL PROPERTIES

Thermal properties will alter as the reduction in size toward nanosize. The melting property of nanomaterials decreases as the function of higher surface energy and change in interatomic spacing. Figure 4.6 shows the melting point of gold nanoparticle decreases as the size reduces.

4.5.3 CHEMICAL PROPERTIES

Ionization energy of nanocluster is more than that of bulk materials. Nanomaterials have an efficiency of chemical reactions, the rate of chemical reaction, very high radical alteration in chemical reactivity and selectivity due to their larger surface area to volume ratio. That is why the catalysis is in the range of nanomaterials. The ionization potential at nanosize is higher than that for the bulk materials.

4.5.4 MECHANICAL PROPERTIES

Nanomaterials contain a large number of defects during their fabrication, which influences the mechanical properties. Some nanostructures have very distinct properties, different from their bulk structure because of their atomic

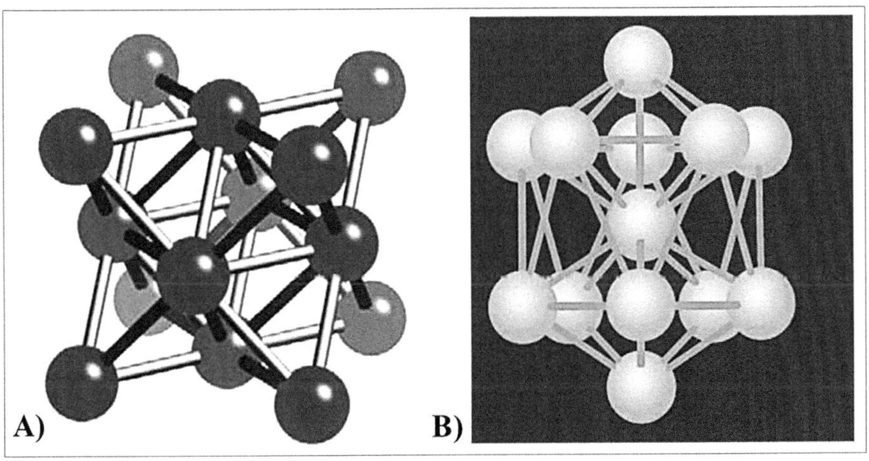

FIGURE 4.5 (See color insert.) Crystal structure. (A) bulk (FCC) and (B) nano (icosahedral) gold.

structural arrangement and may have different mechanical properties like carbon nanotubes (CNT) (single or multi-walled), which achieved the higher mechanical strength, elastic limit, and flexibility. Nanophase ceramic show ductility behavior at elevated temperature, whereas coarse-grained ceramics shows brittle character. In 2002, Qi and Wang[18] found out that the cohesive energy decreases, when the ratio between the atom size and the particle size becomes less than 0.01 to 0.1 and the decrease in cohesive energy influences on the reduction of the melting point. In 2003, Nanda and co-workers[14] investigated on the surface energy of different scale of the materials. He indicated the surface energy is higher in nanoparticle than that of embedded nanoparticles and also higher than that of bulk one.[14]

4.5.5 MAGNETIC PROPERTIES

Large surface to volume of nanomaterials leads to different magnetic coupling with neighboring atoms results in different magnetic property than that of bulk one. For a ferromagnetic material, there are multiple magnetic domains as shown in Figure 4.7, but in case of nanosize of that material, there is one domain, which exhibits superparamagnetic phenomena. The magnetic moment of nanoparticles is very less than its bulk size.[1]

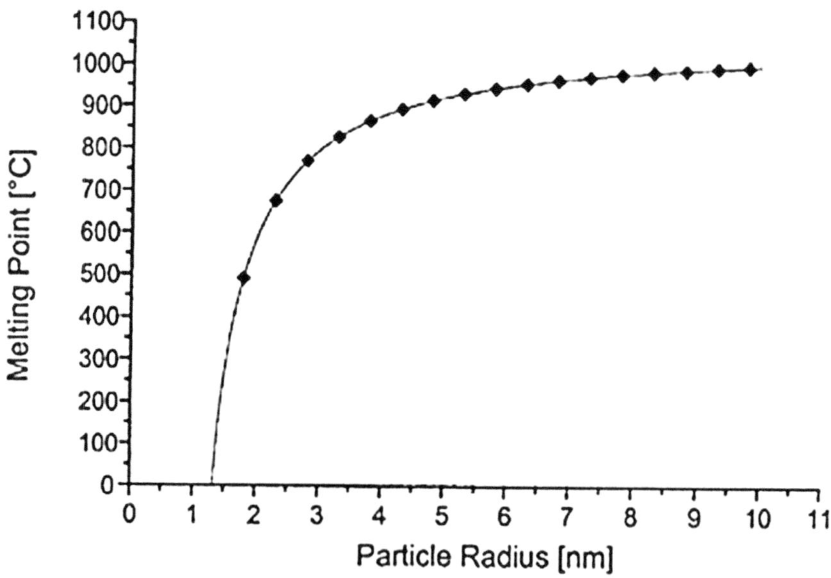

FIGURE 4.6 Particle size versus melting point profile plot of the gold nanoparticle.

4.5.6 OPTICAL PROPERTIES

Optical properties depend on the transition barrier between valence band and conduction band for a nano-sized material. In low-scaled semiconductor and metal, a huge change in an optical property such as color is observed. The gold nanoparticle colloidal solutions have a deep red color, which changes to more yellow with the increment of the particle size (Fig. 4.8) results in surface plasmon resonance in nano-scale materials. The nanosized semiconductor (quantum dot) behaves its size-dependent properties as the function of intensity and frequency of light emission and also improves the nonlinear optical properties as well as enhanced emission energy or wavelength. Other optical properties such as photoconductivity, photocatalysis, photoemission, and electroluminescence are also depended on the size factor.

4.5.7 ELECTRONIC PROPERTIES

The electronic properties are related as wave-like property of the electron for all the small-scaled materials. When size of the materials approaches to the de Broglie wavelength, the limit (discrete nature) of the energy states will be prominent, as depicted in Figure 4.9. The conduction material changes to insulator below the critical length scale, due to the widening of their bandgap. Nanoparticles made of semiconducting materials Germanium (Ge), Silicon (Si), and Cadmium (Cd) are not Semiconductors. The nanoparticles

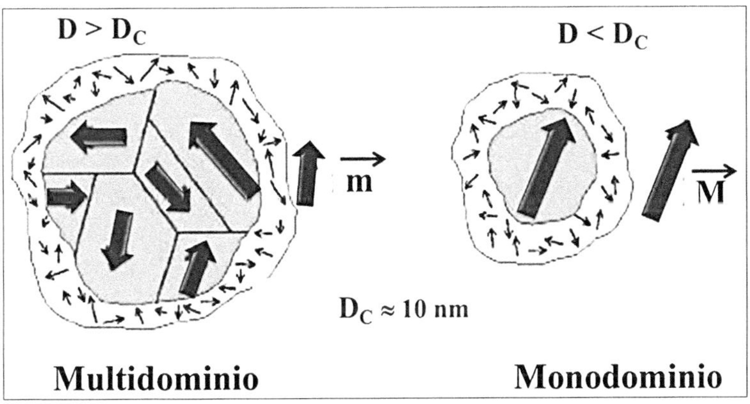

FIGURE 4.7 Multiple magnetic domains for bulk materials and mono-domains for nanomaterials.

of anatase TiO_2 (titanium dioxide), prepared by precipitation method, gives direct bandgap semiconductor behavior, whereas microcrystalline TiO_2 gives indirect bandgap material.[21]

4.6 NANOMATERIALS WITH RESPECT TO THEIR DIMENSION

Nanomaterials can be defined as the materials or structures that have at least one of its dimensions in the range of a nanometer. Thus, the nanomaterials need not be so small that it cannot be visible to human eye; it can be a large surface or a long wire whose thickness is in nanometer scale (depicted in Fig. 4.10). So, nanomaterials can be categorized as:

1. Zero-dimensional nanomaterials,
2. One-dimensional nanomaterials,
3. Two-dimensional nanomaterials, and
4. Three-dimensional nanomaterials.

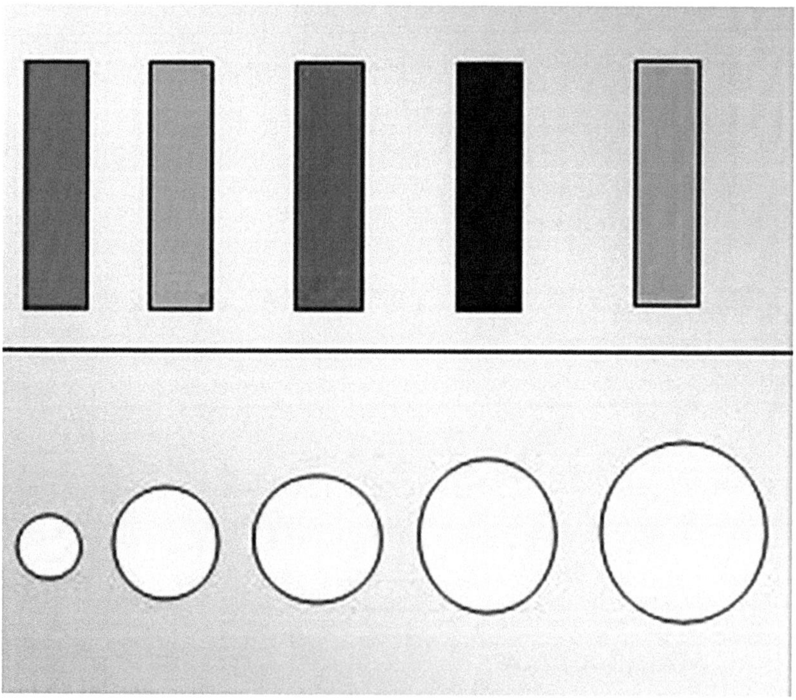

FIGURE 4.8 (See color insert.) The diameter of gold nanoparticle with the respective color.

In zero-dimensional nanomaterials, all the axis are in nanometer range. Nanoparticles are the most common representation of zero-dimensional nanomaterials. In one-dimensional nanomaterials, there are two dimensions at the nanoscale and one dimension at the macroscale. This looks like needle. Some examples of one-dimension materials are nanotubes, nanorods, thin films, platelets or surface coatings etc. In two-dimensional nanomaterials, there is one dimension at nanoscale and two dimensions at macroscale. Two dimension nanomaterials exhibit plate-like shapes. Two-dimensional nanomaterials include nanowires, nanofibers and nanotubes, nanofilms, nanolayers, and nanocoatings. In three-dimensional nanomaterials, there is no dimension in nanoscale, and all the dimensions are the macroscale. Bulk nanomaterials had no dimension in nanoscale but considered dimensions arbitrarily above 100 nm. Bulk nanomaterials are composed of various types of most typically differently orientated nanosize crystals. Three-dimensional nanomaterials are dispersions of nanoparticles, precipitates, colloids and quantum dots, bundles of nanowires, and nanotubes or multinanolayers.

4.7 APPLICATIONS OF NANOMATERIALS

Technology revolutions induce social changes and environmental apprehension. In the words of Peter Grutter, physicist, McGill University one who state their views as "I have a lot of hope that nanotechnology can get governments to put structures in place nationally and internationally that are more adaptable, so that the public could decide it wants the technology and then we could go for it" whereas Thomas Murray (a bioethicist and President at the Hastings center) states that "It is important for the earliest uses of nanotechnology to be thoughtful and attractive. The public is looking for control of their lives. Could you, for example, use nanotechnology to empower people to practice healthy behaviours?" Currently, nanotechnology is mostly used in cosmetics, medicine/drugs, fabrics, defence, energy, and water purification, but within few years all others domains will also be using nanotechnology as a potential tool. Nanotechnology has applications in all conceivable areas, which are discussed below:

4.7.1 IN MEDICINE

Researchers are developing customized nanoparticles in the field of medical application to deliver drugs to specific cells. Due to nanotechnology, there is a great reduction in damage treatment such as chemotherapy. Focus on research that involves the use of nanorobots, which work at the cellular level sometimes referred to as nanomedicine. The maximum drug consumption and their side effects are effectively minimized by the deposition of the active agent in the selected region of morbid. This selective treatment can reduce the cost and also reduces the human agonize. Other application nanotechnologies in the medical field are active agents for the drug, rapid selective tests, antimicrobial agents and coatings treatment, cancer therapy and so on.

4.7.2 IN ENERGY PRODUCTION

Advances in nanotechnology directed to energy saving and energy production by creating storage system, intermediate conversion system, manufacturing development by minimizing materials and process rates, and enhanced renewable energy sources. Nanotechnology reduces the cost of solar cells, producing high-efficiency light bulbs, improving the performance of batteries, reducing power loss in electric transmission wires, and improving the efficiency with reducing the cost of fuel cells. In power plants sector, advanced nanoparticles use sunlight to produce the steam for running the plant. A nanoengineered polymer matrix light bulb has twice the efficiency

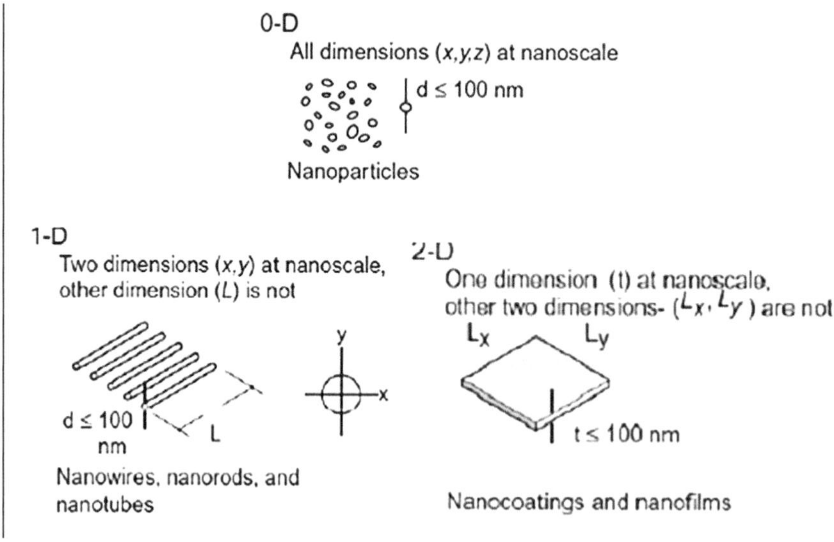

FIGURE 4.10 Different dimension of nanomaterials.

of compact fluorescent light bulbs, and also nanostructured light-emitting diodes (LEDs) gives maximum efficiency nowadays. Nanocrystalline nickel (Ni) and metal hydrides fabricate a Ni-metal hydride battery which can retain the charge to a longer period.

4.7.3 IN ELECTRONICS

Nanotechnology helping to improve data memory, improved high-definition television (HDTV) and LCD monitors, laser diodes, glass fibers, filters, conductive and antistatic coatings. A memory chip that uses CNTs can be applied instead of high-density flash memory chips. Single-atom thick graphene film can build high-speed transistors. Using semiconductor nanowires to build transistors and integrated circuits, silver nanoparticle ink in standard inkjet printers, nanosized cadmium selenide (CdSe) for flexible electronic circuits have highly commercialized.

4.7.4 IN ENVIRONMENT

In propylene oxide used manufacturing plant; the polluting by-products can be checked using a catalyst of silver nanoclusters. Propylene oxide applied to produce plastics, paint, detergents, and brake fluid, clearing volatile organic compounds from air, organic chemicals in groundwater. Nanotechnology developed the efficient and cheap water purification systems, which permits only water molecules through it.

4.7.5 IN DEFENSE

Nanomaterials with nano design provide larger strength for bullet shielding which is much more effective, lightweight, and resistant to chemical and biological attacks. Nanomaterials also reduce the rotor vibration, improving the performance of helicopter rotors. Lightweight solar sails using CNT can use the sunlight to propel the spacecraft up to long range, which is a future solution for fuel energy. Nanosensors are used to detect the levels of remained food or chemicals in spacecraft, and monitor the spacecraft environmental condition for life support systems.

4.7.6 IN MANUFACTURING

Now a day researchers are trying to develop nano steel. Arcelor Mittal is producing nanoparticle content steel to produce thinner gauge, lighter beams, and plates with a lower cost. These steel beams and steel plates are approximately same weight as Al. CNT embedded materials are more interesting to use in many structural devices. Buckyball is the lightweight nanomaterials, used for making badminton racquets which gives greater hitting power and stability. Light-weighting thinner paint coatings, used on aircraft, can able to increase the flight efficiency.

4.7.7 IN CONSUMER PRODUCTS

Nanoparticle minerals used in sunscreen gives longer stability with better biocompatibility, which can strongly block UV rays fall on the skin. Nano-sized TiO_2 and ZnO_2 are applied in most of the sunscreens, as their absorption and reflection capacity is very high. Lipsticks with nanosized iron oxide or ferric oxide (Fe_2O_3), nano vitamins skin care products, nanofabric fitting cloth are the sophisticated applications. Customizing the particles with a few nanometres in diameter can make a better soap. Nanoparticle toothpaste has more ability to clean the teeth. Nanotechnology is successfully used in the processing, developing, maintaining, and packaging of nutritious substance.

4.7.8 IN CONSTRUCTION

Nanotechnology is the pillar to construct strengthen and safer structure much more quickly with lowering the cost. Thin film on the glass is an important material for developing chromogenic-systems (electrochromic, thermochromic, photochromic, and gas chromic) on the walls, doors, and windows. Here nanotechnology utilized to give an effective solution by blocking the incoming light and heat through the glass. Due to sterilizing and anti-fouling property, TiO_2 nanoparticles are used to coat glazing. Nano coating used for corrosion protection as they are hydrophobic in nature. When we replace the macro Silica (SiO_2) with nano Silica, the particle packing can develop mechanical properties. Nano silica added cement-based materials to improve the durability and haematite (Fe_2O_3) nanoparticles added concrete to improve the strength of the construction. The nanosize steel provides more strength steel cables that can be utilized in the large construction.

4.7.9 IN AGRICULTURE

A huge development occurs in agriculture and food industries due to the application of nanotechnology from production, processing to waste treatment. Herbicides, chemicals, or genes contain nanoparticles (magic bullets) can target specific parts of plants to discharge their content. Nanocapsules are used in case of penetration of herbicides in cuticles and tissues, permitting steady secretion of the active substances.[16] Nano-bio-fibers act as a better fertilizer or pesticide absorbent.[5] Nanotechnology expanded its broad area in agriculture, some of the major applications are given below:

1. Bioremediation by nanoparticles.
2. Identification and tracking of agricultural food products by nano-barcodes.
3. Agri-materials nanocapsules are used for agrichemicals delivery that plays an important role in vaccine delivery.
4. Nanotechnology is employed for the defection of nutrients and pathogens with the help of biosensors and quantum dots.
5. Nanosensors are used for detection of crop growth, soil conditions and also for detecting animal and plant pathogens.
6. Single molecule detection (SMD) is enabled with the help of nanotechnology to determine enzyme/substrate interactions.
7. Nano chips are used for identity preservation and tracking in food
8. Quality enhancements of agro-products by nanoparticles.
9. Recycling of agricultural wastes by nanoparticles.
10. Shelf-life enhancement of agricultural produces (like fruits and vegetables) and agro-food products by nanoparticles.
11. Wastewater treatments and disinfection by nanoparticles.
12. Nanotechnology-enabled delivery can regulate hormones in a productive fashion.

4.7.10 IN CHEMICAL

Catalytic activity depends on the contact surface area. As the higher relative surface area of nanomaterials, the catalytic effect will be higher. Nanomaterials have higher chemical reactivity with other chemicals. Platinum nanoparticles are used in automotive catalytic converters due to the higher surface area of nanoparticles can decrease the required platinum amount.

4.8 HOW TO GET AT NANOSCALE?

Synthesis of nanomaterials can have two general approaches: (1) top–down approach and (2) bottom–up approach. The schematic representation of these two methods is given in Figure 4.11.

4.8.1 TOP–DOWN APPROACH

These approaches use larger (macroscopic) structures to produce nanoparticles. Typical top–down approach is attrition or milling for the manufacturing of nanoparticles. This approach is cheaper and quicker to manufacture and the power is only externally controlled during the reduction of size from macroscale to nanoscale structures. The process keys are control energy input and contamination factor. Typical equipment used for this process is high-speed ball milling, etching through the mask, and equipment for severe plastic deformation. There are some significant disadvantages arise in top–down approach are contamination during attrition and crystallographic damage and introduces internal stress. However, top–down processes lead to the bulk production of nanomaterials.

4.8.2 BOTTOM–UP APPROACH

The bottom–up approach includes the atomic level materials to the formation of nanostructures components with further self-assembly process. Fabrication is much less expensive. The main advantage of this process is to get higher purity monosized nanoparticles with better particle size control. Here, the Process keys are control nucleation and growth of the particles. Main disadvantages of this approach are stopping the newly formed particles from the previously formed particle. Production rates of nanoparticle in this method are meager. The physical forces acting at nanoscale during self-assembly are used to construct the basic units up to a larger stable structures. The basic approaches in the bottom-up method are quantum dots by epitaxial growth and nucleation-growth of nanoparticles from colloidal dispersion.

4.9 FABRICATION OF NANOPARTICLES

The most technique of nanoparticle fabrication is classified into three types by their treatment: chemical, biological, and mechanical (Fig. 4.12). Mostly used synthesis route in the industry can be classified as three categories by their phase: solid-state, vapor-phase, and solution precipitation processes.[23]

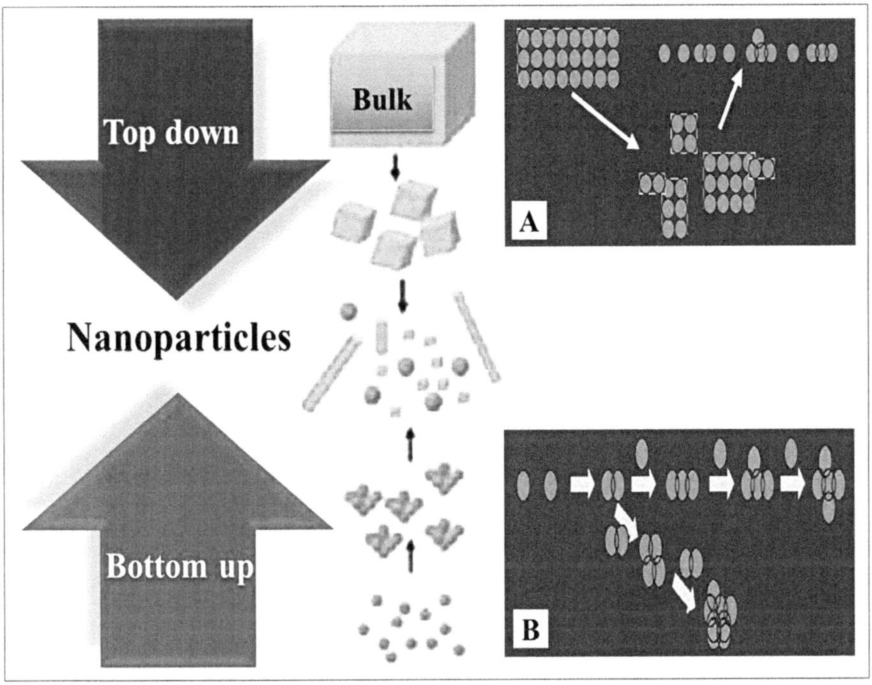

FIGURE 4.11 **(See color insert.)** Representation of nanoparticle fabrication (A) top–down and (B) bottom–up methods.

4.9.1 SOLID STATE SYNTHESIS OF NANOPARTICLES

Solid state synthesis associated with different deformation technique to get the required nanostructure. Solid state synthesis comes under the top–down approach to nanomaterials. The material is synthesized from coarser-grained structures by severe plastic deformation and shear action. There are different techniques like high-speed ball milling; mechanochemical processing, etching through the mask, and so on are involved in reducing the particle size up to nano size.

4.9.1.1 MECHANICAL MILLING

Planetary ball mill, tumbler mills, and high-energy shaker are generally used for mechanical milling operation. During milling, the energy transferred from the refractory ball or steel balls to the coarse-grain powder depends on the rotational speed of the mill as well as the type of ball, size of the balls, ball-to-powder ratio, milling time, and the milling atmosphere. Dry milling induced chemical reactions within a salt matrix, called as the mechanochemical processing of nanoparticles.[9,11,24] The main advantage of this process is the easy scaling up of tonnage quantities of material. Similarly, the serious problem is the contamination from milling media and from their atmosphere.

4.9.1.2 LITHOGRAPHY

The photolithography technique is the most used top–down approach (Fig. 4.13) that can produce the nanostructure lesser than 100 nm. In this technique, Si wafer is coated with a photoresist material, which is to be undergoing photochemical reaction during the exposure to UV light followed by rinsing with a developing solution.

4.9.2 VAPOR-PHASE SYNTHESIS OF NANOPARTICLES

In this technique, atomic particles condensed into the gas phase to form nanoparticles.

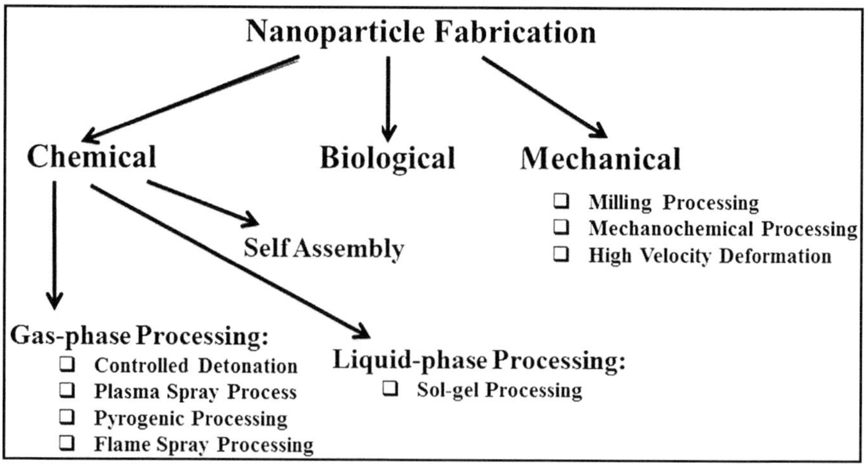

FIGURE 4.12 Classification of fabrication technique of nanoparticle.

4.9.2.1 INERT GAS CONDENSATION OF NANOPARTICLES

Gas condensation processes accomplished in both low and high-pressure atmosphere, and at the end, nanoparticles were quenched rapidly. Metal nanopowder like Al, Co, Cr, Cu, Fe, Ga, Mg, and Ni are processed by this method. Figure 4.14 showing the schematic representation of gas phase synthesis of nanomaterials. At constant temperature with inert-gas pressure, the crucible heated rapidly that form vapor which is a source of nanoparticle nucleation. The production rate was approximately 1 g per run. There are three parameters controlling the cluster size, and their distributions in the atmosphere are: (1) The rate of energy input (evaporation), (2) The rate of energy removal (condensation), and (3) The rate of cluster removal (gas flow).

4.9.2.2 PLASMA-BASED SYNTHESIS OF NANOPARTICLES

This technique utilizes the thermal plasma, that is, the ionized gases as the energy source for melting the materials and consecutively associated with plasma spraying of materials on the substrates. Using thermal plasma both metal and ceramic can be converted to nanoparticle easily.[12,13] Figure 4.15 is showing the plasma-based method. By entering the carrier gas (inert: H/He/Ar and reactive: N_2/O_2, etc.), with the application of electric or magnetic field, there is a formation of high-energetic plasma which is ready to carry the ionized particle toward the substrate. Ultrafine particles of different nanomaterials such as Fe, Al, Co, Cu, Ti, and Ta were produced. The production rate is greater than 50 g/h with about 20 nm diameter.

4.9.2.3 SPUTTERING PLASMA PROCESSING

In the sputtering plasma technique, both DC and RF power has been used to fabricate nanoparticles. The material to be deposited is a sputtering target with carrier gas allowed to agglomerate to produce nanomaterials. To prepare the nano-alloy or nano-oxide particles, multi-target sputtering, mosic target sputtering, and reactive sputtering have been used. The sputtering plasma processing is suitable for the synthesis of contamination-free non-agglomerated nanoparticles.

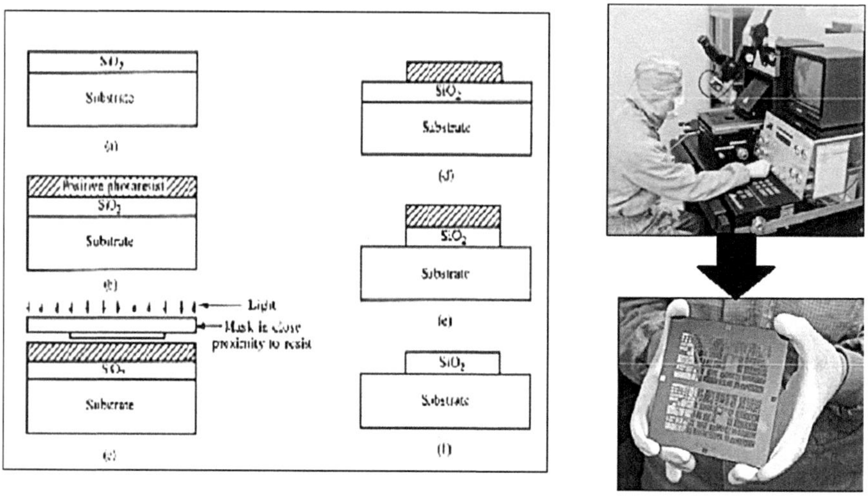

FIGURE 4.13 (See color insert.) Photolithography technique. Processes are coat, protect, expose, etch, repeat…, and the result is multiple patterned layers of different materials.

4.9.2.4 MICROWAVE PLASMA PROCESSING

In this process, microwave is the main factor to produce the plasma. The reaction vessel made of quartz attached with a microwave generator. In this method, there is a less chance of agglomeration due to low-temperature reaction.

4.9.2.5 FLAME-BASED SYNTHESIS OF NANOPARTICLES

In this process, hydrocarbon or hydroxy flame is used to pyrolyze chemical precursor and synthesized the nanoparticles (Fig. 4.16). This technique develops the flames with a flat geometry instead of traditional Bunsen burner conical flames. Many investigations have been carried out on atmospheric flames which produce nanoparticles.[15,17,25] This is a counter-current flow process to collect the nanoparticle. In this process, the reactant was injected through a separate tube into the flame.

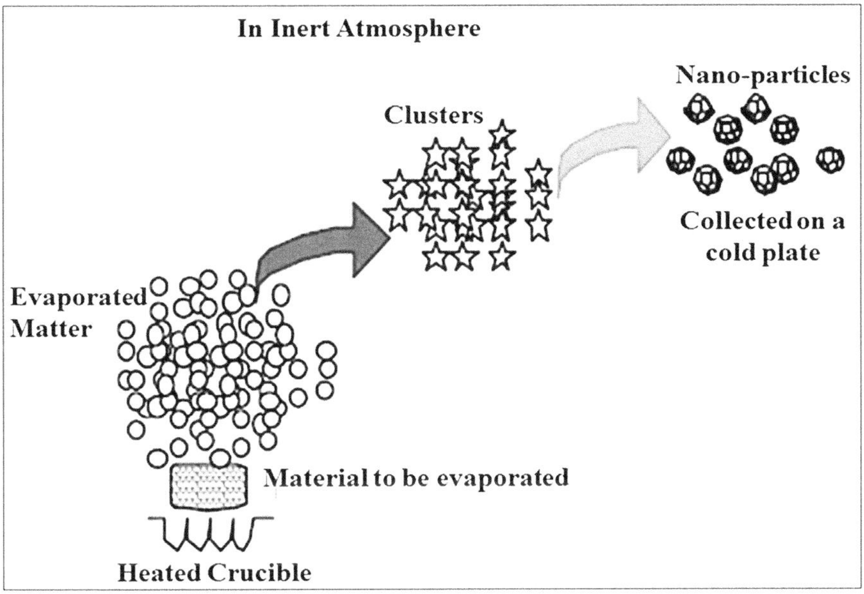

FIGURE 4.14 Schematic representation of single-phase nanomaterials synthesized by gas phase process from a heated crucible.

4.9.3 SOLUTIONS PROCESSING OF NANOPARTICLES

Synthesizing the nanomaterials by precipitating the clusters of particles from a solution has been an attractive technique. A major advantage of this technique is to develop encapsulated nanoparticles, developing the stability of particle in a medium, to control the shape and size of the nanoparticle. Solutions processing can be typically categorized into five major groups: precipitation method, sol–gel method, water–oil microemulsion method, hydrothermal synthesis, and polyol method.

4.9.3.1 SOL–GEL PROCESSING

Most popular commercial method is sol–gel processing technique to produce all type of nanoparticles.[4,22,26] In this process, a colloidal suspension is prepared called sol and then the viscosity of the solution increased to form a continuous network in liquid phase, called gel. In this gelation process,

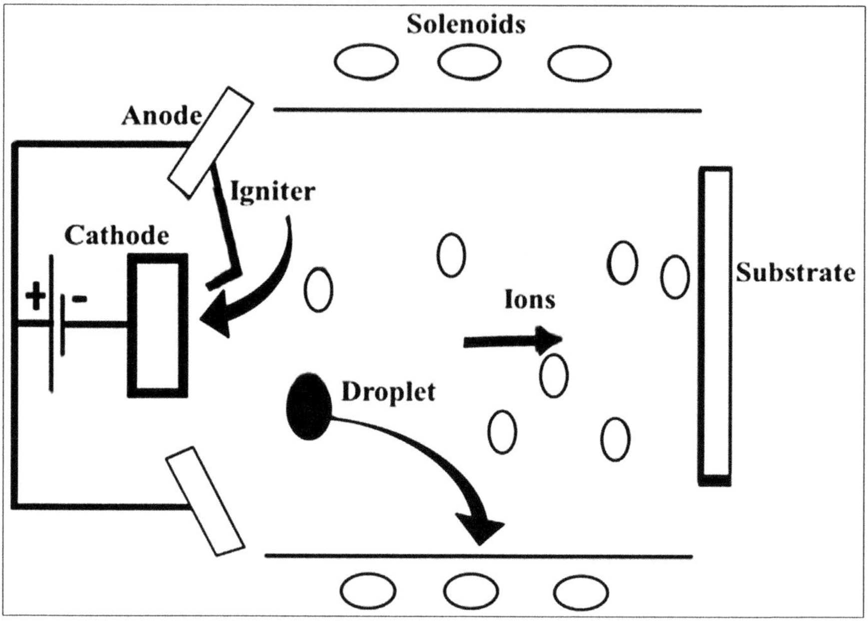

FIGURE 4.15 Plasma synthesis chamber.

the viscosity of the solution increases due to a polycondensation reaction. Hence, the drying of liquid from sol gives a gel. Here the sol/gel transformation can determine about shape and size of the nanoparticle. There are two types of reactions involved that is hydrolysis and condensation in the sol–gel chemistry form metal alkoxides M(OR)z.

$$MOR + H_2O \rightarrow MOH + ROH \text{ (hydrolysis reaction)}$$

$$MOH + ROM \rightarrow M\text{-}O\text{-}M + ROH \text{ (condensation reaction)}$$

By the different procedure of drying, the gel network will change its structure to powder, aerogel, xerogel, coatings and so on (Fig. 4.17). The drying process follows four different periods that is, the constant rate period, the critical point, the falling rate period. Xerogel is formed by thermal evaporation and aerogel is formed by removing the solvent near supercritical conditions.[19]

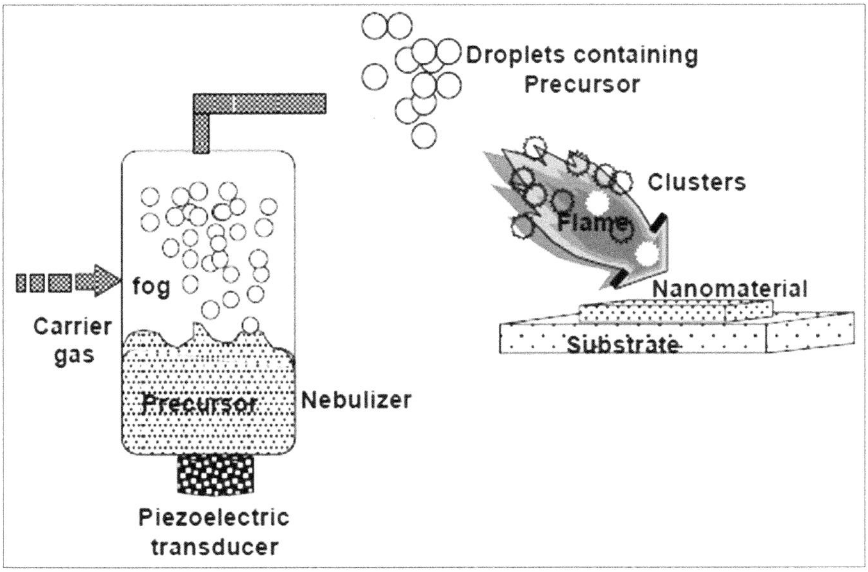

FIGURE 4.16 (See color insert.) Flame-based synthesis of nanoparticles.

4.9.3.2 *SOLUTION PRECIPITATION TECHNIQUE*

Solution precipitation technique, metal salts like chloride (Cl), nitrate (NO^{-3}), acetate ($C_2H_3O_2$), or oxychloride are dispersed in the solvent (water). These cations present in water present as a metal hydrate and are hydrolyzed by addition of a base solution (NaOH or NH_4OH). The hydrolyzed metals condense to form a precipitate of metal hydroxide or hydrous metal oxide. Then the precipitate is cleaned, filtered, and preheated that followed by calcination to get metal oxide phase. By this method, nanocomposites of metal oxides can be prepared by co-precipitation of corresponding metal hydroxide.[10]

4.9.3.3 *SOLUTION PLASMA SPRAY PROCESS*

The solution plasma spray process has normal plasma gun and robotics with a solution precursor storage system act as a feeder. An atomizer used to generate small droplets passes into the plasma plume. Ceramic and superalloy coatings are formed using solution plasma spray process. Figure 4.18 is showing the solution plasma spray system used for making ceramic high-temperature coatings.

4.9.3.4 WATER–OIL MICROEMULSION METHOD

In the water-in-oil microemulsion process, nanoscale water droplet scattered in an oil media. Water droplet becomes stable by the surfactant which is to be absorbed on the surface during formation of the nanoparticle. Uniform nanosized metal, metal oxides, and semiconductor can be synthesized by this method. A notable occurrence of this reverse micelle rout is that the particles are mostly monodispersed and nanosized.[2] The surfactant molecules absorb on the nanoparticle surface to stabilize it. This reverse micelle rout is the mixture of two micro-emulsions with exact reactants. Water droplets of two micro-emulsions react with each other and form nano-particles inside the water droplet. Here the nanoparticle synthesis executed by any of two chemical reactions: (1) hydrolysis of the metal alkoxides or precipitation of the metal salts with a base, (in case of metal oxide) and (2) reduction of metal salts with a reducing agent, such as $NaBH_4$ (in case of metal nanoparticles). Particles are cleaned with acetone and water to ensure the removal of any residual oil and surfactant molecules present on the surface of nanoparticle,[19] and then calcination process gives the final product of powder.

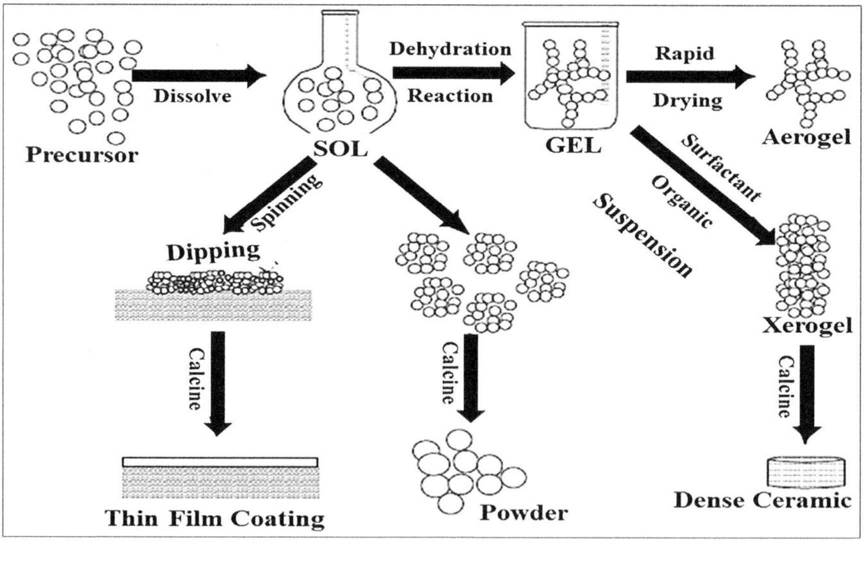

FIGURE 4.17 Schematics of sol–gel method for nanomaterial fabrication.

4.9.3.5 ELECTRODEPOSITION

By electrodeposition, metal layers deposited on a conducting substrate. By applying voltage, the induction of reactions in electrolyte solution is the principle behind the electrodeposition process, for example, in this process which uses electrical current to coat an electrically conductive metal. This technique is favorable to deposit nanostructured materials include metal oxides and chalcogenides. The deposited thickness is a function of current density and time of current flow in electrodeposition.

4.10 CLASSIFICATION OF NANOMATERIALS

Current nanomaterials are catagorized into four types: carbon-based materials, metal-based nanomaterials, polymer-based nanomaterials, and composite nanomaterials.

4.10.1 CARBON-BASED NANOMATERIALS

The main composition of these nanomaterials is carbon and occurs in the form of spherical, ellipsoidal, or tubular structure. Spherical and ellipsoidal structures are termed as fullerenes whereas the cylindrical structures are termed as nanotubes. Carbon-based nanomaterials are stronger, lighter, and

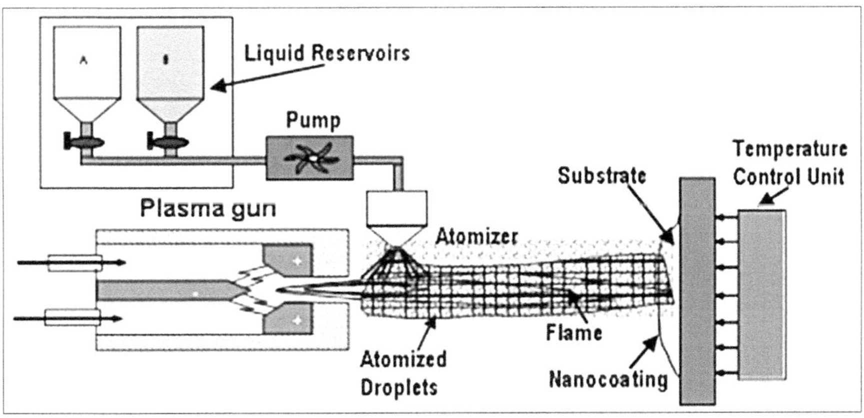

FIGURE 18 Solution plasma spray system.

have broader potential applications. Graphene is the basic structural unit of carbon-based material having a one atom thick planar sheet, in which carbon atoms are present in sp2-bonded results a dense structure. The bond length of carbon–carbon in graphene is around 0.142 nm.

4.10.1.1 BUCKYBALL (FULLERENES, C_{60})

Buckyballs are the first nanoparticles discovered. Buckyballs are also called as fullerenes, composed of covalently bonded carbon atoms, that is, one carbon atom is linked to three nearby carbon atoms and follow a structural pattern of hexagons and pentagons (Fig. 4.19A). The most common buckyball contains 60 number of carbon atoms so often termed as C_{60}. The number of carbon atom in fullerene varies from 20 to more than 100, as shown in Figure 4.19B. Interesting most of the structures of C_{60} neglect the C=C bonds. In C_{60} each C atom has three C–C σ bonds (2s, px, py); the e^- in the 2pz then overlap to form a π system. These fullerenes are black crystalline solid having highest packing density with thermally stable up to 4000°C. Even though each carbon atom is only bonded with three other carbons in a fullerene, shifting a single carbon atom to their adjacent structure will not affect the structure. Due to extremely resilient and sturdy nature and impervious to lasers, buckyballs are used in combat armor and in defenses. Elements can be bonded with the fullerene to create more diverse materials including superconductors and insulators.

4.10.1.2 CARBON NANOTUBES

In 1991, a researcher (Iijima) accidentally observed an unusual structure of carbon in transmission electron microscope that a hollow structure was formed by the rolled and folded sheets of graphene and named it as CNT. CNT composed of several cylinder-shaped macromolecules made by carbon atoms. The walls are made up of a hexagonal lattice of atoms (like a graphite sheet) and are enclosed at their ends by cap like one half of a fullerene-like moiety.[6,20] Nanotubes are classified on the basis of graphene sheet folding: armchair, zigzag and chiral (Fig. 4.20). If an object or system is distinguishable from its mirror image then it is called chiral. Also, that which cannot be distinguished from its mirror image (like sphere) is called an achiral object.

4.10.2 METAL-BASED NANOMATERIALS

Metal-based nanomaterials are quantum dots, nanoparticles of metal and metal oxides, and so on. A quantum dots refers to a number of atoms in a densely packed structure and constitute the building unit of semiconductor crystal. Nanosized aluminum powder or nano-aluminum (n-Al) is commercially used for metal-based nanomaterials. Other nanopowders such as boron (B), Mg, or zirconium (Zr) have also been considered[7] for structural element. Application of n-Al is considered as a potential replacement for the conventional aluminum powders and flakes that are used in explosives, propellants, and pyrotechnics.

4.10.3 POLYMER-BASED NANOMATERIALS

Polymer nanocomposites composed of a polymer or copolymer having nanoparticles or nanofillers embedded in the polymer matrix. Polymer-based materials are called dendrimers and are nanosized polymer built from branched units. Large number of chain-ends present at the surface of dendrimer, which can activate the chemical functions easily and also act as a catalyst. Three-dimensional dendrimers contain hollow interior that can hold other molecules and can be used in drug delivery system. Polymer nanocomposites also applied in gene therapy and medical diagnosis.[8]

4.10.4 NANOCOMPOSITES

Nanocomposites have at least one of the phase in the range of nanometre dimension. On the basis of matrix, nanocomposites are of three types: metal matrix nanocomposites (MMNC), ceramic matrix nanocomposites (CMNC), and polymer matrix nanocomposites (PMNC). Table 4.1 shows the different class of nano-composites with their examples. Generally, in case of MMNC, the matrix material is ductile metal or alloy matrix with nanosized reinforcement. The constituent of MMNC combines the metal and ceramic properties that can provide strength, ductility, toughness, and modulus. In case of CMNC, the higher mechanical properties dominated as comparison to other nanocomposites. PMNC are lightweight, easier fabrication, and often ductile in nature for which it has wide use in industry. Due to low modulus and strength, the demand of PMNC comes after MMNC and CMNC. Henceforth, the nanocomposite covers a wide range of application, such as aerospace industries, automotive industries, and development of any structural materials.

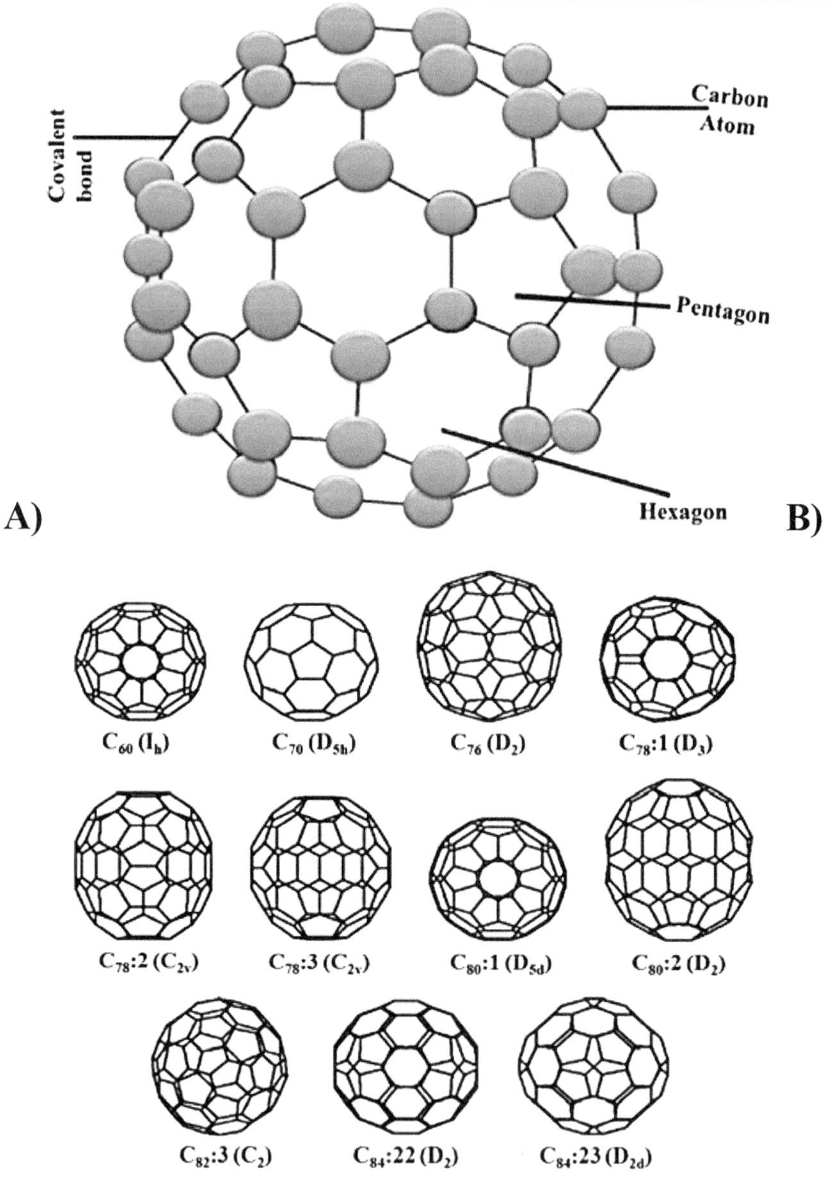

FIGURE 4.19 (A) The structure of fullerene and (B) the fullerene family.

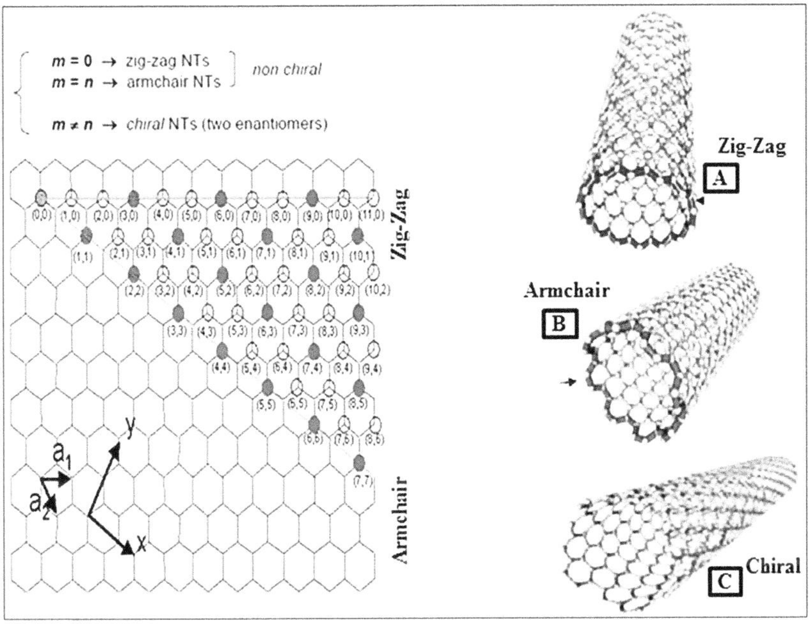

FIGURE 4.20 (See color insert.) Representation of graphene sheet and types of graphene sheet folding (A) zigzag, (B) armchair, and (C) chiral nanotubes patterns.

TABLE 4.1 Classification of Nanocomposites.

Class	Examples
MMNC	Fe-Cr/Al_2O_3, Ni/Al_2O_3, Co/Cr, Fe/MgO, Al/CNT, Mg/CNT
CMNC	Al_2O_3/SiO_2, SiO_2/Ni, Al_2O_3/TiO_2, Al_2O_3/SiC, Al_2O_3/CNT
PMNC	Thermoplastic/thermoset polymer/layered silicate, polyester/TiO_2, polymer/CNT, polymer/layered double hydroxides

4.11 DISADVANTAGES OF NANOMATERIALS

4.11.1 INSTABILITY

Nanomaterials are thermodynamically metastable and the kinetics associated with nanomaterials is very active and faster. So to control the activity is highly challenging and somehow nanoparticles are encapsulated in some other matrix. Relative higher surface area of nanomaterial is the only cause to react greatly and act as strong explosives when there is sudden contact with oxygen

4.11.2 IMPURITY

It is very difficult to nullify the impurity at the time of nanomaterials fabrication and storage, because of their highly reactive nature. In solid phase rout, during mechanical milling, the impurity generated due to cascading and cataracting effect of ball and materials with the lining. In liquid phase rout, encapsulation of nanoparticle is necessary during synthesis in a solution to avoid any foreign atomic contact.

4.11.3 BIO-INCOMPATIBILITY

Toxicity, irritation, and carcinogenic of nanomaterials also appears in cell owing to their high surface area and enhanced surface activity for which most of the cases taken as bio-incompatibility. It is very difficult to separate the nanoparticles from the cell-dermis, due to its transparency nature. During inhale, the lower mass of the nanoparticles entraps inside lungs, and there is no way to expel out of body.

4.11.4 SYNTHESIS AND ISOLATION BEFORE APPLICATION

Immediately after synthesis, it is very difficult to keep the nanoparticles contamination free. so, it is necessary to encapsulate with the stable material. Hence, nanoparticles are extremely difficult to be utilized in isolation. During fabrication, it is very difficult to restrict the grain growth of nanomaterials at exact point.

4.11.5 RECYCLING AND DISPOSAL

There are no fix safe recycling and disposal policies developed for nanoparticle due to their toxicity and reactivity nature.

4.12 SUMMARY

Recent technology focus on how much more can we continue to shrink the existing structural materials? In future, the question may not be as how much smaller the size of the computer to keep on our lap, but how many computers can we fit on our finger. This chapter describes the fundamentals of nanomaterials which is the main interest to squeeze the structural materials. It describes how nanomaterials can be distinctly different in properties than that of bulk one. The size-dependent on physical, electrical, magnetic, and optical properties of nanoparticle and nanostructures materials has been better understood in this chapter. Classification of the nanomaterials has done depending on different aspect. A wide range of applications clearly reviewed in this chapter.

KEYWORDS

- **nanomaterials**
- **nanocomposite materials**
- **encapsulated nanoparticles**
- **gaschromics**
- **metal-based nanomaterials**
- **nanobarcode**
- **nanofibers**

REFERENCES

1. Abolfazl, A.; Samiei, M.; Davaran, S. Magnetic Nanoparticles: Preparation, Physical Properties, and Applications in Biomedicine. *Nanoscale Res. Lett.* **2012,** *7* (1), 144. DOI: 10.1186/1556-276X-7-144
2. Capek, I. Preparation of Metal Nanoparticles in Water-in-oil (w/o) Microemulsions. *Adv. Colloid Interface Sci.* **2004,** *110* (1–2), 49–74. DOI: 10.1016/j.cis.2004.02.003.
3. Schodek, D. L.; Ferreira, P.; Ashby, M. F. *Nanomaterials, Nanotechnologies and Design: An Introduction for Engineers and Architects*. Butterworth-Heinemann, 2009, 560, ISBN: 9780080941530.
4. Dong-Hwang, C.; Xin-Rong, He. Synthesis of Nickel Ferrite Nanoparticles by Sol-gel Method. *Mater. Res. Bull.* **2001,** *36*, 1369–1377. http://dx.doi.org/10.1016/S0025-5408(01)00620-1
5. Eugene, V. D. *Trends in Nanotechnology Research*. Verlag: Nova Publishers, 2004, ISBN: 9781594540912.
6. Gleiter, H. Nanostructured Materials: Basic Concepts and Microstructure. *Acta Materialia*, **2000,** *48* (1), 1–29.
7. Jolivet, J. P.; Tronc, E.; Chaneac, C. Synthesis of Iron Oxide- and Metal-based Nanomaterials. *Eur. Phys. J. Appl. Phys.* **2000,** *10*, 167–172.
8. Jung, K. O.; Jong, M. Park. Iron Oxide-based Superparamagnetic Polymeric Nanomaterials: Design, Preparation, and Biomedical Application. *Prog. Polym. Sci.* **2011,** *36* (1), 168–189.

9. Lee, J.; Zhou, F.; Chung, K.; Kim, N.; Lavernia, E. Grain Growth of Nanocrystalline Ni Powders Prepared by Cryomilling. *Metal. Mater. Trans. A*, **2001**, *32* (12), 3109-3115.
10. Lu, H.; Hsu, C.; Lin, I.; Weng, C. *Method for Preparing ITO Nanometer Powders.* U.S. Patent Application, US 2003/0211032 A1, 2003.
11. Luton, M.; Iyer, R.; Petkovic-Luton, R.; Vallone, J.; Matra, S. *Method of Extruding Oxide Dispersion Strengthened Alloys*. U.S. Patent 4,818,481, 1989.
12. Masaya, S.; Murphy, A. B. Thermal Plasmas for Nanofabrication. *J. Phys. D Appl. Phys*. **2011,** *44* (17):174025.
13. Maurice, G.; Eric, H. J.; Matthew, T.; Cetegen, B. M.; Padture, N. P.; Xie, L.; Chen, D.; Ma, X.; Roth, J. Thermal Barrier Coatings Made by the Solution Precursor Plasma Spray Process. *J. Therm. Spray Technol.* **2008,** *17* (1), 124–135.
14. Nanda, K. K.; Maisels, A.; Kruis, F. E.; Fissan, H.; Strappert, S. Higher Surface Energy of Free Nanoparticles. *Phys. Rev. Lett.* **2003,** *91*, 106102.
15. Nobuhiko, W. Preparation of Fine Metal Particles by the Gas Evaporation Method with Plasma Jet Flame. *Jpn. J. Appl. Phys.* **1969,** *8* (5), 51.
16. Perea-de-Lugue, A.; D. Rubiales. Nanotechnology for Parasitic Plant Control. *Pest Manag. Sci. J.* **2009,** *65* (5), 540–545. DOI: 10.1002/ps.1732.
17. Pratsinis, S. E.; Zhu, W.; Vemury, S. The Role of Gas Mixing in Flame Synthesis of Titania Powders. *Powder Technol.* **1996,** *86* (1), 87–93.
18. Qi, W. H.; Wang, M. P. Size Effects on the Cohesive Energy of Nanoparticles. *J. Mater. Sci. Lett.* **2002,** *21*, 1743–1745.
19. Rachel, A. C.; Markus, A. Sol–Gel Nanocoating: An Approach to the Preparation of Structured Materials. *Chem. Mater.* **2001,** *13* (10), 3272-3282. DOI: 10.1021/cm001257z
20. Rao, C. N. R. *Inorganic Nanotubes, Trends in Chemistry of Materials" Selected Research Papers of C N R Rao*; IISe Press, World Scientific: India, 2008.
21. Reddy, K. M.; Manorama, S. V.; Reddy, A. R. Bandgap Studies on Anatase TiO_2 Nanoparticles. *J. Solid State Chem.* **2001,** *158*, 180–186.
22. Schoofs, B.; Cloet, V.; Vermeir, P.; Schaubroeck, J.; Hoste, S.; Driessche, I. V. A Water-based Sol-gel Technique for Chemical Solution Deposition of (RE)$Ba_2Cu_3O_{7-y}$ (RE=Nd and Y) Superconducting Thin Films. *Superconduct. Sci. Technol.* **2006,** *19* (11), 1178–1184.
23. Skandan, G.; Singhal, A. Perspectives on the Science and Technology of Nanoparticle Synthesis Chapter 2, In *Nanomaterials Handbook*; Gogotsi, Y., Ed.; CRC Press, 2006. DOI: 10.1201/9781420004014.ch2
24. Todaka, Y.; Nakamura, M.; Hattori, S.; Tsuchiya, K.; Umemoto, M. Synthesis of Ferrite Nanoparticles by Mechanochemical Processing Using a Ball Mill. *Metal. Mater. Trans. A*, **2003,** *44*, 277–284.
25. Wooldridge, M. S.; Torek, P. V.; Donovan, M. T.; Hall, D. L.; Miller, T. A.; Palmer, T. R.; Schrock, C. R. An Experimental Investigation of Gas-phase Combustion Synthesis of SiO_2 Nanoparticles Using a Multi-element Diffusion Flame Burner. *Combust. Flame*, **2002,** *131* (1–2), 98–109.
26. Yu, L.; Yadong, Y.; Mayers, B. T.; Xia, Y. Modifying the Surface Properties of Superparamagnetic Iron Oxide Nanoparticles through A Sol-Gel Approach. *Nano Lett.* **2002,** *2* (3), 183–186. DOI: 10.1021/nl015681q.

CHAPTER 5

CHARACTERIZATION TECHNIQUES FOR NANOMATERIALS: RESEARCH AND OPPORTUNITIES FOR POTENTIAL BIOMEDICAL APPLICATIONS

S. P. SURIYARAJ*, DEEPAK KUMAR VERMA*, H. BAVA BAKRUDEEN, Y. ANTONY PRABHU, S. VAIDEVI, B. RAMIYA, V. MONIKA, J. PRASANA MANIKANDA KARTIK, and K. CHANDRARAJ

*Corresponding authors. E-mail: jarayirus.nano@gmail.com

ABSTRACT

Nanomaterials are increasingly used in a wide range of applications in science, engineering, technology, and medicine. Properties of nanomaterials can differ significantly from those of the bulk structured materials. Currently, nanomaterials are used in various fields such as, carbon black particles for wear resistant automobile tires; nanofibers for insulation and reinforcement of composites; iron oxide for magnetic disk drives and audio-video tapes; nano-Zinc oxides and Titanium as sun shields for UV-rays and so on. The products of nanotechnology (dental bonding agents and wound dressing) are available in the markets. Numerous techniques are employed to characterize the material and a systematic application of one or more techniques among them leads to a complete understanding of the nanomaterial. These techniques involve Auger-electron microscopy, dielectric spectrometry, electron microscopy, elemental trace analysis, indentation testing, mass spectrometry, optical techniques, particle counting, scanning probe microscopy, X-ray analysis and so on. In this chapter, authors have explored selected characterization techniques that are useful for the basic nanotechnologist, who work on nanomaterials.

5.1 INTRODUCTION

The characterization of nanomaterials is usually done with a probe, which may be a combination of electrons, photons, neutrons, atoms, or ions. Different types of nanoparticles (NPs) have varying frequencies, ranging from gamma rays to infrared rays according to their relativistic energies. The ensuing information can be processed to retrieve images or spectra, which describe the physical, chemical, structural, geometric, or topographic details of the material. Numerous techniques are available, which are employed to characterize the material and a systematic application of one or more techniques among them leads to a complete understanding of the nanomaterial. Various techniques such as auger-electron microscopy, characterization of powders and dispersions, dielectric spectrometry, electron microscopy, elemental trace analysis (ETA), indentation testing, mass spectrometry, optical techniques, particle counting, scanning probe microscopy (SPM), X-ray analysis, etc., are available to characterize nanomaterials.

In this chapter, authors have explored selected characterization techniques that are useful for the basic nanotechnologists, who work on nanomaterials.

5.2 METHODS FOR NANOMATERIALS CHARACTERIZATION: ELECTRON MICROSCOPY

5.2.1 SCANNING ELECTRON MICROSCOPY

Scanning electron microscope (SEM) is classified under electron microscopy and it forms the images by utilizing the focused beam of electrons. The formation of surface topography of samples and its composition is due to the various signals that are released from the electrons, which collide with the atoms present in the sample. The reason for the image development is the position of the beams merge with the signal that is detected which is generally scanned in a raster pattern. The specimens for SEM can be under different conditions such as in high vacuum, in low vacuum, in wet conditions, and at cryogenic or high temperatures. The resolution capacity of SEM is better than 1 nm. Generally, SEM will detect the secondary electrons to analyze the sample surface. The electron beam makes the atoms in the sample to excite and release the secondary electrons. The emission of secondary electrons, that is discovered, is based on the topography of specimen. There is a special detector in SEM to scan and collect the secondary electrons from the sample surface to present its topographic image.

5.2.1.1 HISTORICAL DEVELOPMENT

1930s The scientists started working on electron microscope. M. Knoll[26] and Manfred von Ardenne developed the electron microscope.[59,60]

1942 Zworykin and his co-worker reported the first SEM to explore the solid surface of specimen.[74] The limitation of this first SEM was: specimen might fell down from the column. About 50 nm of resolution could be attained with this first SEM.

Late 1940s C. W. Oatley fascinated in conducting the experiments in electron optics and to rediscover the SEM in 1940.[39]

1952 C. W. Oatley and Dennis McMullan built the first SEM with 50 nm resolution.[32] This SEM could produce the image with three-dimensional characteristics like modern SEM. Further, McMullan improved the SEM in different ways by developing the electron system and the collection of secondary electron.[32] This was the first time that they could produce the stable image by utilizing the total secondary electrons with low energy.

Early 1953 O. C. Wells under Oatley's supervision constructed the second SEM with electrostatic lenses and electron gun at the bottom of the column.[63] Wells utilized the second SEM for different kinds of specimen and he was the first person to use stereographic pairs with quantifiable depth information for SEM micrographs.[63] It is majorly used in the study of fibers at large scale.

1955 Oatley with his student Everhart developed the secondary electron detector with scintillator, which could convert electrons to photons.[14,39] Further, it was transferred to the photomultiplier tube via light pipe. The combination of scintillator and photomultiplier enhanced the signal and it led to advancement of signal to noise ratio. The contrast of the image was changed according to the voltage that was applied to the specimen. Based on this, Oatley and Wells coined the term "voltage contrast."[39]

1956 Oatley's student, Peter Spreadbury constructed the SEM with cathode rays tube (CRT) as a display unit.[39]

1958 Gary Stewart expands the SEM by fitting the ion gun in specimen chamber. He went to Cambridge Instrument Company to explore the construction of stereoscan.

1960 Alec Broers developed the ion beam optics by inserting magnetic objective lens for greater resolution. Fabien Pease constructed the first SEM by utilizing all magnetic lenses at the resolution of 10 nm in 1960.

5.2.1.2 PRINCIPLE/THEORY/CONCEPT OF OPERATION OF SEM

The electron gun present in the SEM releases the electrons to focus the specimen surface. These electrons by passing through the magnetic lens interact across the specimen surface and produce accelerated electrons, which have significant kinetic energy to release various signals. These signals include secondary electrons, backscattered electrons, photons of characteristics X-rays and light, absorbed current, and transmitted electrons. The two major signals that are involved in the formation of images are secondary electrons and back-scattered electrons.[19] The formation of morphology and topography of sample is due to the secondary electrons while the back-scattered electrons illustrate the sample composition. The shape and chemical composition of the sample entirely depends on the strength of secondary electrons. The detector collects these electrons to generate electronic signals and these signals produce an image. The image is formed in a CRT tube by scanning the signals from the specimen. This image can be taken from the CRT tube and nowadays, the modern SEM has the facility to take picture utilizing the digital camera. SEM image magnification depends on the ratio of the screen size to the surface area scanned by the specimen.[43] For example: if the screen size is 300 mm and the specimen scanned area is 3 mm, then the magnification will be 100× (= 300 ÷ 3). To get a greater magnification, the specimen is scanned by decreasing the area to 0.3 mm so that the magnification will be 1000×. Unlike light microscope, SEM does not form the real image of specimen. It develops the electronic image by utilizing the stream of electron data.[13]

For list of components and instrumentation, reader is advised to consult SEM manual or SEM reference book.

5.2.1.3 SAMPLE PREPARATION IN SEM FOR NANOMATERIALS CHARACTERIZATION

Sample is prepared according to the information required by the researcher. The following considerations should be taken into account while preparing the sample: sample size, geometry, state, and conductivity. To prepare the specimen, the scientist can follow the steps as outlined below:

- ***Drying:*** Drying sample is the preliminary step before viewing under the microscope. For wet sample, there is a SEM with low vacuum environment. But the limitations under low vacuum are hindrance at high resolution and magnification of images. The most common drying methods are air drying, critical point drying, and freeze drying.

- ***Air drying***: In this technique, the organic compounds such as hexamethyldisilazane (HMDS) and tetramethylsilane (TMS) with high volatile are used to dry the sample. Before using this organic compound, the samples are treated with 100% ethanol in order to replace the water in sample.
- ***Critical point drying***: **This technique** involves increasing the temperature and pressure of a wet sample. Therefore, the sample gets dried at the critical point of the liquid. The critical point of water is 228.5 bar and 374°C; and ethanol is 60.8 bar and 241°C. Both of them are not suitable for this drying method. Since they will destroy the sample at high critical point. The critical point of CO_2 liquid is 72.8 bar and 31°C and it can be used for drying. It is used in the mixture of acetone, which acts as intermediate since CO_2 is insoluble in water. Before this method preparation starts, sample must be treated with liquid dehydration process in order to avoid any presence of water.
- ***Freeze drying***: Here, the solid phase is directly converted to gaseous phase under vacuum without the formation of liquid. Therefore, it does not affect the surface tension of sample. It would be a problem only if the change occurs during liquid to gaseous phase.

- ***Mounting***: The sample must be mounted in the sample stub by utilizing stereo-microscope. The tools for mounting the samples in stub are: Leit-C, double side carbon adhesives, fine spatulas, stork bill forceps or insect forceps, and sharpened softwood sticks. These tools are used in order to avoid any damage that can be caused to the sample. The important parameters while mounting the sample are its stability, orientation, and adherence of sample to SEM stub.
- ***Coating and conductivity***: The electrons from the electron beam react with the biological sample and form the electron clouds of negative charge. This will create the charging problem in the SEM. However, due to this charging effect, it may cause image contort, devoid of contrast between bright and dark areas. To avoid these problems, the

samples are laminated with conductive materials. Though the sample is conductive if it is mounted in epoxy or Bakelite then it is necessary to coat with conductive materials. Mostly carbon is applied for conductivity because it is cheap and it does not appear in X-rays. For some sample, which is irregular in shape, one can use gold or palladium. The problem with this coating is that sample composition analysis could not be found (since it is difficult to remove the gold or palladium from the sample for further analysis). The coating thickness must be 20 nm so that it does not affect the sample topography.

- ***Particulate sample***: The particles—such as NPs, subtle crystals, dry powders and spores—do not require any special preparation. The prime factor to be considered is: the particle must avoid loose stacking in one over the other. Further, the particle must be strongly attached in the specimen stub.
- ***Large and bulk samples***: Large samples (such as plastic, metal, wood, rock, and electronic component) are easy to handle and it does not require any processing. They should be placed in suitable large holder and the sample height should fit in the SEM. To elude any vibrations and movement, the base and across the sample must be sturdy in the stub. It should not affect the area being observed. The materials—such as carbon paint, silver paint, Leit-C, and double-sided carbon tapes and tabs—are used for fixing the sample in the stub. Among these, carbon tapes and tabs are more efficient in the use. The coating of sample is necessary for non-conductive samples.

5.2.1.4 CASE STUDY ON APPLICATION OF SEM FOR NANOMATERIALS CHARACTERIZATION

5.2.1.4.1 Nanomaterials as Absorbents for Water Treatments

Wastewater treatments utilizing the NPs are emerging as an alternative method for removing pollutants. These NPs are made of either organic or inorganic substances. The properties such as active surface, small size, and high porosity make the particles not only to absorb contaminants but are also effective in manufacturing process. The advantage of nanosize particle is their large surface area and the ratio of area to volume. In addition to this, it also exhibits potential characteristics such as high reactivity and catalysis.

Due to large surface area, it has many active sites for absorbing different chemical compounds in wastewater.

In recent years, synthesis of NPs by green synthesis method is considered to be eco-friendly and cheap. Devatha et al.[11] synthesized iron NPs from leaf extracts (from *Mangifera indica, Murraya koenigii, Azadiracta indica, Magnolia champaca*) to study the wastewater treatment efficiency. These NPs can be characterized by the techniques such as SEM (Fig. 5.1), UV-visible spectrophotometer, Fourier-transform infrared spectroscopy (FTIR). Characterization by these methods is helpful for confirmation of nanomaterial existence and formation. These NPs are used to remove phosphates, chemical oxygen demand (COD) and ammonia nitrogen in wastewater. Among the leaf extracts, *Azadiractia indica* showed maximum removal of pollutant from wastewater. It showed 98% of phosphate, 84% of ammonia nitrogen and 82% of COD removal.

5.2.1.4.2 Nanomaterials as Photocatalysts

The biological treatment is not effective in degradation of organic pollutants. Therefore, degradation utilizes photocatalysts these days. Nanomaterials are becoming more imminent. To remove the pollutant in wastewater the photocatalystic particle of TiO_2 is used to degrade dyes in presence of UV or visible light. There are many reasons for this degradation reaction mechanism. One mechanism includes that the free radicals initiate the oxidation of organic substance due to the induction of electron–hole pairs at the photocatalytic surface.

5.2.1.4.3 Nanomaterials as Antibacterial Agents

Nowadays, the most of the infection causing bacteria are resistant to antibiotics. Therefore, use of nano antimicrobials offers benefits to prevent the infections. The antibacterial activity due to membrane and oxidative stress have been studied by utilizing graphite (Gt), graphite oxide (GtO), graphene oxide (GO) and graphene oxide (rGO).[29] These graphene materials with *E. coli* are incubated in an isotonic saline solution to study the antibacterial activity. Among these, GO showed maximum antibacterial activity of 69.3% against the control without the graphene material. Liu et al.[29] studied the antibacterial activity of graphene coated silver NPs for two bacterial strains of *E. coli* and *S. aureus*. GO with silver NPs showed the highest antibacterial activity of 98.3% and 96.1% in *E. coli* and *S. aureus*, respectively.

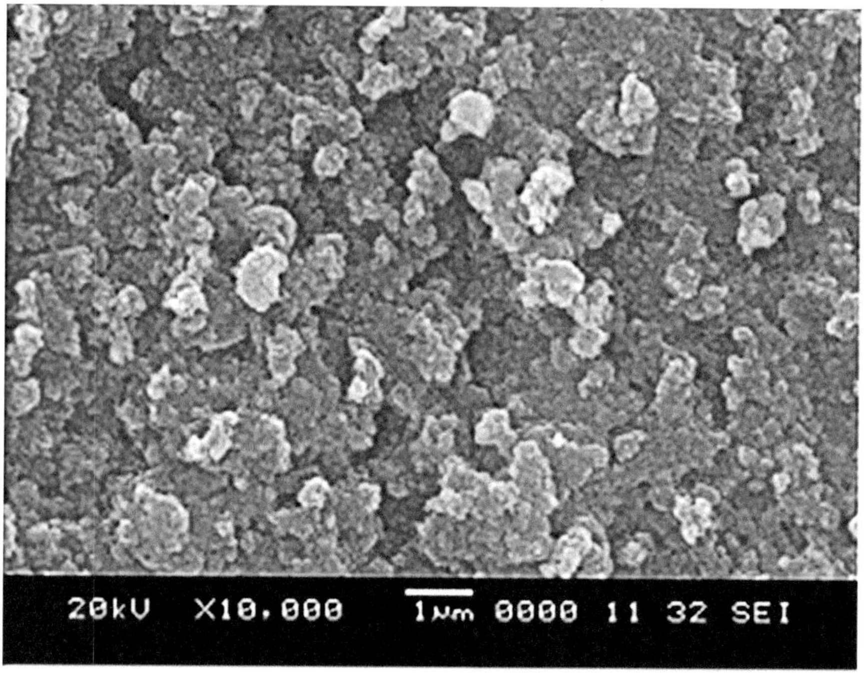

FIGURE 5.1 Scanning electron microscopy (SEM) image for *Azadiracta indica* iron nanoparticles (AI-FeNPs). Reprinted with permission from Ref [11]. © 2016 Elsevier.

In other study, Staneva et al.[51] prepared magnetic nanoparticles (MNPs) to study the antibacterial activity. The MNPs of barium hexaferrite ($BaFe_{12}O_{19}$) were produced in presence of ultrasound by co-precipitation of Ba^{2+} and Fe^{3+} with NaOH. Hydrogel was used as precursor in dispersing the NPs in aqueous solution. Then, the solution was saturated with dyed fabricated cotton in presence of photoinitiator. Finally, authors made the $BaFe_{12}O_{19}$ material of hydrogel cotton fabricated NPs (Fig. 5.2). These NPs were analyzed by SEM, infrared spectroscopy and X-ray diffraction (XRD).

5.2.1.4.4 Temperature Monitoring of Nanoparticles on Membrane Surfaces

The temperature is the key factor to affect the chemical process and membrane. Today, thermocouples are used to monitor the temperature. Although thermocouples are cost-effective and accurate for temperature detection, yet the only limitation is to miniaturize the sensors. These sensors have major applications in nanomedicine, biotechnology, and microfluidics. Currently, luminescent probe is being used to detect the temperature at nanoscale because their fluorescence emissions affect the temperature. The silica NPs are immobilized with tris(phenantroline)ruthenium(II) chloride ($Ru(phen)_3$).

5.2.1.4.5 Immobilization of Lipase in Magnetic Nanoparticles

Nowadays, the research is going on for alternative fuels from renewable sources. Biodiesel can be used as an alternative fuel because it is climate-friendly and is produced from renewable sources. Both edible and non-edible oils are utilized as feedstock for biodiesel engenderment. The enzyme lipase is used to catalyze the reaction for biodiesel production from oils. The

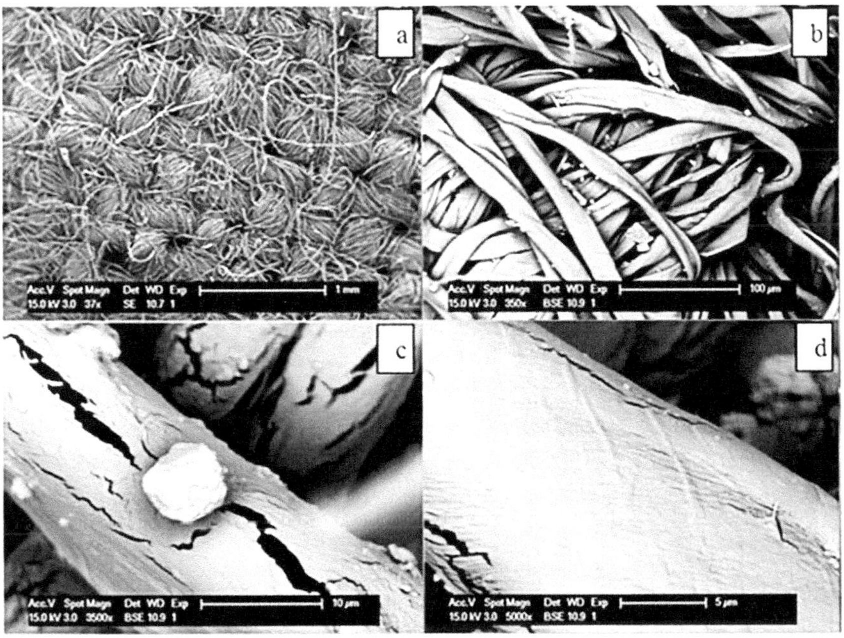

FIGURE 5.2 SEM micrographs of modified cotton fabric: (a) at magnification ×37, (b) at magnification ×350, (c) at magnification ×3500, and (d) at magnification ×5000. Reprinted with permission from Ref [52]. © 2017 Elsevier.

factors such as stability, activity, and temperature affect the biodiesel production. The other limitation is cost of enzyme production. To overcome this problem, enzyme immobilization can be followed by recycling the enzymes. The different immobilization methods are followed such as cross-linking, adsorption, entrapment, and encapsulation. Nowadays, there is great attention toward MNPs since these have high surface area and effective physical properties. This makes the enzyme to immobilize easily.

Mehrasbi el al.[34] covalently immobilized the lipase from *Candida antarctica* with MNPs for biodiesel production. The MNPs were prepared by coating Fe_3O_4 core with silica shell (Fe_3O_4@SiO_2). This particle was further used as immobilization matrix with (3-glycidoxypropyl) trimethoxylsilane (GPTMS) to make it as functional. The specific activity of enzyme was maintained at 97% by interacting lipase with functionable Fe_3O_4@SiO_2 particle. Later, physical and chemical properties of the NPs and immobilized enzymes were characterized by using the techniques such as TEM, SEM, IR, TGA, XRD, and DSL. These immobilized NPs were used for conversion of oil to methanol and it is highly stable. It retained 100% of enzyme activity even after recycling for six times. The SEM image of Fe_3O_4@SiO_2 showed the morphology and size of the particles.

5.2.2 TRANSMISSION ELECTRON MICROSCOPY

Transmission electron microscope (TEM) is an advanced microscope for analysis of NPs.[47] Nanotechnologist, Physicist and Material scientists in nanoscale research are always dependent on TEM.[22] Scrutinizing of particles—in the nanoscale ranges from 1 nm to 100 nm—requires first-grade mechanisms for obvious ultra-high exaggerations.[1] Initially, electron beam gets passed and transmitted on ultra-thin specimen, promptly interacts with the objects, micro-secondly or nano-secondly on the specimen of sample. Easily attainable configuration makes the focus ultra-visible through an imaging device such as charge coupled device, florescent screen, and photographic films.[69]

TEMs are useable for higher resolution of imaging and better than a light microscope. Electron de Broglie wavelengths permit very tiny objects to wider-range up to thousand times greater. Consequently, it provides diverse applications in various fields such as physics, chemistry, biophysics, green nanotechnology, material science, virology, cancer biology, and microbiology.[18,35,49] TEM imaging up to ultra-nanoscale is absolutely possible.

TEMs have many features for nanomaterials such as crystal orientation, microelectronics structure, identity of chemicals, particle sizing and adsorption imaging.[2,27]

5.2.2.1 HISTORICAL DEVELOPMENT

1858 Deflection of cathode rays was feasible via usage of magnetic systems by Plucker.

1891 Ferdinand Braun built the cathode ray oscilloscopes (CROs) measurement device. Riecke described that the cathode rays could be focused by magnetic fields, allowing for simple electromagnetic lens designs.

1926 Hans Busch showed that the lens maker's equation could, with congruous postulations, be applied to electrons.

1928 Max Knoll and his students Ernst Ruska and Bodo von Borriesto worked on designing of electromagnetic lens and CRO column placement, to optimize parameters for the construction of better CROs performance, and make electron optical components to generate low magnification (nearly 1:1) images.

1931 Max Knoll and his group used two magnetic lenses to achieve higher magnifications, arguably creating the first electron microscope; and designed TEM with better resolution than a light microscope in 1933. First electrostatic lens of the electron microscope was patented by the Siemens Company in 1931.

1938 Paul Anderson and Kenneth Fitzsimmons constructed the first TEM.

1939 First commercial TEM was installed in the Physics department of IG Farben-Werke.

1949 The first International conference in electron microscopy was held in Delft.

1970s Albert Crewe developed the field emission gun.

1986 Ruska was honored with the Nobel Prize for the improvement in TEM.[44]

2008 Albert Crewe and his team developed the cold field electron emission source and built a STEM able to visualize single heavy atoms on thin carbon substrates. Jannick Meyer and his team described the direct visualization of light atoms such as carbon, hydrogen utilizing TEM and a clean single-layer graphene substrate.

The functioning parts of TEM are given in Figure 5.3.

5.2.2.2 SAMPLE PREPARATION IN TEM FOR NANOMATERIALS CHARACTERIZATION

Various complex methods are available for the synthesis of nanomaterials such as chemical preparation, precipitation method, vacuum deposition, vaporization, chemical vapor deposition, chemical vapor condensation, mechanical attrition, vibrating ball mill, attrition ball mill, low energy tumbling mill, planetary ball mill, high energy ball mill, electro-deposition, and sol–gel techniques.[21,42] Diverse types of nanomaterial preparing methods are given in Table 5.1. TEM has numerous advantages in many areas such as biology, material science, and electronics[5,23,24,52,53,70]:

- Electron beam interacts the sample instantly, an effect that mainly upsurges atomic numbers.
- Tissue sectioning microtome is fully automated method and it can be used to cut ultra-tiny sections in TEM.
- Sample staining in TEM is accessible through the organisms.
- Like bacteria, fungi effortlessly absorb the stain; reflects the contracts at atomic numbers thus it can be described by imaging system.
- Lead, uranium, osmium heavy metals may be used to visualize the dense electron atoms in the biological sample.

Tissue sample coating in heavy film of the conducting material is achieved by a sputter device. Mechanical improvement like a polishing method can be utilized to formulate a sample but the ion etching is inevitable for sample thinning. Rod-like (44.6–88 nm range of negatively charged nano-hydroxyapatite) can expose great intake by Bel-7402 cells. Developed nanomaterials were entered into the cells via endocytosis process. Incessant uptake of nano-hydroxyapatite into the hepatoma cell, radically affects its adhere and they ultimately die. Integrin blocking efficiency was proved the anti-cell adhesion in Bel-7402 cell.[69]

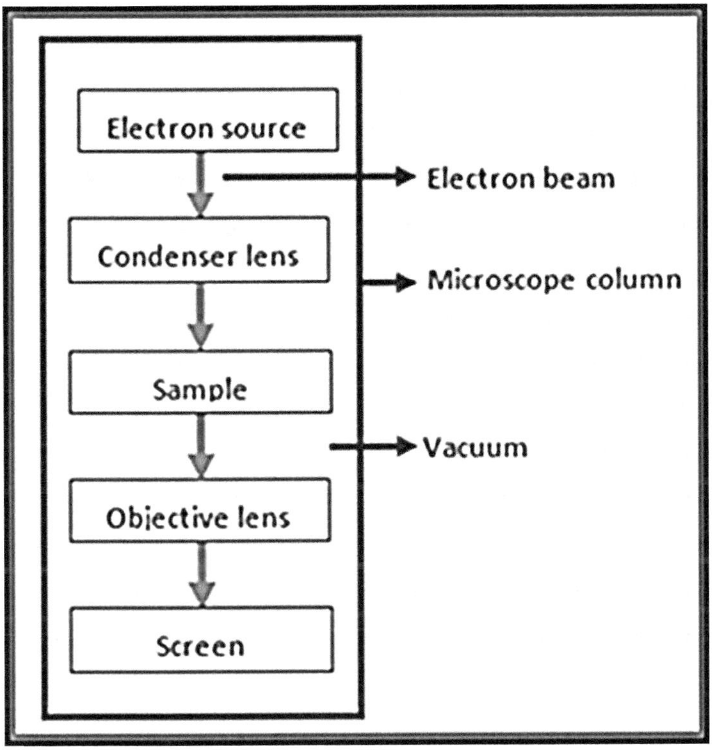

FIGURE 5.3 Flowchart showing the sequence of functioning parts of TEM.

TABLE 5.1 Diverse Types of Nanomaterial Preparing Methods.

Nanomaterials	Characterized particle size	Method	Applications
Gallium nitride	50, 100 nm	Chemical vapor deposition	Semiconductor
Lead sulfide NPs and nanowires	5 μ (5000 nm)	Chemical	Ammonia gas sensing agent
Lithiophilite ($LiMnPO_4$)	20 nm	Sol–gel	Electrochemical performance
Manganese ferrite	50 nm	Chemical	Apoptotic and necrotic activity in Murine breast cancer cell (4T1)
Nanodiamond and chitosan	1–10 μm (1000–10,000 nm)	Solution casting	Mechanical properties
Sulfur NPs	10 nm	Precipitation	Antimicrobial activity

5.2.2.3 CASE STUDY ON APPLICATION OF TEM FOR NANOMATERIALS CHARACTERIZATION

TEM is becoming a significant tool for a nanomaterial characterization.[30] It contains various features like a material transformation; structural dynamics properties.[4] TEM comprises four-dimensional analysis that is most powerful. Specifically, as it can be seen in fascinating area from nanotechnology to material sciences and biology. In situ TEM characterization techniques provide abundant prospects to describe dynamic changes in electronic state, shape, chemical composition, structure, and size in materials exactly at the nanoscale.[71] Characterization of catalytic gold (Au) NPs supported on cerium oxide (CeO_2) is given in Figure 5.4, where:

- Test is under high vacuum environment.
- A gas comprises 1% volume of CO air at 45 Pa and at room temperature. Both of these higher magnifications regions are exposed in the beneath panels. The distance of Au NPs at 1 Torris exactly at (~133 Pa to ~1.3 mbar) CO upsurges the contrast at high vacuum (Fig. 5.4).
- Ultra-high exaggerated image of Au NPs with adsorbed CO/air.
- Simulated image established on a forcefully favorable pattern.

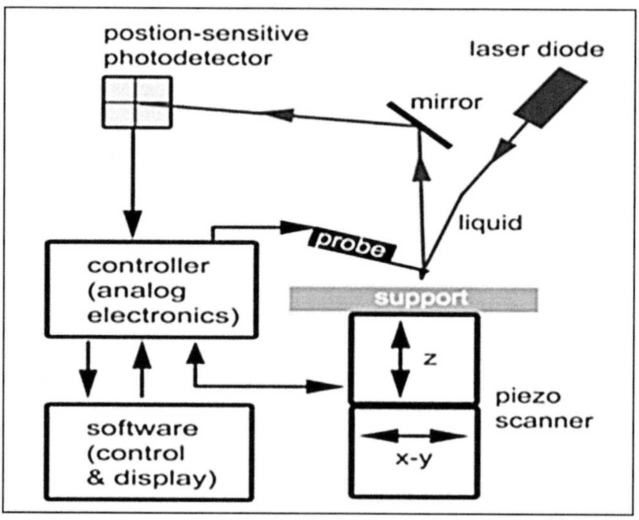

FIGURE 5.4 Characterization of catalytic gold (Au) nanoparticles supported on cerium oxide (CeO_2). Reprinted with permission from Ref [72]. © 2015 Cambridge University Press.

Currently, spatial resolution of imaging up to sub-electron-volt range and single atoms are possible. Different types of TEM holder have been developed to visualize imaging under various conditions like optical excitation, electric fields, stress, magnetic and heat. Developments of TEM at higher-resolution imagining have enabled the numerous discoveries at atomic level.[45] Construction of environmental TEM (ETEM) provided several breakthroughs for imaging of liquid samples. With high vacuum of TEM in microfabricated imaging, it is possible to achieve colloidal dynamic growths. Rapid controllable, ultra-nanomaterial synthesis includes wide applications for many fields. Well-known chemical compound such as sodium borohydride, citric acid, and oxalic acid can feasibly synthesize a nanomaterial through coupled interactions. Various types of nanomaterials such as cerium, titanium, copper, zinc, and gold have been synthesized in wet nanolaboratory. Hydroxyapatite functionalized nanowires have been utilized for different materials in tissue engineering on MG63 cells, and it promotes high bioactivity and osteoblast differentiation.[73] Well-characterized, cobalt-doped hydroxyapatite has been vigorously used against *Shigella flexneri* and *Micrococcus luteus*.[54] Quercetin-doped well-ameliorated hydroxyapatite is characterized in TEM, and it was effectively able to overhaul human bone, therefore new osteoblast cells were grown.[16]

Nanomaterial characterization of hydroxyapatites comprises several applications that are described in Table 5.2. Nanomaterials-doped hydroxyapatites express numerous advantages in biology.[6] It is becoming indispensable for activating apoptosis mechanisms, anti-cell adheres, anti-radial activity, and repairing bone and teeth.[33,40,69]

TABLE 5.2 HR TEM Characterization of Hydroxyapatites Nanomaterials.

Nanomaterials	Characterized particle size	Morphology	Applications
Arginine-doped hydroxyapatite and Europium-doped hydroxyapatite	100 nm	Nanocrystal	Internalization of cell binding, DNA binding with the side effects and lack of cell cytotoxicity.
Hydroxyapatite NPs	100 nm	Bar size	Destruction of cell organs, apoptosis and anticancer activity.
Hydroxyapatite NPs	44.6–88 nm	Rod shape	Anticell adhere and anti-cancer activity.
Quercetin-doped hydroxyapatite	200 nm	Nanocrystal	Antiradical activity
Silver-doped hydroxyapatite	30–50 nm	Rod shape	Antibacterial activity

5.2.3 SCANNING PROBE MICROSCOPY

SPM is a significant branch of microscopy for the surface analysis. Originated in 1981, it forms an illustration of surface through the physical probe.[31] Feasibly, it reveals the specimen images up to atomic level instantaneously. Piezoelectric actuators are responsible for the extraordinary atomic resolution.[65] Various Illustrations of ultra-high resolution images taken by SPM are given in Figure 5.5.

SPM comprises ultra-high resolutions better than an optical microscope, TEM, and SEM.[66] SPM does not have a lens for clear observation, conversely, it detects the specimen surface via cantilever. However, it analyzes the vertical resolution of the sample visibly up to 0.01 nm. Horizontal imagining visualization can possibly go up to 0.2 nm. Obviously, it can measure the various specimens in diverse surroundings, for example, insulating dried and wet.[15]

5.2.3.1 PRINCIPLE/THEORY/CONCEPT OF SPM

The Figure 5.5 depicts principle of operation of SPM. It uses the ultra-sharp needle-like tip on the termination of a cantilever. Scanner reads the specimen horizontally (X and Y) and precisely, whereas the surface profile analysis many differs in the sample. Continuous controlling of (Z) tool will retain the constant amplitude in cantilever, therefore, it produces the 3-D image.[41]

Optical microscope rendering software is vital to generate an image. Such software is embedding on the SPM instrument by the manufactures, however, it is also accessible from the workgroups. Few of the freeware SPM topographic images generating software are Gwyddion, WSxM, and FemtoScan online.

5.2.3.2 SAMPLE PREPARATION IN SPM FOR NANOMATERIALS CHARACTERIZATION

Different types of samples can be used for SPM imaging under ambient conditions. Different methods can be used for sample preparation. Atomic force microscopy (AFM) imaging essentially requires:

- The substrate should be clean, flat and very low roughness,
- The sample must be properly placed and attached with the substrate,
- The substrate must be fixed in a stationary position, and

- For SPM sample analysis, the substrate must be flatter/smoother than the substrate.

In other words, the roughness of the sample should be greater than the topographical features of the substrate. The most commonly used substrates include Si, glass, ITO, FTO, mica, and HOPG. The adhesives used are generally double sided tape, carbon tape and silver paste. SPM can measure various specimens in diverse environments, for example, wood, insulating, dried, and wet.

5.2.3.3 *DIFFERENT TYPES OF SPM*

- **AFM**—Atomic force microscopy
  - Atomic force microscopy—infrared
  - Contact atomic force microscopy
  - Dynamic contact atomic force microscopy
  - Non-contact atomic force microscopy
  - Tapping atomic force microscopy

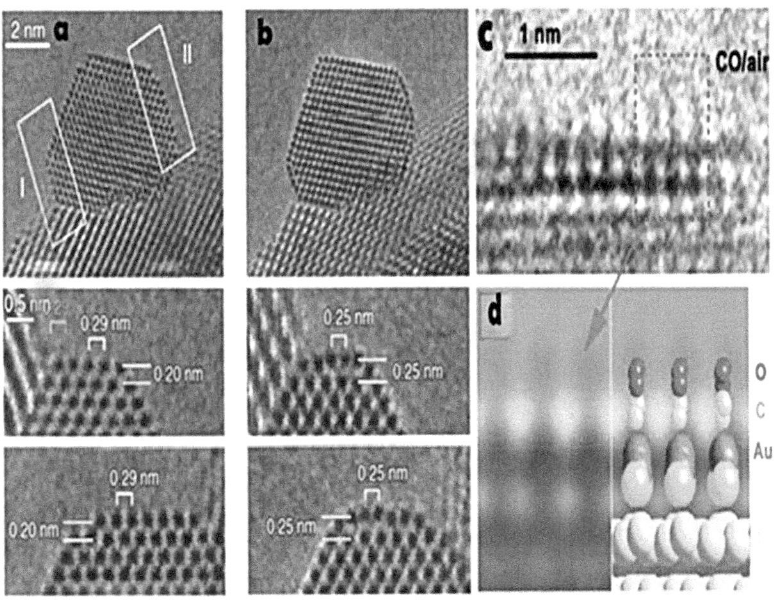

FIGURE 5.5 (See color insert.) Schematic diagram of scanning probe microscopy. Reprinted with permission from Ref [73]. © 2015 Royal Society of Chemistry.

- **BEEM**—ballistic electron emission microscopy
- **C-AFM**—conductive atomic force microscopy
- **CFM**—chemical atomic force microscopy
- **Fluid FM**—fluidic force microscopy
- **FMM**—force modulation microscopy
- **KPFM**—Kelvin probe force microscopy
- **MFM**—magnetic force microscopy
- **MFM**—magnetic probe microscopy
- **MRFM**—magnetic resonance force microscopy
- **NSOM**—near-field scanning optical microscopy
- **PTMS**—photothermal micro spectroscopy/ microscopy
- **SGM**—scanning gate microscopy
- **SICM**—scanning ion-conductance microscopy
- **SPSM**—Spin-polarized scanning tunneling microscope
- **STM**—scanning tunneling microscopy

All of these microscopic techniques of STM and AFM are unique microscopy for precise atomic surface imaging; and it goes beyond the diffraction limits and delivers ultrahigh image of sample.[17,50]

5.2.3.3.1 Atomic Force Microscopy

AFM is an ultrahigh resolution microscope under the category of SPM. The AFM is an ultra-explorer meticulously and inspects the specimen on the basis of single atom. Unraveling of the nanomaterial atomic properties at nanoscale can provide a better understanding of the NPs.[7,64] Atom petite bonding contact, at the sharp tip of AFM to specimen, divulges the interatomic forces by the order of picoscale to nanonewtons. Of these, closest tip separation will tremendously shorter to 5 A. Cantilever surface frequency of AFM can be accurately quantified by the functioning dynamic mode of detecting method.[10,55] AFM technique is a multifunctional tool that can be used to reveal cellular dynamics, and protein charges up to unprecedented range. AFM allows single molecules assessments into higher resolution in various fields such as closeness of cell, biochemical, physical, and chemical properties.[37]

5.2.3.4 SAMPLE PREPARATION IN AFM/STM FOR NANOMATERIALS CHARACTERIZATION

The AFM method (Fig. 5.6) includes different methods, which are used in the preparation of samples for AFM/SPM analysis. Each technique can deliver the unique nanorod and nanocrystal coverage for the experiments. Nanocrystal can isolate robustly the substrate by binding.

Dip casting or simple drop method can be used to form an array (Fig. 5.6). Thus, the nanocrystals are unperturbed, albeit the interaction that may vary in postdeposition technique, chemical treatments, and annealing. Spacing can be reduced because of the utilization of the tightly packed arrays. Chemical treatments can also change the native ligands, thus forming attached array of nanocrystals.[38] It is necessary for a specimen to be consistent. Pre-heating action gives the smooth surface of grease, under various temperatures. However, the lubrication type of grease, thickener type, base oil, and chemical concentration may provide varying results.[66] Scanning tip moves toward the sample that is softly interconnected to the surface. DNA imaging of AFM has been investigated in the research laboratory of C. Bustamante. Earlier sample was prepared via ionic treatment. Principally, this mica exterior is treated through the Mg^{2+} to upsurge the affinity of negatively charged DNA sample. Physical and chemical analysis of the DNA studies delivers the structural surface distortions.[65]

5.2.3.5 CASE STUDY FOR APPLICATION OF AFM FOR NANOMATERIALS CHARACTERIZATION

AFM technique can be used to characterize the biological systems on a nanometer scale. Molecular and cell machinery include myosin, hemolysin, ATP synthase, kinesin, and RNA polymerase. Rapid scanning mechanisms exactly expose the unprecedented range of sensitivity up to the piconewton levels.[37] Accordingly, it became most powerful surface observing tool in the research and for the molecular biology. AFM allows investigators a thorough analysis of a single molecule up from the biochemical and physical or chemical[10] point of view.

The vertical interchange exploring in atomic manipulation is given in Figure 5.7 as follows:

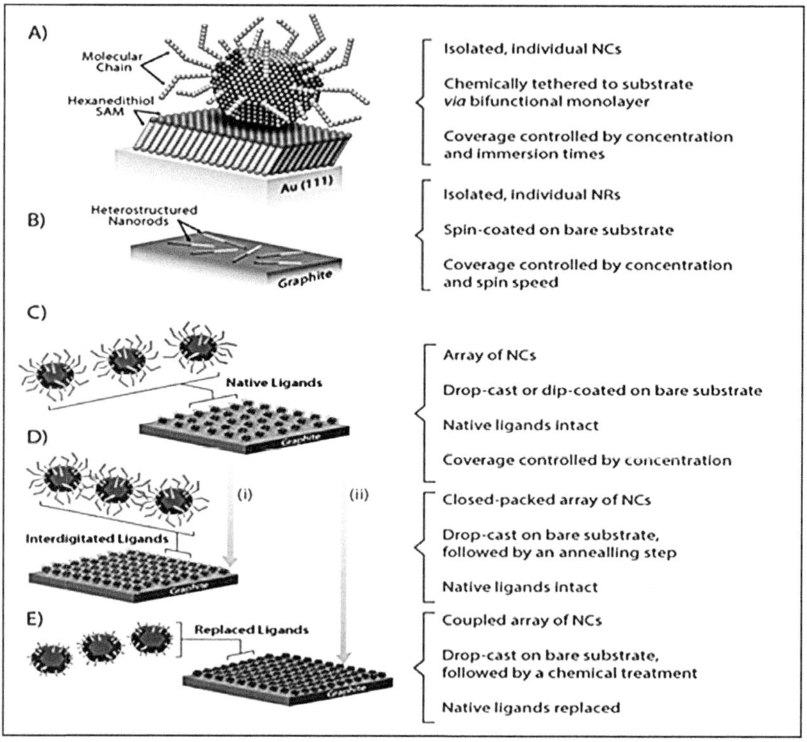

FIGURE 5.6 (See color insert.) Sample preparation methods for AFM analysis[38]: (A) Isolated and tethered nanocrystals (NCs) were assembled on Au; (B) Spin coating of nanorods; (C) Easiest way for array construction is by drop casting system; (D) Drop casting, and (E) NCs are prepared via chemical treatment.

- Meticulous bonding interaction analysis of repulsive foremost atom forces at probe's surface is achieved. Si atom (white circle) is substituted by the Sn atom via AFM tip. Recently developed (dark circle) Sn atoms alternate the Si atom by AFM tip.
- Erase.
- Si atom deposition and elimination in minor concentrations (Fig. 5.7b and c).
- Of these, vertical interchanging method permits a new way to design an atomic surface at room temperature by AFM.
- In Figure 5.7e, hydrogen, tin and silicon atoms are illustrated in various colors (white sphere, red and yellow) individually. Apex tip and the surface models were arranged. It can be used to analyze the atomic topography image of surface.[10]

Meticulous observing of cellular functions,[37] and signaling at atomic nano level are precise and advantageous for better understanding of gap junctions for signaling in human. Exploring a chaperone complex in AFM is utterly attainable in real time.[21,56] Figure 5.8 shows the meticulous observing of individual cellular activities at ultrahigh surface resolution as below:

- Human signaling channels are well-known as gap junctions,
- Bovine rhodopsin is essentially a pigment of eye (encircled in white),
- Gathering of light harvesting I (large size) and II (small size) can absorb the photons and convert simultaneously into electric energy,
- Utilization of (anti-cancer) taxol drug modifies the proto-filaments diameters, and
- Entry of amino-sulphonate compounds in human cell can modify the closed state of gap channels to open state from pH < 6 to > 7.5.

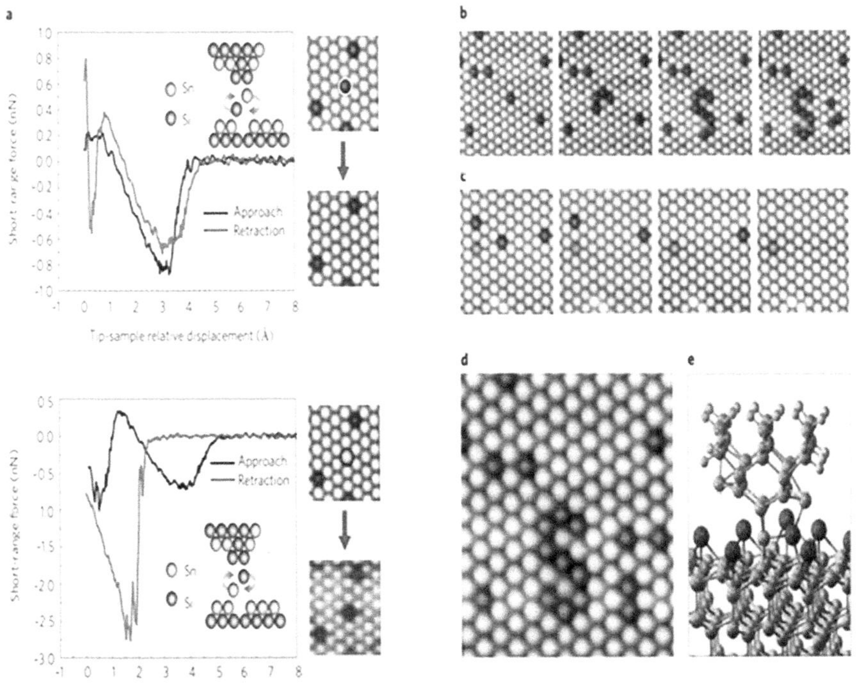

FIGURE 5.7 (See color insert.) Vertical interchange exploring in atomic manipulation. Reprinted with permission from Ref [10].

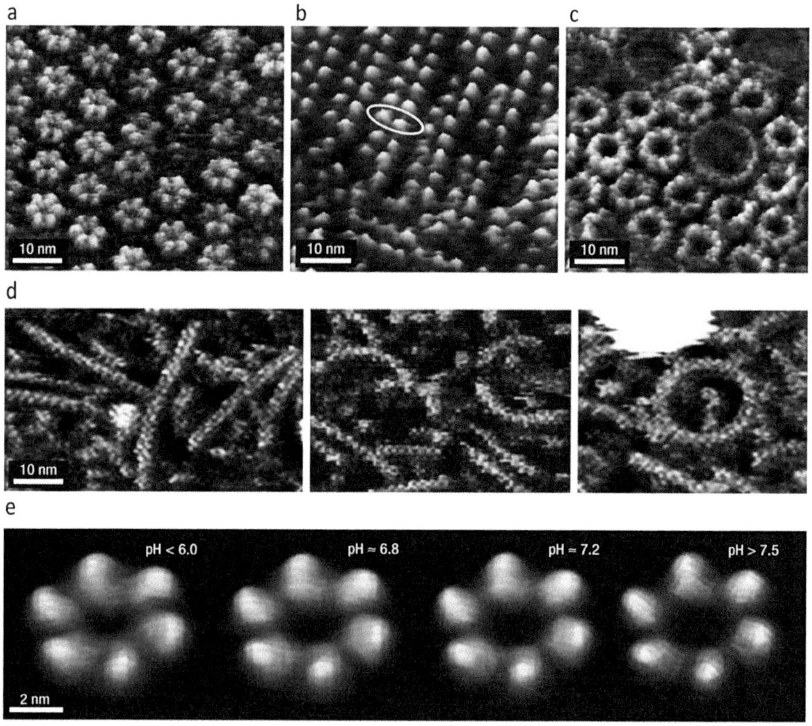

FIGURE 5.8 (See color insert.) Meticulous observations of individual cellular activities at ultrahigh surface resolution. Reprinted with permission from Ref [37]. © 2008 Nature Publishing Group.

5.2.4 X-RAY DIFFRACTION

X-ray is a form of electromagnetic radiation having wavelength range from 0.01 to 10 nm. It was first discovered by German physicist Wilhelm Conrad Rontgen in 1895. They have wavelengths shorter than ultra-violet radiation but longer than gamma rays. Diffraction is a phenomenon where a wave meets an obstacle or a slit and bends around the corner of an interference or aperture.[61] Huygens-Fresnel principle states that diffraction describes the interference of waves like light waves, sound waves, electromagnetic waves, X-rays, radio waves and water waves. The 90% of solids available on earth are crystalline in nature and their structure can be interpreted by observing diffraction pattern when X-rays are made incident on them. The electromagnetic spectrum of different scattered wavelength of lights is shown in Figure 5.9.

5.2.4.1 HISTORICAL DEVELOPMENT OF XRD

1891 E. Fedorov (a Russian crystallographer, mineralogist, petrographer, and chemist) worked on 230 space groups of crystal symmetry and proved that 127 could occur in nature, whereas 103 were unreal.

1912 Max von Laue announced his discovery of XRD on June 8. He used copper sulfate crystals as diffraction grating. Later on, W. H. Bragg and W. L. Bragg confirmed Barlow's hypothetic model of rock salt (NaCl). This proved to be the first XRD analysis of a single crystal. P. Debay and P. Scherrer analyzed polycrystalline samples and study of numerous objects began afterwards.

1930s Desmond Bernal and Dorothy Crowfoot determined structure of sterols and later on structures of penicillin and vitamin B12.

1956 Crick and Watson worked on the structure of DNA and received Nobel prize for the same along with Maurice Wilkins in 1962.[9,62]

1969 Dorothy Hodgkin obtained the structure of insulin (a very well-known biological compound).[9,62]

5.2.4.2 PRINCIPLE/THEORY/CONCEPT OF XRD

XRD can help to determine atomic and molecular structure of a crystal. The crystalline atoms diffract the incident X-rays in different directions and this proves it to be a primary use of to identify and characterize compounds. XRD methodology finds an advantage over other techniques such as IR-, UV-, FTIR spectroscopy, NMR or MS due to its non-destructive nature. The theory of XRD was first proposed in the year 1913 by William Lawrence Bragg and his father William Henry Bragg. The peaks of scattered intensity can be observed when X-rays are incident on a crystal lattice following two conditions: (1) angle of incidence = angle of scattering; (2) path difference = integer number of wavelengths. X-ray scattering pattern is only produced when the incident beam of rays follow Bragg's law of diffraction (Fig. 5.10) and satisfies the following equation:

$$n\lambda = 2d \sin\theta \tag{5.1}$$

where, n = positive integer, λ = wavelength of incident wave, θ = scattering angle, and d = interplanar distance.

Using eq (5.1), one can get details about the crystal structure and also determine the wavelength of X-rays incident on the crystal. The principle of operation of XRD is based on the fact that characteristic X-rays are produced where the electrons gain high energy to remove inner shell electrons of the molecule. These rays are emitted from heavy elements when their electrons make transition between lower atomic energy levels. On emission, they showed two sharp peaks that are left, when vacancies are produced in n =1 or K shell of the atom. The electrons fall back from outer shell to inner shell to fill the gap and X-rays are produced by transitions from $n = 2$ to $n = 1$ levels is called "Kα X-rays" and those for $n = 3$ to $n = 1$ transition are called "Kβ X-rays". The Kα- and Kβ X-rays are the two most common components of the spectra. Kα consists of two parts: Kα1 and Kα2 where the former has slightly shorter wavelength and the twice the intensity of the later. To produce monochromatic X-rays, filters such as foils and crystal monochromators are used. The X-rays generated are collimated and directed on the sample; as the sample and detector are rotated, intensities of the diffracted X-rays are noted.[58]

A typical XRD instrumentation consists of three main components[3]: X-ray source, specimen stage, and X-ray detector (Fig. 5.11). Generally, two types of detectors are used in XRD that is, Scintillation and gas-filled. The XRD instrument in the IITM is shown in Figure 5.12. The parafocusing (or Bragg–Brentano) diffractometer is the most common geometry for diffraction instruments. This has several advantages such as high resolution and as well as high beam intensity at low cost using carefully prepared samples.

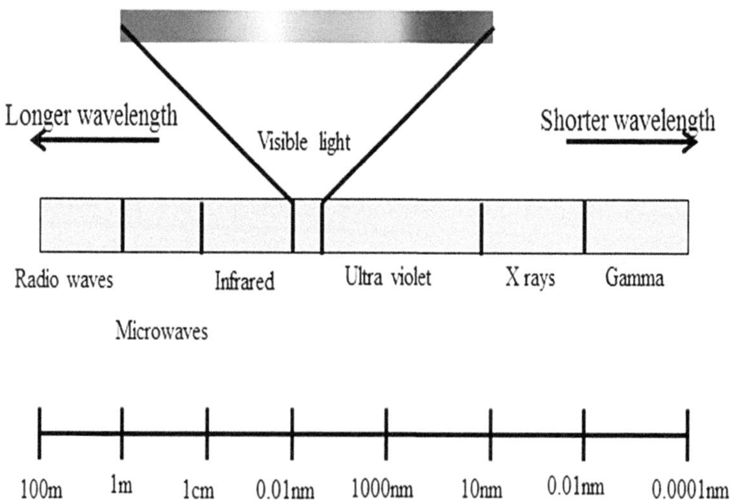

FIGURE 5.9 **(See color insert.)** Electromagnetic spectrum for different types of waves.

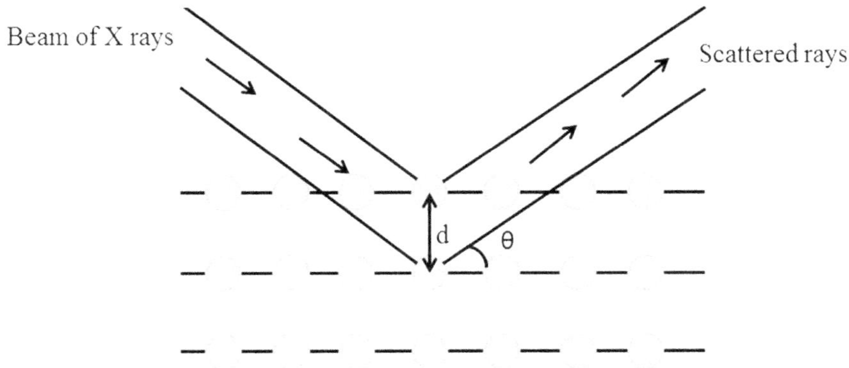

FIGURE 5.10 Bragg's law.

5.2.4.3 SAMPLE PREPARATION IN XRD FOR NANOMATERIALS CHARACTERIZATION

The analysis in XRD may be done for number of purposes such as cell refinement, indexing, phase identification, polyphasic mixture qualitative and quantitative analyses, structural determination, and so on. The materials to be analyzed are usually in random forms and so they have to be grounded to the powder form. Except for this initial milling down of materials, the rest of procedure is non-destructive. Generally, the samples are distinguished as single crystalline form and powder or polycrystalline forms. Single crystal form is the perfect crystal form with the cross-section of about 0.3 mm; whereas the materials are ground down to 0.002 mm to 0.005 mm cross section[48] in the powder form. Both solid and liquid samples can be used for XRD and the processing methods vary based on the state.

5.2.4.3.1 Solid Samples

The sample can be presented in the form of powder, briquettes or fusion products. The diffraction of X-rays depends upon the arrangement of atoms, hence it is necessary to produce a smooth surface so that the diffraction happens only from the atoms. Any surface defects will lead to decrease in the precision of data obtained. Bulk composition of the sample can be analyzed if the sample contained homogenous arrangement of atoms in three-dimensional spaces. Sample deposition can be done by various methods,[25] such as:

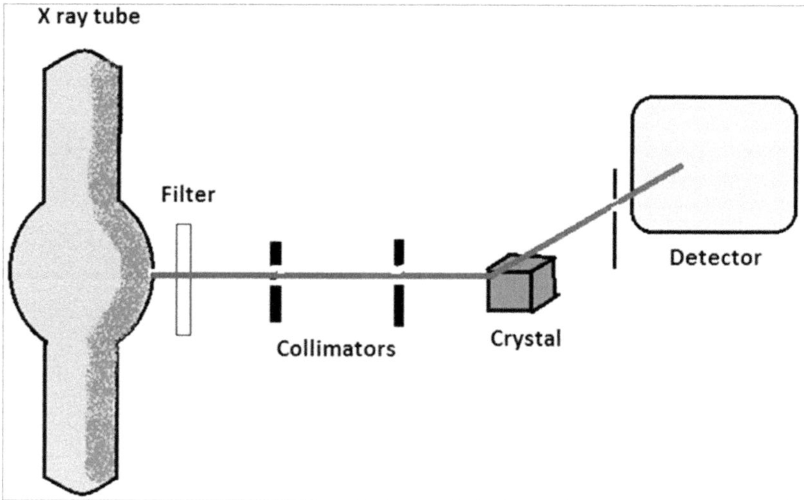

FIGURE 5.11 Components of XRD.

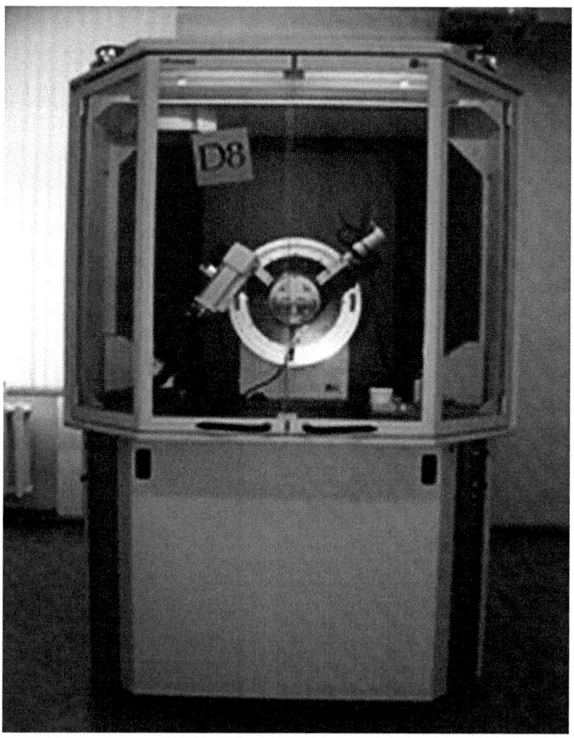

FIGURE 5.12 **(See color insert.)** XRD instrument. Courtesy: Chemistry, IITM.

- Dry dusting
- Dusting in oil, grease, silicones, and so on.
- Backfilling or side loading
- Mixing with inert powder (wheat, cabosil)

Volatile inert liquids (acetone, ethanol, etc.)

- Non-volatile suspending inert liquids (amyl acetate, + 5% collodion)
- Spray drying
- Thin film deposition (for transparent materials)

A sample collection method is considered efficient when it is able to collect all the selected particles on the tip of the glass fiber for introduction into the X-ray beam. Common techniques incur loss of more particles in the process of grinding and transfer of the particles to the fiber. This can be overcome by collecting the specimen particles directly on the glass fiber. Powder samples are preferred when the original material is more heterogeneous in nature. Powders are formed from variety of materials ranging from metals to biological samples. Some of them are already present as powders but they should be pulverized by grinding or milling. They are used as such in the powder form for the analysis or packed in cells or spread on as a thin film.

5.2.4.3.2 Liquid Samples

Liquid specimens are homogenous, particle effects are minimized and the results can be represented for the whole bulk regime. Sensitivity increases as the beam can penetrate to higher depths and also the emitted secondary radiation is less absorbed. Calibration standards can be prepared easily for liquid samples, whereas internal standardization can also be applied.[57] Sometimes, the elemental concentrations of the analytes in the original sample are less (e.g., environmental samples). For such samples, preconcentration is needed as like other analytical methods. The ideal method is the presentation of samples in the form of thin films. The intensity of diffracted X-ray is directly proportional to the mass of the compound to be analyzed. The thin films are prepared by filtering the solution through the filter preferably Millipore or Nucleopore and then the retained materials are dried. The major drawback of liquids is the highly scattered background.

This makes it difficult to work with low concentrations of light elements hence limiting the detection range.

5.2.4.4 CASE STUDY ON APPLICATION OF XRD FOR NANOMATERIALS CHARACTERIZATION

5.2.4.4.1 Characterization of Cobalt Oxide Nanomaterials

Li et al.[28] synthesized cobalt oxide (Co_3O_4) nanotubes for application in the batteries and gas sensors, using anodic aluminum oxide membranes as templates and chemical thermal decomposition method. The samples were characterized by powder XRD for phase purity and crystallinity. For comparison, NPs and nanorods were also prepared by similar procedures. XRD plot showed consistent positions of the characteristic peaks in all three structures. However, the peak intensity of the NPs was the strongest, indicating its highly crystalline structure. No other significant peaks were observed implying the high purity and absence of any other phases.

In another research carried out by Yang et al.,[67] Cobalt oxide NPs were synthesized by calcination of the precursor at 300°C; and the precursor was obtained by mechanochemical reaction of $Co(NO_3)_2 \cdot 6H_2O$ with NH_4HCO_3. The cobalt oxide crystals were characterized by XRD technique and crystal size was calculated to be 13 nm. Further, the crystals were characterized using TEM, differential thermal analysis (DTA) and thermogravimetric analysis (TGA) and the effect of calcination temperature were studied. Salavati-Niasari et al.[46] synthesized cobalt oxide NPs using N-N`-bis (salicylaldehyde)-1, 2-phenylenediimino cobalt(II); Co (salophen) precursor was formed by a two-step process. The NPs were characterized using XRD. Sample was prepared by thermal decomposition and oxidation at 773 K for analysis and the XRD pattern was observed. The data produced by XRD was incorporated in Scherrer formula to calculate diameter of cobalt oxide crystal and it was found to be 30 nm. These results elaborate on the synthesis of pure single crystalline cobalt oxide NPs, as no peaks of impurities were noted.

5.2.4.4.2 Characterization of ZnS Nanomaterials

Synthesis of zinc sulfide (ZnS) NPs and its characterization are carried out widely. Based on the method adopted for synthesis, the NPs vary in properties. Yang et al.[68] synthesized ZnS NPs using microwave irradiation. After 10 min of irradiation, ZnS powder was synthesized and characterized using XRD method; and the pattern revealed the size of the crystal as 6.5 nm using Scherrer formula. Mu et al.[36] performed synthesis and embedding of ZnS NPs in silica nanospheres by seeded growth procedure. The XRD pattern of non-coated and silica-coated ZnS nanocomposites were analyzed. It was found that XRD peaks were quite similar for both, showing little shift due to coating effect. The size of the NPs was found to be 10 nm and 6 nm for ZnS and silica-coated ZnS particles, respectively.

5.2.4.4.3 Determination of Nanoparticle Sizes by X-ray Diffraction

XRD was used to analyze nickel and iron (3+) oxide nanomaterials of different particle sizes. Various methods for analysis of XRD line profile were performed such as Scherrer, Williamson–Hall, Warren–Averbach, and modified Warren–Averbach methods based on nanostructure formation. XRD reveals size of crystallite, microstrain of a lattice, dislocation structure etc., based on width and shape of reflections. By studying the diffraction patterns of Ni NPs, it was observed that lines of nanocrystalline Ni has characteristic shape of Lorentz function (a sharp peak and extended tails). Crystallite dispersity factor contributes mainly to the physical broadening of lines which is confirmed by rigorous analysis. Lines of annealed NPs were found to be narrower than the initial NPs.[17] Calculations were performed by different methods using following Scherrer equation that expresses the volume-weighted size of crystallites:

$$(L)v = \frac{K\lambda}{B\cos\theta o} \tag{5.2}$$

where, λ is radiation wavelength, B is physical width of a reflection (in 2θ); θ_0 is the diffraction angle of line maximum; and K is a constant close to unity.

In this method broadening of lines is caused by high dispersity of crystallites.

5.2.4.5 WILLIAMSON–HALL METHOD

This method considers that line width is a function of the diffraction angle or diffraction vector in the presence of microstrains. The dependence on angle makes it differentiate between effects of dispersity and microdistortions by considering several reflections. The Willaimson–Hall equation is given below:

$$\beta = 2eQ + [1/Lv), \tag{5.3}$$

where β is physical width in units of diffraction vector, e is maximum micro-deformation of a lattice, and L_v = volume-weighted size of crystallites. For $e = 0$, eq (5.3) reduces to Scherrer equation.

5.2.4.6 *WARREN–AVERBACH METHOD*

This is more exact method based on consistent physical model of X-ray scattering by small distorted crystals. Here, a physical profile is presented as convolution of two bell-shaped functions, where the broadening of one is due to crystalline dispersity and the other is due to lattice microstrains. This is given in the form of Fourier coefficients of the function of diffraction line. The Fourier coefficients of dispersity are related to size distribution of crystallites and independent of reflection order, as shown in the following equation:

$$A^S L_n = \pi r^2 = [1/La]\int_{Ln}^{\infty}[-Ln][L_n']d\,L_n', \tag{5.4}$$

where $p_a(L_n)dL_n$ is the relative number of columns with a length from L_n to $L_n + dL_n$.

Voigt function is a universal approximation of which Fourier transformation is given in an analytical form yielding stable solution of incorrectly formulated problem for estimation of substructure parameters in Warren–Averbach method. Thus, this method is considered more appropriate physical model and has consistent mathematical apparatus. However, in practice, Fourier coefficients seem to be more sensitive towards experimental errors.[61]

5.3 SUMMARY

Characterization of the nanomaterials is a primary process in the field of materials science and nanoscience, without which no scientific properties of engineering materials could be determined. The characterization terms are used to study the microscopic and nanoscopic structure and physical, chemical and biological properties of materials. Although there are various techniques available to characterize the nanomaterials, yet there is a need for the development of advanced characterization tools to characterize the nanoscale level objects precisely and effectively in complex environmental media.

KEYWORDS

- **atomic force microscopy**
- **differential thermal analysis**
- **Fourier-transform infrared spectroscopy**
- **scanning transmission electron microscopy**
- **transmission electron microscope**

REFERENCES

1. Alaqad, K.; Saleh, T. A. Gold and Silver Nanoparticles: Synthesis Methods, Characterization Routes and Applications Towards Drugs. *J. Environ. Anal. Toxicol.* **2016,** *6* (4), 1–10.
2. Anjum, D. H. Characterization of Nanomaterials with Transmission Electron Microscopy. *Mater. Sci. Eng.* **2016,** *146*, 1–10.
3. Bish, D. L.; Post, J. E. *Modern Powder Diffraction. Reviews in Mineralogy*; Mineralogical Society of America: USA, 1989; p 369.
4. Brongersma, M. L.; Halas, N. J.; Nordlander, P. Plasmon-induced Hot Carrier Science and Technology. *Nature Nanotechnol.* **2015,** *10*, 25–34.
5. Cariaga, G. G. C.; Hernandez, K. V. C.; Castro, N. C.; Zaldivar, M. H.; Gutierrez, R. G.; Lopez, O. E. C. Synthesis and Characterization of GaN Rods Prepared by Ammono-chemical Vapor Deposition. *Adv. Chem. Eng. Sci.* **2012,** *2*, 292–299.
6. Ciobanu, C. S.; Iconaru, S. L.; Coustumer, P. L.; Constantin L. V.; Predoi, D. Antibacterial Activity of Silver-doped Hydroxyapatite Nanoparticles Against Gram-positive and Gram-Negative Bacteria. *Nanoscale Res. Lett.* **2012,** *7*, 1–9.
7. Cordova, G.; Lee. B. Y.; Leonenko, Z. Magnetic Force Microscopy for Nanoparticle Characterization. *Nano World J.* **2016,** *2* (1), 10–14.
8. Crewe, A. V.; Isaacson, M.; Johnson, D. A Simple Scanning Electron Microscope. *Rev. Sci. Instrum.* **1969,** *40* (2), 241–46.
9. Crick, F. *What Mad Pursuit: A Personal View Of Scientific Discovery*?; Weidenfeld and Nicolson: London, 1989; p 182
10. Custance, O.; Ruben Perez, R.; Morita, S. Atomic Force Microscopy as a Tool for Atom Manipulation. *Nature Nanotechnol.* **2009,** *4*, 803–810.

11. Devatha, C. P.; Thalla, A. K.; Katte, S. Y. Green Synthesis of Iron Nanoparticles Using Different Leaf Extracts for Treatment of Domestic Waste Water. *J. Cleaner Prod.* **2016,** *139*, 1425–1435.
12. Dorofeeva, G. A.; Streletskiib, A. N.; Povstugara, I. V.; Protasova, A. V; Elsukova, E. P. Determination of Nanoparticle Sizes by Xray Diffraction, *Coll. J.* **2012,** *74* (6), 675–685.
13. Egerton, R. F. *Physical Principles of Electron Microscopy: An Introduction to TEM, SEM, and AEM*; Springer: US, 2005, 202.
14. Everhart, T. E; Thornley, R. F. M. Wide-band Detector for Micro-microampere Low-energy Electron Currents. *J. Sci. Instrum.* **1960,** *37* (7), 246–248.
15. Fardi, M.; Hassani, S. S. Scanning Impedance Microscopy (SIM): A Novel Approach for AC Transport Imaging. *Int. J. Nano Dimen.* **2016,** *7* (4), 278–283.
16. Forte, L.; Torricelli, P.; Boanini, E.; Gazzano, M.; Rubini, K.; Fini, M.; Bigi, A. Antioxidant and Bone Repair Properties of Quercetin-Functionalized Hydroxyapatite: An in Vitro Osteoblast–Osteoclast–Endothelial Cell Co-Culture Study. *Acta Biomater.* **2016,** *1* (32), 298–308.
17. Friedrich, L.; Rohrbach, A. Surface Imaging Beyond the Diffraction Limit With Optically Trapped Spheres. *Nature Nanotechnol.* **2015,** *10*, 1064–1069.
18. Golding, C. G.; Lamboo, L. L; Beniac D. R.; Booth T. F. The Scanning Electron Microscope in Microbiology and Diagnosis of Infectious Disease. *Sci. Rep.* **2016,** *23* (6), 1–8.
19. Goldstein, J.; Newbury, D.E.; Joy, D.C.; Lyman, C.E.; Echlin, P.; Lifshin, E.; Sawyer, L.; Michael, J.R. *Scanning Electron Microscopy and X-Ray Microanalysis;* Springer: Boston, MA, 2003; p 689.
20. Guz, N. V.; Patel, S. J.; Dokukin, M. E.; Clarkson, B.; Sokolov, I. AFM Study Shows Prominent Physical Changes in Elasticity and Pericellular Layer in Human Acute Leukemic Cells Due to Inadequate Cell–cell Communication. *Nanotechnology* **2016,** *27,* 1–12.
21. Jaiswal, J.; Chauhan, S.; Chandra, R. Influence of Sputtering Parameters on Structural, Optical and Thermal Properties of Copper Nanoparticles Synthesized by DC Magnetron Sputtering. *Int. J. Sci. Technol. Manag.* **2015,** *4* (1), 678–688.
22. Jonge, N. D.; Ross, F. M. Electron Microscopy of Specimens in Liquid. *Nature Nanotechnol.* **2011,** *6*, 695–704.
23. Kanagesan, S.; Aziz, S. B. A.; Hashim, M.; Ismail, I.; Tamilselvan, S.; Alitheen, N. B. B. M.; Swamy, M. K.; Rao, B. P. C. Synthesis, Characterization and *In Vitro* Evaluation of Manganese Ferrite ($MnFe_2O_4$) Nanoparticles for Their Biocompatibility with Murine Breast Cancer Cells (4T1). *Molecules* **2016,** *21* (312), 1–9.
24. Karami, H.; Ghasemi, M.; Matini, S. Synthesis, Characterization and Application of Lead Sulfide Nanostructures as Ammonia Gas Sensing Agent. *Int. J. Electrochem. Sci.* **2013,** *8*, 11661–11679.
25. Kleeberg, R.; Monecke, T.; Hillier, S. Preferred Orientation of Mineral Grains in Sample Mounts for Quantitative XRD Measurements: How Random are Powder Samples? *Clays Clay Miner.* **2008,** *56*, 404–415.
26. Knoll, M. Charging Potential and Secondary Emission of Electron-irradiated Bodies (*Aufladepotentiel und Sekundäremission elektronenbestrahlter Körper*). *Z. Tech. Phys.* **1935,** *16,* 467–475.

27. Krueger, A. B.; Carnell, P.; Carpenter, J. F. Characterization of Factors Affecting Nanoparticle Tracking Analysis Results with Synthetic and Protein Nanoparticles. *J. Pharm. Sci.* **2016,** *105*, 1434–1443.
28. Li, W. Y.; Xu, L. N.; Chen, J. Co_3O_4 Nanomaterials in Lithium-ion Batteries and Gas Sensors. *Adv. Funct. Mater.* **2005,** *15* (5), 851–857.
29. Liu, Z.; Guo, W.; Guo, C.; Liu. S. Fabrication of AgBr Nanomaterials as Excellent Antibacterial Agents. *RSC Adv.* **2015,** *89*, 72872–72880.
30. Lu, J.; Chen, Z.; Ma, Z.; Pan, F.; Curtiss, L. A; Amine, K. The Role of Nanotechnology in the Development of Battery Materials for Electric Vehicles. *Nature Nanotechnol.* **2016,** *11*, 1031–1038.
31. Ma, Z.; Zhao, M.; Qu, Z.; Gao, J.; Wang, F.; Shi, Y.; Qin, L.; Liu, J. Precise Nanoscale Measurements with Scanning Probe Microscopy (SPM): A Review. *J. Nanosci. Nanotechnol.* **2017,** *17*, 2213–2234.
32. McMullan, D. An Improved Scanning Electron Microscope for Opaque Specimens, *Proc. IEE Part II: Power Eng.* **1953,** *100* (75), 245–256.
33. Meena, R.; Kumar, K.; Rani, K. M.; Paulraj, R. Effects of Hydroxyapatite Nanoparticles on Proliferation and Apoptosis of Human Breast Cancer Cells (MCF-7). *J. Nanopart. Res.* **2012,** *14* (712), 1–11.
34. Mehrasbi, M. R.; Mohammadi, J.; Peyda, M.; Mohammadi, M. Covalent Immobilization of *Candida antarctica* Lipase on Core-shell Magnetic Nanoparticles for Production of Biodiesel from Waste Cooking Oil. *Renew. Energy* **2017,** *101*, 593–602.
35. Mishra, M.; Chauhan, P. Applications of Microscopy in Bacteriology. *Microscop. Res.* **2016,** *4*, 1–9.
36. Mu, J.; Gu, D.; Xu, Z. Synthesis and Stabilization of ZnS Nanoparticles Embedded in Silica Nanospheres, *Appl. Phys. A* **2005,** *80* (7), 1425–1429.
37. Muller, D. J.; Dufrene, Y. F. Atomic Force Microscopy as a Multifunctional Molecular Toolbox in Nanobiotechnology. *Nature Nanotechnol.* **2008,** *3*, 261–269.
38. Nanayakkara, S. U., Lagemaat, J. V. D.; Luther, J. M. Scanning Probe Characterization of Heterostructured Colloidal Nanomaterials. *Chem. Rev.* **2015,** *115* (16), 8157–8181.
39. Oatley, C. W.; Everhart, T. E. The Examination of p–n Junctions in the Scanning Electron Microscope. *J. Electron. Control* **1957,** *2* (6), 568–570.
40. Prakasam, M.; Locs, J.; Ancane, K. S.; Loca, D.; Largeteau, A.; Cimdina, L. B. Fabrication, Properties and Applications of Dense Hydroxyapatite: A Review. *J. Funct. Biomater.* **2015,** *6*, 1099–1140.
41. Raigoza, A. F.; Dugger, J. W.; Webb, L. J. Review: Recent Advances and Current Challenges in Scanning Probe Microscopy of Biomolecular Surfaces and Interfaces. *ACS Appl. Mater. Interfaces* **2013,** *5* (19), 9249–9261.
42. Rajput, N. Methods of Preparation of Nanoparticles - A Review. *Int. J. Adv. Eng. Technol.* **2015,** *7* (4), 1806–1811.
43. Reimer, L. *Scanning Electron Microscopy: Physics of Image Formation and Microanalysis*; Springer-Verlag Berlin Heidelberg, 1998; p 527.
44. Ruska, E.; Binnig, G.; Rohrer, H. *The Nobel Prize in Physics, Life Through a Lens: perspectives*,<u/c> 1986.
45. https://www.nobelprize.org/nobel_prizes/physics/laureates/1986/perspectives.html (accessed Apr 27, 2017).
46. Ruzmetov, D.; Oleshko, V. P.; Haney, P. M.; Lezec, H. J.; Karki, K.; Baloch, K. H.; Agrawal, A. K.; Davydov, A. V.; Krylyuk, S.; Liu, Y.; Huang, J. Y.; Tanase, M.;

Cumings, J.; Talin, A. A. Electrolyte Stability Determines Scaling Limits for Solid-state 3D-li Ion Batteries. *Nano Lett.* **2011,** *12*, 505–511.

47. Salavati-Niasari, M.; Khansari, A; Davar, F. Synthesis and Characterization of Cobalt Oxide Nanoparticles by Thermal Treatment Process, *Inorganica Chimica Acta* **2009,** *362* (14), 4937–4942.
48. Segets, D. Analysis of Particle Size Distributions of Quantum Dots: From Theory to Application. *Kona Powder Particle J.* **2016,** *33*, 48–62.
49. Sharma, R.; Bisen, D. P.; Shukla, U.; Sharma, B. G. X-ray Diffraction: A Powerful Method of Characterizing Nanomaterials. *Recent Res. Sci. Technol.* **2012,** *4* (8), 77–79.
50. Soanpet, P.; Samatha. D.; Babu, A. S.; Bharathi, D.; Raj, R. Gold Nanoparticles for the Treatment of Cancer: Review. *World J. Pharm. Pharm. Sci.* **2015,** *5*, 381–392.
51. Somnath, S.; Collins, L.; Matheson, M. A.; Sukumar, S. R.; Kalinin, S. V.; Jesse, S. Imaging via Complete Cantilever Dynamic Detection: General Dynamic Mode Imaging and Spectroscopy in Scanning Probe Microscopy. *Nanotechnology* **2016,** *27*, 1–11.
52. Staneva, D.; Koutzarova, T.; Vertruyen, B.; Tonkova, E. V.; Grabchev, I. Synthesis, Structural Characterization and Antibacterial Activity of Cotton Fabric Modified with a Hydrogel Containing Barium Hexaferrite Nanoparticles. *J. Mol. Struct.* **2017,** *1127*, 74–80.
53. Suleiman, M.; Masri, M. A.; Ali, A. A.; Aref, D.; Saadeddin, A. H. I.; Warad, I. Synthesis of Nano-sized Sulfur Nanoparticles and Their Antibacterial Activities. *J. Mater. Environ. Sci.* **2015,** *6*, 513–518.
54. Sun, Y.; Yang, Q; Wang, H. Synthesis and Characterization of Nanodiamond Reinforced Chitosan for Bone Tissue Engineering. *J. Funct. Biomater.* **2016,** *7* (27), 1–16.
55. Tank, K. P.; Chudasama, K. S.; Thaker, V. S.; Joshi M. J. Cobalt-doped Nanohydroxyapatite: Synthesis, Characterization, Antimicrobial and Hemolytic Studies. *J. Nanopart. Res.* **2013,** *15,* 1–11.
56. Tikhomirov, G.; Petersen, P.; Qian, L. Programmable Disorder in Random DNA Tilings. *Nature Nanotechnol.* **2017,** *12*, 251–259.
57. Umemura, K.; Izumi, K.; Oura, S. Probe Microscopic Studies of DNA Molecules on Carbon Nanotubes. *Nanomaterials* **2016,** *6* (180), 1–14.
58. Vahvaselka, K. S. Temperature Dependence of the Liquid Structure of Ga. *Physica Scripta* **1981,** *22*, 647–652.
59. Van Grieken, R.; Markowicz, A. (Eds.). *Handbook of X-ray Spectrometry*; CRC Press, 2001, 1016.
60. Von Ardenne, M. The Electron Microscope: Theoretical Basics (*Das Elektronen-Rastermikroskop, Theoretische Grundlagen*). *Z. Tech. Phys.* **1938,** 553–572.
61. Von Ardenne, M. The Electron Scanning Microscope: Practical Design (*Das Elektronen-Rastermikroskop. Praktische Ausführung*). *Z. Technical Phys.* **1938,** 407–416.
62. Warren, B. E. X Ray Diffraction. *Acta Crystallographica* (New York: Addison Wesley) **1968,** 1–14.
63. Watson, J. *The Double Helix: A Personal Account of the Discovery of the Double Helix*; Weidenfeld & Nicolson: New York, Atheneum and London, 1968, pp 1–10.
64. Wells, O. C. Correction of Errors in Electron Stereomicroscopy. *Br. J. Appl. Phys.* **1960,** *11* (5), 119–201.
65. Wilson, N. R.; Macpherson, J. V. Carbon Nanotube Tips for Atomic Force Microscopy. *Nature Nanotechnol.* **2009,** *4*, 483–491.

66. Wutscher, T.; Niebauer, J.; Giessibl, F. J. Scanning Probe Microscope Simulator for the Assessment of Noise in Scanning Probe Microscopy Controllers. *Rev. Sci. Instrum.* **2013,** *84*, 1–4.
67. Yamashita, M.; Yoshida, M.; Matsuo, M.; Sato, S.; Yamamoto, H. Observations of Wood Cell Walls With a Scanning Probe Microscope. *Mater. Sci. Appl.* **2016,** *7*, 644–653.
68. Yang, H. S.; Holloway, P. H.; Ratna. B. B. Photo Luminescent and Electroluminescent Properties of Mn-doped ZnS Nanocrystals. *J. Appl. Phys.* **2003,** *93*, 586–592.
69. Yang, H.; Huang, C.; Su, X.; Tang, A. Microwave-assisted Synthesis and Luminescent Properties of Pure and Doped ZnS Nanoparticle. *J. Alloys Compd.* **2005,** *402* (1–2), 274–277.
70. Yin, M.; Yin, Y.; Han, Y.; Dai, H.; Li, S. Effects of Uptake of Hydroxyapatite Nanoparticles into Hepatoma Cells on Cell Adhesion and Proliferation. *J. Nanomater.* **2014,** 1–7.
71. Zhang, S.; Meng, F. L.; Wu, Q.; Liu, F. L.; Gao, H.; Zhang, M.; Deng, C. Synthesis and Characterization of $LiMnPO_4$ Nanoparticles Prepared by a Citric Acid Assisted Sol–gel Method. *Int. J. Electrochem. Sci.* **2013,** *8*, 6603–6609.
72. Zheng, H.; Meng, Y. S.; Zhu, Y.; Frontiers of *in situ* Electron Microscopy. *Mrs Bull.* 2015, *40*, 12–18.
73. Zhong, J.; Yan, J. Seeing is Believing: Atomic Force Microscopy Imaging for Nanomaterial Research. *RSC Adv.* **2016,** *6*, 1103–1121.
74. Zhong, Z. Y.; Yan, H. Y.; Jun, Z.; Hong, Z. S.; You, L. Z.; Kechao, Z. Characteristics of Functionalized Nanohydroxyapatite and Internalization by Human Epithelial Cell. *Nanoscale Res. Lett.* **2011,** *6*, 1–8.
75. Zworykin, V. A.; Hillier, J.; Snyder, R. L. A Scanning Electron Microscope. *ASTM Bull.* **1942,** *117*, 15–23.

PART III

Potential Applications of Nanotechnology in Food Safety

CHAPTER 6

ROLE OF ENCAPSULATION IN FOOD SYSTEMS: A REVIEW

FARHAN SAEED, HUMA BADER-UL-AIN, MUHAMMAD AFZAAL, NAZIR AHMAD, MUNAWAR ABBAS, and HAFIZ ANSAR RASUL SULERIA*

*Corresponding author. E-mail: hafiz.suleria@uqconnect.edu.au

ABSTRACT

Encapsulation is a process to entrap solids, liquids, and sometimes gases within a thin coating. Encapsulation techniques are primarily utilized to deliver certain functional ingredients for example, antimicrobials, antioxidants, dyes, vitamins, hormones, and drugs. In this manuscript, microencapsulation techniques that is, complex coacervation, polymer–polymer incompatibility, spray-drying, and centrifugal extrusions are primarily focused. Instead, nanotechnology involves enclosing a bioactive ingredient in nano-scale capsule. Nanoencapsulation holds potential for its application in nutritional additives and specialized nano-packagings. It has vast importance for its intentional creation, manipulation, and characterization. Certain toxicological issues associated with the consumption of nano-particles exist in food formulations. Conclusively, the contemporary chapter focuses on the application of encapsulation techniques in pharmaceutical and nutraceutical industry with the key purpose to avoid the drug and nutrient loss during the passage in digestive tract. Yet the investigations are required to explore the encapsulation techniques with respect to their losses and toxicological aspects.

6.1 INTRODUCTION

Principally, microencapsulation is related to holding an interior material, for example, solid, liquid, or even gas, within a shielding layer.

The resulting microcapsules increase the effectiveness of the enveloped compound and offer a broad range of applications in different industries. Microcapsules exist in different shapes, depending upon their sizes and diameter that is, simple, irregular, multi-core, multi-wall, and matrix. For example, in pharmaceutical industry, this technique has been applied for the preparation of "state of the art capsule," systems for controlled release of drugs.[26] In food processing, this technique had been previously applied to combat the intolerable flavor and taste of a variety of food ingredients, and to convert liquids into solids. In the modern era, the food industry predicts more and more intricate characteristics from food constituents which can easily be attained through microencapsulation. Moreover, the perception of controlled release of encapsulated ingredient at the correct place and time is becoming more and more attractive. This character can increase the effectiveness of food additives thus assure optimal dose level.[28]

Microencapsulation had been explored widely in the fields of biomedicine and biopharmaceutics for rehabilitation of cells to the transportation of drug/medicines. Unique properties of encapsulation have made it appropriate for food industry, particularly for the development of functional foods and nutraceuticals against various ailments. With the passage of time, the encapsulation of bioactive molecules with certain benefits, for example, antioxidants and probiotics, is escalating.[8] Apart from food and pharmaceutics, microcapsules also find numerous applications in other industries like cosmetics, textile industry, and agriculture.[39] Various types of organic (polysaccharides, lipids, proteins, polymers, etc.) and inorganic materials have been used as coating material for encapsulation. Similarly, the selection becomes even more complex due to subsequent processing and storage conditions. In general, providing a good protection to the internal coated material, the coating material should have flexibility for application in several microencapsulation techniques. However, not all techniques are suitable for all types of ingredients. Besides, the nature of coating and to be coated material, the flexibility and cost of operation are among the decisive factors to be considered for the selection of a microencapsulation technique. This chapter focuses on available microencapsulation techniques in relation to their potential applications in the domain of food science and technology.

6.2 MICROENCAPSULATION TECHNIQUES

Microencapsulation techniques can be categorized into two classes, that is, chemical processes and physical/mechanical processes. In first category (chemical processes), the microcapsules are produced in a tank/reactor containing liquid (as in a chemical process). Most commonly employed techniques included in this category are five different classes. In the second category (physical/ mechanical processes), a gas phase is employed for encapsulation and various commercially available equipment/devices are used in a physical process. Spray drying, fluidized bed, centrifugal extrusion, co-extrusion, and submerged nozzle extrusion are some commonly used techniques in this category.[16] These techniques have been described one by one in coming sub-sections. To give an overview, these techniques are listed in Table 6.1 together with their important process steps, loading efficiency, particle size range, and specific advantages.

TABLE 6.1 Encapsulation Techniques and Their Advantages.

Encapsulation technique	Application	Reference
Centrifugal extrusion	Ingredients protection techniques which offers a process for formulation	[36]
Complex coacervation	Water purification, adhesives, coatings, and biotechnology	[33]
Interfacial polymerization and in situ polymerization	Rapid development of capsule	[10]
Polymer–polymer incompatibility	Protect insoluble material	[11]
Spray drying process	To entrap the flavors and related compounds	[20]

6.2.1 CHEMICAL PROCESSES

6.2.1.1 COMPLEX COACERVATION

This technique combines both electron attracting and proton attracting hydrophilic polymers for the formation of liquid-rich polymers. This results in development of different of totally different and unique coacervate of polymers and water. A droplet of microcapsule of the material being entrapped is instinctively coated with coacervate thin film on the dispersion of water-insoluble core material.[14] The liquid film is then solidified (through crosslinking or desolation techniques) to attain harvestable capsules. This technique is used to entrap the wide range of hydrophobic or water-repelling compounds[31] The size of capsules produced by this technique generally ranges from 20 to 1000 μm. However, capsules out of this range can also be prepared.

6.2.1.2 POLYMER–POLYMER INCOMPATIBILITY

Two chemically dissimilar polymers, which have ability to dissolve in an ordinary solvent, are mixed together in this technique. Consequently, these essential chemicals react with each other and develop two distinct liquid phases with each phase containing predominantly one polymer. When an insoluble core material is dispersed in such a system, it gets coated spontaneously with a thin layer of the compatible polymer (designed to be the core–shell material). A small amount of the incompatible polymer phase can be entrapped in the final capsules as an impurity. The whole approach is generally carried out in organic solvents that encapsulate the solids with restricted degree of water solubility.[37]

6.2.1.3 INTERFACIAL POLYMERIZATION AND IN SITU POLYMERIZATION

In interfacial polymerization, a capsule shell is developed on exterior of tiny drops of different coated matter by rapid by the collection of chemically active mono-molecules. In the liquid core material, a versatile functional singlet is liquefied and the resultant. A co-reactant, generally a multi-functional amine is then incorporated to the aqueous phase. Finally, capsule generation occurs by rapid polymerization reaction at the crossing point. This technique can be applied for both liquids and solids but the polymerization chemistry is characteristically different.[49,36]

6.2.1.4 SOLVENT EXTRACTION/EVAPORATION

An aqueous solution of core material is added to the organic phase, containing a polymer solution in solvents (e.g., dichloromethane and chloroform). This emulsion is then dispersed in aqueous phase containing emulsifier to make water-in-oil-in-water (w/o/w) emulsion. The process is suitable for encapsulation of both the hydrophobic (e.g., lidocaine and progesterone) and hydrophilic (e.g., proteins and peptides) materials.

6.2.2 PHYSICAL PROCESSES

6.2.2.1 SPRAY DRYING

Core particles, dispersed in a polymer solution are sprayed into a hot chamber and subsequently evaporation of the solvent results in solidification of the shell material on core particles.[57] Typical shell materials take an account of gum acacia, maltodextrins.[7] However, strength of encapsulated ingredients can be enhanced by incorporation of a minute quantity of some hydro-colloids having stumpy solubility.[32] It is important to address that hydrophobically-modified octyl-substituted starches can be used as shell material for encapsulation of almost 50% flavor oils. A variety of other innate gums have also been discussed regarding their emulsification and shell characteristics. For example, the mesquite gum has enhanced emulsions stability and superior encapsulation efficiency in contrast to gum acacia.[21]

6.2.2.2 EXTRUSION

the extrusion microencapsulation was early developed by Izquierdo-Barrientos et al.[23] and was refined by improving basis by Izquierdo-Barrientos et al.[24] This technique can be used exclusively for the volatile flavors encapsulated in glassy carbohydrate matrices. The glycerol, potato starch and water are subjected to the process of low temperature and then gelatinized in extruder with 100°C. The mass is introduced to low temperature of 50°C at last barrel of extruder where bioactive formulation is injected. Finally, the extruded threads are cut into required size and dried[53] As a result of hydrophilic matric, permitting very limited gases to interior viscous material, the shelf life of extruded capsule is five times more in contrast to spray dried microcapsules. The glassy carbohydrate matrics have proven to be better impermeable barrier for the flavors and oils that are prone to oxidation. The citrus oil and volatile flavors are encapsulated in the carbohydrate shell material like fluidize bed coating and extrusion is the most compatible technique for such shell materials.

6.2.2.3 CENTRIFUGAL EXTRUSION

In centrifugal extrusion, immiscible core and shell materials are pushed through a spinning, two-fluid nozzle, resulting in the formation of an unbroken rod. Adequate capsules are formed when the extruded rod breaks up into droplet.[33]

6.2.2.4 CO-EXTRUSION

In co-extrusion, a double fluid stream having liquid core and shell materials is siphoned from the concentric tubes and then droplets are formed in the way of vibration. This operation can be performed by using numerous types of extrusion nozzles.[9]

6.2.2.5 SUBMERGED NOZZLE EXTRUSION AND FLUIDIZED BED

In submerged nozzle extrusion, the core material is extruded with the water-soluble polymer (e.g., gelatin) through a two-fluid nozzle resulting in the entrapment of the core material with the polymer. The capsules are then cooled to gel followed by drying. This is one of the most efficient techniques for the formation of solid capsules.[56]

6.3 APPLICATIONS OF MICROENCAPSULATION IN FOOD

Variety of coating materials is used for encapsulation of food ingredients. In foods, microencapsulation of functional ingredients is performed either to improve or enhance nutritional value or as a preservative.[50] In the following subsections, some of the applications of encapsulation of food ingredients are discussed.

6.3.1 ENCAPSULATION OF FLAVORS

Encapsulation of flavors is one of the most important applications of micro-encapsulation in food industry.[15] Flavors are encapsulated for their excellent chemical stability, retention of their volatiles, undesirable interaction with food components, and restricted release.[15,30] Countless studies have been done on the effect of wall material composition and the working conditions for the preservation and restricted discharge of encapsulated flavors.[4] Currently, the wall systems of skimmed milk powder and whey protein concentrate are being used for encapsulation of flavors of oregano, citronella, and marjoram[15,40] found that microencapsulation of flavors in two-layered emulsions (stabilized by protein isolate/pectin complex) increases the retention of volatile compounds during spray drying. Carnauba wax has been reported as an attractive encapsulation material for liquid flavors and aroma owing to the easy handling of wax-aroma forms.[48]

Monoterpene rich essential oils are generally used as flavor ingredients and their high polarity can provide superior solubility to the compound.[11] Volatiles transmission via wall matrix during the process of SD can be elaborated by a mean of discerning permeability. This presumes principally that the preservation of the compound properties is a purpose of the foundation molar volume.[44] In another study, SD technique was lucratively used to encapsulate sumac flavor in wall material sodium chloride.[41]

6.3.2 ENCAPSULATION OF COLORS

There has been an increasing interest in the replacement of synthetic dyes with natural food colorants due to both the legislative actions and increasing consumer concerns. These colorants obtained from natural sources can become unstable under various physical and chemical conditions and can be affected by various factors including pH, temperature, light, presence of co-pigments, enzymes, oxygen and so on. The best way to preserve them against degradation is through microencapsulation.

Food colors have been encapsulated using various techniques and wall materials. Researchers successfully encapsulated anthocyanin pigments of black carrot through spray drying using a variety of maltodextrins as wall material. Resulting pigment powder showed excellent stability when stored at 4°C. Astaxanthin (ketocarotenoid responsible for red–orange color) was microencapsulated through solvent extraction using chitosan cross-linked with glutaraldehyde. Stability of the red rose pigments encapsulated in Bee's-wax stearic acid mixture (1:1) was significantly enhanced against pH, light, and heat. In another investigation, anthocyanins extracted from Jaboticaba (a fruit native to southeastern Brazil) were encapsulated through spray drying using, maltodextrin (30%), maltodextrin + Arabic gum (5% + 25%), and Capsul + maltodextrin (25% + 5%), as wall material. Maximum desirability of the product was achieved while using maltodextrin (30%).[45]

6.3.3 ENCAPSULATION OF LIPIDS

Microencapsulation of both the essential oils and oils rich in unsaturated fatty acids has been well documented. Essential oils are those extracted from the plants and contain volatile compounds which are used for flavoring foods and drinks. Unsaturated fatty acids (USFA) are claimed to have positive effects on human health and prevent the cardiovascular diseases.[45] These oils can be easily oxidized leading to unpleasant taste and smell or can undergo degradation reactions and loss of volatiles.[2] These have claimed to reduce these oxidation and degradation reactions together with reduction in loss of volatiles.

In an attempt to encapsulate flaxseed oil through spray drying using gum Arabic as wall material, lipid oxidation was reported to reduce with increasing encapsulation efficiency. Similar effects were observed when linseed oil was encapsulated using gum Arabic, maltodextrin, methylcellulose, and whey protein isolate. Researchers used gum Arabic alone and its blends with depolymerized (radiations or ezymatically) guar gum to encapsulate mint oil.[29] It was found that the microcapsules prepared with radiation depolymerized guar gum could better retain mint oil compounds such as menthol and isomenthol during storage period of 8 weeks.[3]

6.3.4 ENCAPSULATION OF VITAMINS

Vitamins are the micronutrients essential for the normal growth and maintenance of body health.[20] It has been confirmed through numerous research investigations that water-soluble vitamins including ascorbic acid, thiamin, riboflavin, vitamin B6, and folic acid are lost during food processing. In addition to these processing losses, vitamins can also be subjected to degradation by light, heat, oxygen, and pH in the presence of metal ions like Cu^{+2}, Fe^{+2} and so on.[19] These degradations can be minimized through encapsulation of vitamins.

Among water-soluble vitamins, encapsulation of ascorbic acid (vitamin C) has been most extensively investigated. Some of the researchers have demonstrated the potential of ascorbic acid encapsulation by spray drying as a mean to enhance the retention of the vitamin C. Researchers encapsulated the ascorbic acid and they found that not only the stability of the encapsulated vitamin has been improved but it also helped to mask the acidic taste of ascorbic acid. It has been investigated the release of vitamin C from chitosan microcapsules by varying the cross-linked density through glutaraldehyde (GA) treatment. Release rates of vitamin C decreased by increasing crosslink density. Heat treatment of liquid foods containing folic acid and 5-methyltetrahydrofolic acid (MTHFA) resulted in thermal degradation of the vitamin. Microencapsulation of MTHFA in pectin: alginate (80:20) provides the stability to the vitamin especially at elevated temperatures, for example, in extrusion.[18] Microencapsulated ferrous fumarate plus ascorbic acid sprinkled on some complimentary food can help in the treatment of anemia. Other investigations related to encapsulation of water-soluble vitamins (vitamin C) can be found in Ref. [19].

Li et al.[36] entrapped vitamin D_2 in chitosan micro cores using spray drying and then microencapsulated by ethyl cellulose coating. The resulting microcapsules showed sustained release in intestinal juice. Manna et al.[38] prepared lipid microcapsules containing α-tocopherol (vitamin E) in lipid matrices through spray chilling. Encapsulation efficiency and level of retention of active compound were high with good stability over storage time and temperature. Encapsulation of vitamin A has been reported by a group of researchers through gelatin-acacia complex Coacervation. Spherical microcapsules with excellent appearance were obtained by drying the microcapsules through freeze drying.

6.3.5 ENCAPSULATION OF BIOACTIVE COMPOUNDS

These are the compounds that are beneficial for consumer beyond the nutrition and they express themselves in small quantities in foods. Addition of bioactive compounds in functional foods encounters plenty of challenges related to their stability during processing and storage, and undesirable interactions with carrier foods.[46] These challenges can be overcome through encapsulation of these bioactive compounds. Bioactive compounds extracted from cactus pear (*Opuntia ficus-indica*), when encapsulated with maltodextrin, showed slow degradation at storage temperature of 60°C. Microencapsulation of polyphenolic extracts from bayberry using ethyl cellulose as a coating material not only protected the antioxidant capacity but also improved the storage stability under adverse environmental conditions. Blackcurrant polyphenols (anthocyanins and flavonols) when microencapsulated show high storage stability at 8°C and 25°C. Researchers reported that the microencapsulation of bioactive compounds extracted from red pepper results in improved thermal (200°C) and storage stability.[60]

6.3.6 ENCAPSULATION OF PROBIOTICS

Some other techniques can also be combined to improve the survival of these beneficial organisms (probiotics), that is, strain selection in adverse environment, addition of prebiotics and buffering of yoghurt mixes with whey proteins.[10,51] Certain strains of lactic acid bacteria, such as *Bifidobacterium bifidum* and *Lactobacillus gasseri*; prevent some diseases linked to the gastrointestinal tract.[1,50] The microencapsulation techniques improve the viability of *Bifidobacterium pseudolongum* in the gastrointestinal tract when incorporated in dairy products. Microencapsulation can help to maintain the survival of *Bifidiobacterium* BB-12 during stress conditions and the resistance of encapsulated *Bifidiobacterium* bacteria can be increased by the addition of prebiotic inulin and oligofructose-enriched inuline. Probiotics encapsulated in solid lipid microcapsules produced through spray chilling can efficiently protect them against passage through gastric and intestinal fluids. Microcapsules prepared with aluminum carboxymethyl cellulose-rice bran composites have potential to protect the probiotics against heat.[59,34]

6.3.7 ENCAPSULATION OF MISCELLANEOUS FOOD INGREDIENTS AND NANO-ENCAPSULATION

Spray-drying is believed to be a most valuable technique for iodine encapsulation and the most excellent consequences have been attained by dextrin when use as an encapsulating agent. The resultant microparticles comprise 1% iodine, which is stable for a period of up to 12 months below a temperature of 40°C and at elevated relative humidity. Spray dried microencapsulation of bixin with gum Arabic can provide higher stability to photodegradation.[40]

It has been studied that the Maillard reaction products (MRPs) exhibit antioxidant characteristics along with the formation of stable and vigorous shell around the oil phase. Aqueous two-phase systems are (ATPS) product of phase severance in a mixture of soluble polymers in ordinary solvent as a result of squat entropy of mixing of polymer mixtures. This can be used to develop a double-encapsulated ingredient in a single spray drying step. The proteins exhibited the detachment between the two phases but the entire cells have a tendency to contemplate in one of the polymer phases, which make them perfect contender for ATPS spray drying. The microcapsule configuration can be controlled by adjustment of two polymer ratio and concentration.[27]

There is an increasing interest in the preparation of nanoparticles/nanocapsules in the area of food processing due to their easy incorporation in food systems and increased bioavailability. The preparation of such nanoparticles can be either through downsizing of capsules produced by previously described "classic" microencapsulation techniques or by using some new techniques.[8] Some examples can be solvent evaporation, salting out, coacervation, ionic gelation, and polymerization.

Various active ingredients like vitamins, fatty acids, co-enzymes, isoflavones, flavonoids, carotenoids, phyto-extracts, essential oils, food colorants, and bioactive molecules are encapsulated at nanoscale for their commercial applications in a variety of products including beverages, meats, cheese, and other foods preparation. Liposomal entrapment of vitamin C retained its antioxidant activity for 50 days of refrigerated storage as compared to un-encapsulated vitamin which lost all of its activity[52] Similarly, encapsulation of antioxidant glutathione (GSH) in nanoliposome delivery system increases the protection and residency period of GSH in the body. Nanosized milk protein "casein" has been employed as a delivery system for vitamin D_2. Beta carotene is encapsulated in maize protein zein through electrospinning technique for its nutraceutical purposes.[6] Thus, in future the success of nanoencapsulation has a great glimpse target delivery of nutraceutical and function foods.

6.4 SUMMARY

Microencapsulation is a fruitful expertise for viable uses in the medicinal and agro-alimentary industries. Generally, the techniques remain primarily different, especially in the field of practical paint where the chance of getting functional surface with microcapsules is almost indefinite. The technology increases the functions of different properties combinations of various materials that functioning is impossible through other technologies. Regarding the food and pharmaceutical industry, microcapsules have been proven as a unique carrier system for target delivery of desired molecules. There are still many challenges of microencapsulation besides its acceptable evaluation. Therefore, in future, the in-depth research is needed for its safety and efficiency from biological and technological aspects. However, the slogan of encapsulation technology can become the rising trend in the 21st century demands the coherent and systematic studies involving researchers from various domains to bring the meticulousness of the issue.

KEYWORDS

- emulsion
- encapsulation
- microencapsulation
- nanoencapsulation
- spray drying

REFERENCES

1. Abed, N.; Couvreur, P. Nanocarriers for Antibiotics: A Promising Solution to Treat Intracellular Bacterial Infections. *Int. J. Antimicrob. Agents* **2014,** *43*, 485–496.
2. Adabi, M.; Saber, R.; Faridi-Majidi, R.; Faridbod, F. Performance of Electrodes Synthesized with Polyacrylonitrile Based Carbon Nanofibers for Application in Electrochemical Sensors and Biosensors. *Mater. Sci. Eng.* **2015,** *48*, 673–678.
3. Ahmed, E. M. Hydrogel: Preparation, Characterization, and Applications: A Review. *J. Adv. Res.* **2015,** *6*, 105–121.
4. Arrazola, G.; Herazo, I.; Alvis, A. Micro Encapsulation of Artificial Tangerine Eggplant Anthocyanins Obtained by the Spray Drying Technique. M.Sc. Thesis, Polytechnic University of Valencia, Spain, 2014, pp 152–162.
5. Asensio, C. M.; Grosso, N. R.; Juliani, H. R. Quality Preservation of Organic Cottage Cheese Using Oregano Essential Oils. *LWT-Food Sci. Technol.* **2015,** *60*, 664–71.
6. A-sun, K.; Thumthanaruk, B.; Lekhavat, S.; Jumnongpon, R. Effect of Spray Drying Conditions on Physical Characteristics of Coconut Sugar Powder. *Int. Food Res. J.* **2011,** *23* (3), 1315–1319.
7. Azagheswari, Kuriokase, B.; Padma, S.; Priya, S. P. A Review on Microcapsules. *Global J. Pharmacol.* **2015,** *9* (1), 28–39.
8. Barrasso, D.; Eppinger, T.; Pereira, F. E.; Aglave, R.; Debus, K.; Bermingham, S. K.; Ramachandran, R. A. Multi-scale, Mechanistic Model of a Wet Granulation Process Using a Novel Bi-directional PBM–DEM Coupling Algorithm. *Chem. Eng. Sci.* **2015,** *123*, 500–513.
9. Battal, D.; Celik, A.; Güler, G.; Aktas, A.; Yildirimcan, S.; Ocakoglu, K.; Çomeleko, glu, U. SiO2 Nanoparticule-induced Size-dependent Genotoxicity—An In vitro Study Using Sister Chromatid Exchange, Micronucleus and Comet Assay. *Drug Chem. Toxicol.* **2015,** *38*, 196–204.
10. Bazzarelli, F.; Piacentini, E.; Poerio, T.; Mazzei, R.; Cassano, A.; Giorno, L. Advances in Membrane Operations for Water Purification and Biophenol Recovery/Valorization from OMWWs. *J. Membr. Sci.* **2016,** *497*, 402–409.
11. Bhargava, K.; Conti, D. S.; da-Rocha, S. R. P.; Zhang, Y. F. Application of an Oregano Oil Nanoemulsion to the Control of Foodborne Bacteria on Fresh Lettuce. *Food Microbiol.* **2015,** *47*, 69–73.
12. Bhusari, S. N.; Muzaffat, K.; Kumar, P. Effect of Carrier Agents on Physical and Microstructural Properties of Spray Dried Tamarind Pulp Powder. *Powder Technol.* **2014,** *266*, 354–364.

13. Capablanca, L.; Bonet, M.; Bou, E.; Ferrándiz, M.; Franco, E.; Dolçà, C. *Aplicación de técnicas Biotecnológicas y de Microencapsulación Para la Funcionalización de Agrotextiles*. Presented at the II Congreso I+D+i, Campus de Alcoy, Creando Sinergias: Alcoy, 2014.
14. Castro, N.; Durrieu, V.; Raynaud, C.; Rouilly, A.; Rigal, L.; Quellet, C. Melt Extrusion Encapsulation of Flavors: A Review. *J. Polym. Rev.* **2016,** *56* (1), 151–159.
15. Comunian, T. A.; Abbaspourrad, R.; Favaro-Trindade, C. S.; Weitz, D. A. Fabrication of Solid Lipid Microcapsules Containing Ascorbic Acid Using a Microfluidic Technique. *Food Chem.* **2014,** *152*, 271–275.
16. Fernandes, R. V. B.; Marques, G. R.; Borges, S. V.; Botrel, D. A. Effect of Solids Content and Oil Load on the Microencapsulation Process of Rosemary Essential Oil. *Indus. Crops Prod.* **2014,** *58*, 173–181.
17. Ghayempour, S.; Montazer, M.; Mahmoudi, Rad, M. Tragacanth Gum as a Natural Polymeric Wall for Producing Antimicrobial Nanocapsules Loaded with Plant Extract. *Int. J. Biol. Macromol.* **2015,** *81*, 514–520.
18. Ghayempour, S.; Montazer, M.; Mahmoudi, Rad, M. Tragacanth Gum Biopolymer as Reducing and Stabilizing Agent in Biosonosynthesis of Urchin-like ZnO Nanorod Arrays: A Low Cytotoxic Photocatalyst With Antibacterial and Antifungal Properties. *Carbohydr. Polym.* **2016,** *136*, 232–241. doi:10.1016/j.carbpol.2015.09.001
19. Ghayempour, S.; Mortazavi, S. M. Microwave Curing for Applying Polymeric Nanocapsules Containing Essential Oils on Cotton Fabric to Produce Antimicrobial and Fragrant Textiles. *Cellulose* ***2015,*** *22*, 4065–4075.
20. Giro-Paloma, J.; Martinez, M.; Cabeza, L. F.; Fernandez, A. I. Types, Methods, Techniques, and Applications for Microencapsulated Phase Change Materials (MPCM): A Review. *Renew. Sustain. Energy Rev.* **2016,** *53*, 1059–1075.
21. Grimard, J.; Dewasme, L.; Wouwer, A. V. A Review of Dynamic Models of Hot-Melt Extrusion. *Processes* **2016,** *4*, 19. doi:10.3390/pr4020019
22. Hassabo, A. G.; Mohamed, A. L.; Wang, H.; Popescu, C.; Moller, M. Metal Salts Rented in Silica Microcapsules as Inorganic Phase Change Materials for Textile Usage. *Inorg. Chem. Ind. J.* **2015,** *10*, 59–65.
23. Izquierdo-Barrientos, M. A.; Sobrino, C.; Almendros-Ibáñez, J. A. Experimental Heat Transfer Coefficients Between a Surface and Fixed and Fluidized Beds with PCM. *Appl. Therm. Eng.* **2015,** *78*, 373–379.
24. Izquierdo-Barrientos, M. A.; Sobrino, C.; Almendros-Ibáñez, J. A. Energy Storage with PCM in Fluidized Beds: Modeling and Experiments. *Chem. Eng. J.* **2015,** *264,* 497–505.
25. Jin, H.; Mangun, C. L.; Griffin, A. S.; Moore, J. S.; Sottos, N. R.; White, S. R. Thermally Stable Autonomic Healing in Epoxy using a Dual-Microcapsule System. *Adv. Mater.* **2014,** *26*, 282–287.
26. Jouki, M.; Yazdi, F. T.; Mortazavi, S. A.; Koocheki, A. *Food Hydrocolloid*, **2014,** *36*, 9.
27. Kang, S., Baginska, M., White, S., Sottos, N. R. Core–Shell Polymeric Microcapsules with Superior Thermal and Solvent Stability, *ACS Appl. Mater. Interfaces* **2015**. DOI: 10.1021/acsami.5b02169
28. Karimi, M. A.; Pourhakkak, P.; Adabi, M.; Firoozi, P.; Adabi, M.; Naghibzadeh, M. Using an Artificial Neural Network for the Evaluation of the Parameters Controlling PVA/chitosan Electrospun Nanofibers Diameter. *e-Polymers* **2015,** *15* (2), 127–138.
29. Keshani, S.; Wan, W. R.; Nourouzi, M. M.; Namvar, F.; Ghasemi, M. Spray Drying: An Overview on Wall Deposition, Process and Modeling. *J. Food Eng.* **2015,** *146*, 152–162.

30. Kha, T. C.; Nguyen, M. H.; Roach, P. D.; Stathopoulos, C. E. Microencapsulation of Gac Oil: Optimization of Spray Drying Conditions Using Response Surface Methodology. *Powder Technol.* **2014,** *264*, 298–309.
31. Khadiran, T.; Hussein, M. Z.; Zainal, Z.; Rusli, R. Encapsulation Techniques for Organic Phase Change Materials as Thermal Energy Storage Medium: A Review. *Solar Energy Mater. Solar Cells* **2015,** *143*, 78–98.
32. Kumar, A.; Vercruysse, J.; Vanhoorne, V.; Toiviainen, M.; Panouillot, P. E.; Juuti, M.; Vervaet, C.; Remon, J. P.; Gernaey, K. V.; De-Beer, T. Conceptual Framework for Model-based Analysis of Residence Time Distribution in Twin-screw Granulation. *Eur. J. Pharm. Sci.* **2015,** *71*, 25–34.
33. Li, P.; Yang, Z.; Wang, Y.; Peng, Z.; Li, S.; Kong, L.; Wang, Q. Microencapsulation of Coupled Folate and Chitosan Nanoparticles for Targeted Delivery of Combination Drugs to Colon. *J. Microencapsul.* **2015,** *32*, 40–45.
34. Li, Z. et al. Sonochemical synthesis of hydrophilic drug loaded multifunctional bovine serum albumin nanocapsules. *ACS Appl. Mater. Interfaces* **2015,** *7*, 19390–19397.
35. Liu, C.; Rao, Z.; Zhao, J.; Huo, Y.; Li, Y. Review on Nanoencapsulated Phase Change Materials: Preparation, Characterization and Heat Transfer Enhancement. *Nano Energy* **2015,** *13*, 814–826.
36. Malleswari, K.; Desi, R. R. B.; Swathi, M. Microencapsulation: A Review A Novel Approach In Drug Delivery. *Eur. J. Pharm. Med. Res.* **2016,** *3* (6), 186–194.
37. Manna, J.; Goswami, S.; Shilpa, N.; Sahu, N.; Rana, R. K. Biomimetic Method to Assemble Nanostructured Ag@ZnO on Cotton Fabrics: Application as Self-cleaning Flexible Materials with Visible-light Photocatalysis and Antibacterial Activities. *ACS Appl. Mater Interfaces* **2015,** *7*, 8076–8082.
38. Marcela, F.; Lucía, C.; Esther, F.; Elena, M. Microencapsulation of L-Ascorbic Acid by Spray Drying Using Sodium Alginate as Wall Material. *J. Encapsul. Adsorp. Sci.* **2016,** *6*, 1–8.
39. Martins, M. I.; Barreiro, M. F.; Coelho, M.; Rodrigues, A. E. *Chemical Eng. J.* **2014,** *245*, 191.
40. McClements, D. J. Encapsulation, Protection, and Release of Hydrophilic Active Components: Potential and Limitations of Colloidal Delivery Systems. *Adv. Colloid Interface Sci.* **2015,** *219*, 27–53.
41. Montserrat-dela, Paz, S.; Fernandez-Arche, M. A.; Bermudez, B.; Garcıa-Gimenez, M. D. The Sterols Isolated From Evening Primrose Oil Inhibit Human Colon Adenocarcinoma Cell Proliferation and Induce Cell Cycle Arrest Through Upregulation of LXR. *J. Funct. Foods* **2015,** *12*, 64–69.
42. Muriel-Galet, V.; Cran, M. J.; Bigger, S. W.; Hernandez-Munoz, P.; Gavara, R. Antioxidant and Antimicrobial Properties of Ethylene Vinyl Alcohol Copolymer Films Based on the Release of Oregano Essential Oil and Green Tea Extract Components. *J. Food Eng.* **2015,** *149*, 9–16.
43. Navarro, M.; Fiore, A.; Fogliano, V.; Morales, F. J. Carbonyl Trapping and Antiglycative Activities of Olive Oil Mill Wastewater. *Food Funct.* **2015,** *11*, 574–583.
44. Pinho, E.; Soares, G.; Henriques, M. Cyclodextrin Modulation of Gallic Acid in vitro Antibacterial Activity. *J. Inclus. Phenomena Macrocyc. Chem.* **2015,** *81*, 205–214.
45. Rahmani, V.; Shams, K.; Rahmani, H. Nanoencapsulation of Insulin Using Blends of Biodegradable Polymers and *In Vitro* Controlled Release of Insulin. *Chem. Eng. Proc. Technol.* **2015,** *6*, 228.

46. Rodriguez-Garcia, I.; Silva-Espinoza, B. A.; Ortega-Ramirez, L. A.; Leyva, J. M.; Siddiqui, M. W.; Cruz-Valenzuela, M. R.; Gonzalez-Aguilar, G. A.; Ayala-Zavala, J. F. Oregano Essential Oil as an Antimicrobial and Antioxidant Additive in Food Products. *Crit. Rev. Food Sci. Nutr.* **2015**. http://dx.doi.org/10.1080/10408398.2013.800832.
47. Shen, Q.; Quek, S. Y. Microencapsulation of Astaxanthin with Blends of Milk Protein and Fiber by Spray Drying. *J. Food Eng.* **2014,** *123*, 165–71.
48. Su, W.; Darkwa, J.; Kokogiannakis, G. Review of Solid–Liquid Phase Change Materials and Their Encapsulation Technologies. *Renew. Sustain. Energy Rev.* **2015,** *48*, 373–391.
49. Subramanian, S. S.; Durga, S., Loshni, K. R., Dinesh, V. K. A Review on Control of Plastic Extrusion Process. *Int. J. Adv. Res. Electr. Electron. Instrum. Eng.* **2016,** *5* (1), 167–171.
50. Tatiana, A.; Leonardo, M.; Esteban, N. L.; Ana, L. R.; de-Souza, Charlene, P. Effect of Mucoadhesive Polymers on the In Vitro Performance of Insulin-loaded Silica Nanoparticles: Interactions with Mucin and Biomembrane Models. *Eur. J. Pharm. Biopharm.* **2015,** *93*, 118–126.
51. Thangaraj, S.; Seethalakshmi, M. Microencapsulation of Vitamin C Through an Extrusion Process. *Int. J. Adv. Res. Biol. Sci.* **2014,** *7*, 16–21.
52. Thiry, J.; Krier, F.; Evrard, B. A. Review of Pharmaceutical Extrusion: Critical Process Parameters and Scaling-up. *Int. J. Pharm.* **2015,** *479*, 227–240.
53. Turchiuli, C.; Jimenez-Munguia, M. T.; Hernandez, S. M.; Cortes, F. H.; Dumoulin, E. Use of Different Supports for Oil Encapsulation in Powder by Spray Drying. *Powder Technol.* **2014,** *255*, 103–108.
54. Umesha, S. S.; Manohar, R. S.; Indiramma, A. R.; Akshitha, S.; Naidu, K. A. Enrichment of Biscuits with Microencapsulated Omega-3 Fatty Acid (Alpha-linolenic Acid) Rich Garden Cress (Lepidium sativum) Seed Oil: Physical, Sensory and Storage Quality Characteristics of Biscuits. *LWT-Food Sci. Technol.* **2015,** *62*, 654–661.
55. Ushak, S.; Gutierrez, A.; Galleguillos, H.; Fernandez, A. G.; Cabeza, L. F., Grageda, M. Thermophysical Characterization of a By-product From the Non-metallic Industry as Inorganic PCM. *Solar Energy Mater. Solar Cells* **2015,** *132*, 385–391.
56. Vidovic, S. S., Vladic, J. Z., Vaštag, Z. G., Zeković, Z. P., Popović, L. M. Maltodextrin as a Carrier of Health Benefit Compounds in Satureja montana Dry Powder Extract Obtained by Spray Drying Technique. *Powder Technol.* **2014,** *258*, 209–215.
57. Wang, X.; Yuan, Y.; Yue, T. The Application of Starch-based Ingredients in Flavor Encapsulation. *Starch* **2015,** *67* (3–4), 225–236.
58. Wang, Y.; Li, P.; Chen, L.; Gao, W.; Zeng, F.; Kong, L. X. Targeted Delivery of 5-Fluorouracil to HT-29 Cells Using High Efficient Folic Acid-Conjugated Nanoparticles. *Drug Deliv.* **2015,** *22*, 191–198.
59. Zabihi, F.; Yang, M.; Leng, Y.; Zhao, Y. PLGA–HPMC Nanoparticles Prepared by a Modified Supercritical Anti-solvent Technique for the Controlled Release of Insulin. *J. Supercrit. Fluids* **2015,** *99*, 15–22.
60. Zhao, X.; Meng, Q.; Liu, J.; Li, Q. Hydrophobic Dye/Polymer Composite Colorants Synthesized by Mini-emulsion Solvent Evaporation Technique. *Dyes Pigments* **2014,** *100*, 41–49.

CHAPTER 7

NANOSENSORS IN FOOD SAFETY: CURRENT STATUS, ROLE, AND FUTURE PERSPECTIVES

BARKHA SINGHAL* and SWATI RANA

*Corresponding author. E-mail: barkha@gbu.ac.in

ABSTRACT

Nanotechnology has revolutionized the food industry and confers paradigmatic shift through enrichment in nutritional value, innovations in nutrient delivery systems for enhancement of bioavailability, nano-formulated agrochemicals, intelligent packaging materials, and biosensors for the detection of foodborne pathogens and intoxicants. With an emphasis on toxicity and health concerns, food safety is recognized as a burning problem worldwide and responsibility for ensuring and enhancing the food safety is collectively shared by producers to consumers. The unceasingly growing public concern has incited research interest in the use of nanoparticles to develop nanosensors for detection of various contaminants and pathogens in food system. The elegant amalgamation of nanotechnology with different scientific disciplines leads to refurbished impetus to the development of a novel sensing platform for enhancement in the safety of food materials. Therefore, the present disclosure addresses the various aspects and embodiments related to the importance and types of nanosensors and its fabrication as well as comprehensive applications in food safety. Though the lucrative and innovative applications of nanotechnology have seen exponential growth in the last decade but the risk assessment and safety issues associated with this emerging technology have withdrawn ample attention among the scientific communities. Therefore, this chapter also covers the regulatory framework capable of managing any risks associated with implementation of nanoparticles in food safety. This chapter highlights the potential role of nanosensors in the field of enhancing the food safety. The food safety has been ensured by the prompt preventive action manifested by the rapid detection of food contaminants by the nanosensing devices. This is accomplished due to the remarkable properties of nanosensors in which the selective binding with the biological components with target analytes and their assessment through the transformation of detectable signals by the suitable transducers. This magical phenomenon confers various advantages of fast, sensitive, and portable detection to the nanosensors.

7.1 INTRODUCTION

The availability and consumption of the food are not only concerned with the nutritional aspects but also associated with the health of consumers. The unprecedented rise in the economy due to the rapidly growing food industries brought great concern of food safety to the forefront of public debate nowadays. The importance of food safety was firstly published in "The Jungle" by the scientist Upton Sinclair. Successively, the increased awareness, as well as the public pressure, led to the formulation of "Pure Food and Drug Act," in 1906 that ushered in a new era of the improvement and enhancement of food safety. The dramatic improvements have been seen in the area of food safety that is directly correlated to the enhancement of public health over the century. Despite the tremendous success in improvement of food safety still, it is considered as a critical issue worldwide and the responsibility for ensuring and enhancing safety is the collective efforts shared by producers to consumers.

Therefore, the rapidly growing populace and deleterious impacts of the compromised food safety procedures which have a significant effect on the public health strongly emphasized the need to develop and implement novel methods to improve food safety throughout the food chain. The popular technologies for ensuring the food safety include the chemical methods, enzyme measurements methods, immunological methods, various sensing methods, and chromatographic methodologies. These technologies require tedious and time-consuming processes for the detection but still face major challenges to provide the feasibility and functionalities needed for state-of-the-art in the area of food safety. These challenges have been overcome by the continuous inflow of new technologies and nanotechnology has emerged as a potential aid to revolutionize this ever-growing sector.

The word "nano" refers to the micro, small and atomic by nature. The thriving applications of nanotechnology have tremendous potential in various research areas as well as impacting our day-to-day life. Therefore, amazingly nowadays from scientific perspective it has become the "call of the century."[88] Recent research has highlighted nanotechnology as a potential tool to use in various food applications, including improving food textures, enhancement of the nutrients bioavailability, novel food packaging materials, increasing aroma and flavor, improvement in catalytic properties of cooking oil, and targeted crop pesticides.[70] The extended potentiality of nanotechnology has been viewed in the area of food safety that includes, enhancement of shelf-life of food, detection of pathogens and intoxicants, tracking changes in the extrinsic as well as intrinsic factors of food, sensing for various fertilizers, herbicide, pesticide, insecticide, and carcinogens. Therefore, significant contribution of nanotechnology has been envisaged for the enhancement of food quality and safety.[20]

Researchers believe that analyzing these complex processes affecting the food safety with minimized cost and time-consumption as well as non-invasive manner may become possible due to advancements in the development of nanosensors. For the accomplishment of this, the elegant amalgamation of nanosciences, electronics, computers, and biology have been used for the development of biosensors having exceptional sensing capabilities. That resulted in unprecedented spatial and temporal resolution and reliability, fast response time. The nanosensors open the plethora of opportunities for conducting the basic research as well its translational applications in the food industry. The elegant integration of this technology with lab-on-a-chip technique has facilitated the analysis of food contaminants at molecular level in the food industry.[40,47]

Therefore, the present chapter addresses the various aspects and embodiments related to the importance and types of nanosensors and its fabrication as well as comprehensive applications in food safety. Though the lucrative and innovative applications of nanotechnology have seen exponential growth in the last decade but the risk assessment and safety issues associated with this emerging technology has withdrawn ample attention among the scientific communities. Therefore, this chapter also covers the regulatory framework capable of managing any risks associated with implementation of nanoparticles (NPs) in food safety.

7.2 MECHANISM OF ACTION OF NANOSENSORS

The advent of nanosensors revolutionized the sensing field and various advances have been visualized in different applications such as in the medical diagnosis, national security, aerospace, integrated circuits, drug monitoring, drug delivery and efficacy, and many more. The rapid transformation has been seen in the food industry as well from every niche area from production, to processing, packaging, and consumption.[84] The atomic scale manufacturing of nanosensors confers remarkable ability for the measurements in nanometers. The application of nanotechnology for the advancement of biosensor leads to an efficient miniaturization of nano-biosensors as compared to conventional biosensors. The precise identification of the living cells and their parts is due to different bio-receptive layers that can either contain bio-recognition element or be made of bio-recognition elements covalently attached to the transducer.[49] The calculation and measurement of various properties such as dislocations, acceleration, concentration, displacement, volume, external forces pressure, or temperature, redox of each cell in the living body made significant contribution in sensing capability of nanosensors. Apart from that they also have capability to detect the macroscopic changes such as electrochemical, thermal, optical, and magnetic behavior and affect the technological properties of food.

7.3 FABRICATION OF NANOSENSORS

Nowadays there is a growing demand for adopting stringent food regulations and safety measures for ensuring food safety. The inadequacy of conventional methods necessitates the development of novel and sophisticated technologies to enable rapid, sensitive, and affordable analysis of food contaminants. Modern biotechnology paved the way for employing nanosensing mechanisms based on specific target-binding associated with effective transducing systems. Recent research highlighted the various ways of fabrication of nanosensors.[17] There are four methods broadly employed for the fabrication that are "top–down" lithography, "bottom–up" assembly, molecular self-assembly, and "easy-to-make" method based on nanowires.

7.3.1 "TOP–DOWN" LITHOGRAPHY

This is the simplest method of fabrication of nanosensors. This technique involves breaking down a larger block of material into smaller particles of nanosized scale by physical or chemical means. These nanosized particles, in turn, can be utilized as the components to use in specific microelectronic systems and the creation of many integrated circuits for the nanosensors. The best example has been viewed in the case of nanosensors, is the utilization of the silicon wafer as starting material. The silicon wafer was coated with photo-resistant layer and their parts were created by imparting the light using lithography. That generated parts are nanosize in dimension and further modified with other materials for the development of nanosensors.

7.3.2 "BOTTOM–UP" ASSEMBLY

The method of bottom–up assembly is a complex process but have a quality of conceptual simplicity. This method utilized atomic-scale components as starting material and processed one by one into various positions to create the sensor. For the accomplishment of this, the different methods are used for the development of nanosensors include layer-by-layer deposition, crystallization, solvent extraction/evaporation, self-assembly. Though this technique is having simple concept but found to be an extremely difficult process to be utilized as massive scale production. Till date, the atomic force microscopy has been applied for the development of these nanoparticles at laboratory scale.

7.3.3 MOLECULAR SELF-ASSEMBLY

This method is conceptualized on the basis of "growing" nanostructures and there are two ways for the development of nanosensors based on molecular self-assembly. The first method utilizes a part of previously known or even naturally formed nanostructure as the starting material and interacting with free atoms of their similar scale. The macroscale conversion of this material will be done by attaining the shape with irregular surface having the capacity of attracting more molecules and creating a larger component of the nanosensor. The method is a complex one and it involves automated assembly of the complete set of components into the finished product. The best example for this category has been visualized in the manufacturing of micro-sized computer chips and has yet to be accomplished at the nanoscale.

7.3.4 NANOWIRES

This method of sensor fabrication is based on the implications of semiconductors. For the fabrication, the top of an insulating material was doped with silicon film and then lay down patterns of lines known as masks. The location of nanowires was determined by the masks. Then the etching of silicon wire was done which is not covered by masks. The sensing devices will be developed with the modification of the nanowires that are designed to bind with the targeted biological entity. The conductivity of the nanowire has shown variation in the presence of the target analyte, which will create a signal that can be detected and analyzed. This meticulous process has an advantage of multiple nanosensors on the same chip and rendering them portable.

7.4 TYPES OF NANOSENSORS

Food safety lays an important foundation in the socio-economic perspective of the society. This encompasses manufacturing, processing, and storage of food products that can minimize the outbreak of deadly foodborne illness. The well-established instrumental protocols based on high-performance liquid chromatography (HPLC), gas chromatography (GC) have been explicitly used for detection of food contaminants but still suffers from the tedious and time-consuming preparation of sampling and analysis. Therefore, there is a pressing need for the development of those techniques that can detect the safety threats and quality attributes in the various food matrices. Nowadays the advent of microfluidics, photonics, optics, and their integration with other techniques and modified approaches including micro-processing and biosensing based technology confers paradigmatic shift in the area of food quality and safety. Nanosensors represent a sustainable alternative that may address various analytical challenges with a great potential in food safety. The leveraged growth and excitement in nanomaterial development have revolutionized the research in nanosensors field. The various multidisciplinary aspects need to be considered for the nanosensor development such as transduction, signal generation, fluidics design, surface immobilization chemistry, detection format, and data analysis that leads to the miniaturization of sensing device with less sample preparation and fast response time. The classification of nanosensors is as follows:

7.4.1 OPTICAL NANOSENSORS

In the last decade, the advancements in optical techniques gave impetus to the prolific growth of optical biosensors which gains popularity in food safety and quality evaluation. The appealing capability of detecting analytes in complex food matrices with minimized pre-treatment of samples holds great promises in the current scenario.[15] The basic principle of optical biosensors relies on binding of an analyte to the sensing layer by sorption or complex formation resulting change in the *surface characteristics* of the sensor that can be measured. The installation into the deeper part of cells with minimal alterations in the physical properties of the cell makes them very suitable devices for the comprehensive understanding of the biological process. The nanomaterials utilized for the fabrication of optical sensors such as gold NRs (nanorods), metal NPs, metal oxide NPs, and quantum dots (QDs) possess excellent optical properties and preferentially utilized as an optical label for enhancement of the sensitivity of their transducing surfaces.[28] The various optic phenomenons have been utilized for the development of sensor transducing platform. They are based on fluorescence and plasmonics. The biological activity can be monitored by sensing the optical signal generated between bio-recognition element and a functionalized nanomaterial.

7.4.1.1 FLUORESCENCE-BASED NANOSENSORS

In this sensor, the sensitivity of fluorescence is utilized for making quantitative measurements in the intracellular environment. The most prototype optical nanosensor is molecular dye probe inside a cell that consists of direct loading of fluorescent dyes into the cell. In this method, there is less physical perturbation of the cell, however, the cell sequestration, protein binding, and toxicity has been visualized through the inherent dye–cell chemical interference which makes these sensors unsuitable for the application. Therefore, freely flowing reported attached and labeled nanoparticles can be used for intracellular sensing. In the fluorescent methods, the fluorescence signal of excited molecules, that are insensitive to pH changes such as Alexa Fluor dyes, cyanines, and CdSe/ZnS quantum dots (QDs), red and infrared wavelengths dyes such as Hilyte dyes, have been used for the detection of target analyte.[58] The disadvantage associated with above-mentioned methods has been overcome by using the optical fiber probe which physically separates the cellular environment with sensing surface. These nanosensors proved as minimal invasive method to analyze important cellular processes in vivo. In the current scenario, photonic explorers for bio-analysis with biologically localized embedding (PEBBLE) is being utilized as fluorescent nanosensors. These consist of an analyte-specific dye and a reference dye inside a biologically inert matrix[14] that can be measured at nano-scale. These sensors are found to be less disruptive to the cellular environment as well as prevent encapsulation of the dyes within biological matrix.

7.4.1.2 PLASMONICS-BASED NANOSENSORS

These sensing devices are more sophisticated and sensitive than fluorescence-based sensors. The plasmonics is derived from "plasmons", means the quanta propagating through the collective motion of large numbers of electrons through longitudinal waves present in matter. The sensing surface irradiated with incident light excites the conduction of electrons in the metal and followed by excitation of surface plasmons that leads to enormous electromagnetic enhancement for highly sensitive detection of spectral signatures. They are further divided into two categories, they are (1) surface plasmon resonance (SPR) and (2) surface-enhanced Raman scattering (SERS).

7.4.1.2.1 Surface Plasmon Resonance (SPR)-Based Nanosensors

This is a technique of the measurement of fluctuations in refractive index of the surface that is mediated by the evanescent wave.[3] These waves are produced when an incident, monochromatic light beam at a special angle (a) is able to interact with free electrons (plasmons) in the metal film. The sensing devices incorporating SPR as transducing surface has an ability to measure chemical and biochemical interactions occurring at the interface between a thin gold film and a dielectric interface or transparent material. The basic design of SPR based sensor is depicted in Figure 7.1.

7.4.1.2.2 Surface-Enhanced Raman Scattering (SERS)-Based Nanosensors

SERS is a Raman spectroscopic (RS) technique in which Raman-active analyte molecules get adsorbed onto specialized metal surfaces leads to the enhanced detection by generation of Raman signals. The intriguing interest has been visualized for the application of SERS in nanosensors because there has been 10^{14}–10^{15} fold enhancement in the Raman scattering by the incorporation of nanoparticles for the single analyte. The most preferred

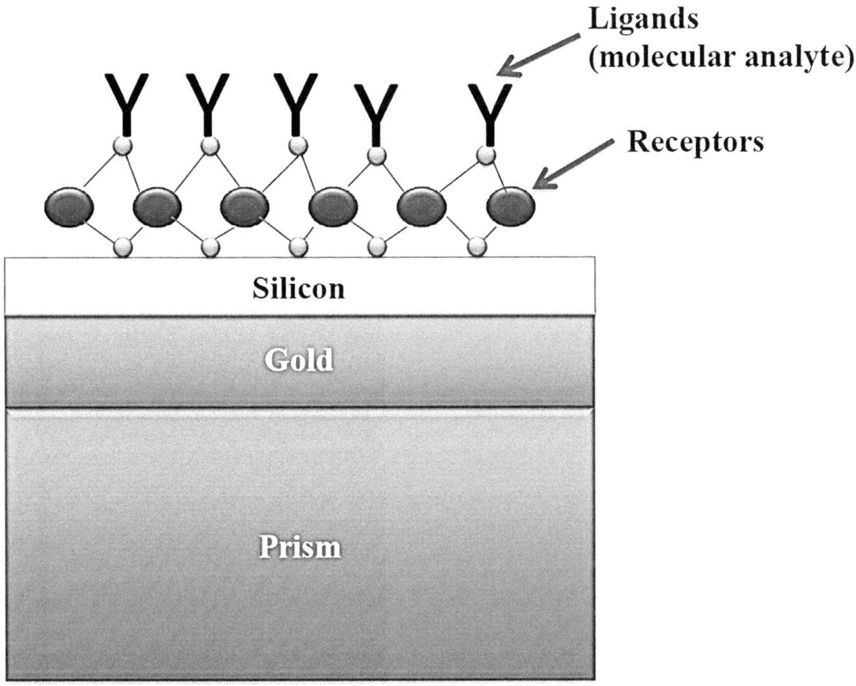

FIGURE 7.1 **(See color insert.)** Design of surface plasmon resonance (SPR) based sensor.

metal structures, such as gold or silver has proven to be a desirable platform for giving the enhanced signals at nanoscale, therefore successfully applied for characterization and detection of food-related analytes.

7.4.2 ELECTROCHEMICAL NANOSENSORS

In the current scenario, the electrochemical sensors hold a leading position among the presently available commercialized sensors constituting the nanomaterials as bio-recognition element and its analysis through the changes to the material's current, voltage, and conductivity.[18,63,65] These sensors are broadly divided into three categories are potentiometric, amperometric, and conductometric. In the potentiometric sensors, the concentration of the ionic species has been determined by the measurement of potential difference between two electrodes.[66] In the amperometric sensors, the current is measured by the oxidation or reduction of an electro-active species through the applied potential between a reference and a working electrode.[48] In the conductometric sensors, the series of frequencies have been applied to measure the conductivity of nanomaterial for the quantification of target analyte.[10] In the current scenario, the advancements in microfabrication technologies laid significant improvement in the area of fabrication of electrochemical nanobiosensors through the development of electrochemical impedance spectroscopy (EIS)[46] in the field of food safety. It consists of a measurement of the ac electrical current, $I^{\char"5E}(v)$, at a certain angular frequency, v, when a certain ac voltage, $V^{\char"5E}$ (v), is applied to the system.

7.4.3 MECHANICAL NANOSENSORS

These nanosensors are based on the measurement of mechanical forces at the molecular scale that occurs in the biological matrices. They have remarkable ability to measure the sub-cellular processes that includes various parameters like transport and affinity forces, mass changes displacements at the molecular scale. The assessment has been done by analyzing the proportionality with respect to resolution of mass in the mechanical devices with respect to the total mass of the device. These sensors can narrow down at nanoscale level, resulting in the enhancement of mass resolution. Research studies reported the various nanomaterials as a sensing platform have been utilized for the fabrication of mechanical nanosensors. The most commonly utilized nanomaterials are as follows.

7.4.3.1 CARBON NANOTUBE-BASED NANOSENSORS

The development of carbon nanotube (CNTs) leads to the substantial progress in the electrochemical sensor area and their great demand and growing interest in analytical chemistry have laid significant applications in the food analysis. CNTs consist of concentric cylinders a few nanometres in diameter and up to hundreds of micrometers in length. Recent research speculated that CNTs are more advantageous with respect to silicon-based circuits due to their attractive electrical properties. These are considered as one of the major contributing factors for the development of nanoelectromechanical systems (NEMS) which are the upcoming sensing devices integrating electrical and mechanical components with critical dimensions of 100 nm. The layout of CNTs is shown in Figure 7.2.

7.4.3.2 NANOCANTILEVER-BASED NANOSENSORS

These are one of the novel innovations of mechanical nanosensors and recognized as a "miniature diving board." Nanocantilevers work on the basis of their ability to detect biological interactions by physical and electrochemical signaling.[73] The various biological entities such as enzyme and substrate, antigen and antibody, receptor, and ligand have been successfully applied for the analysis. The nanocantilever undergoes two responses to measure interactions: firstly, due to the molecular interaction change in mass leads to shifts in resonance frequency and secondly, adsorption of the molecule generates surface stress that in turn leads to bending of the cantilever. A variety of optoelectrical techniques including, capacitance, piezoresistivity, piezoelectricity, optical beam deflection has been utilized for assessing the above-mentioned bending and resonance frequency shifts. These devices have excellent capability of recognizing pathogenic bacteria and viruses as well as proteins through their silicon-based material counterpart which are used for their designing. Currently, nanocantilevers have been explicitly utilized for detection of antibiotic residues, contaminant chemicals, toxins in food products as well as molecular interactions in the food matrices.

The conceptual designing of nanocantilever is depicted in Figure 7.3. The commercialized cantilever based sensing device has been developed by

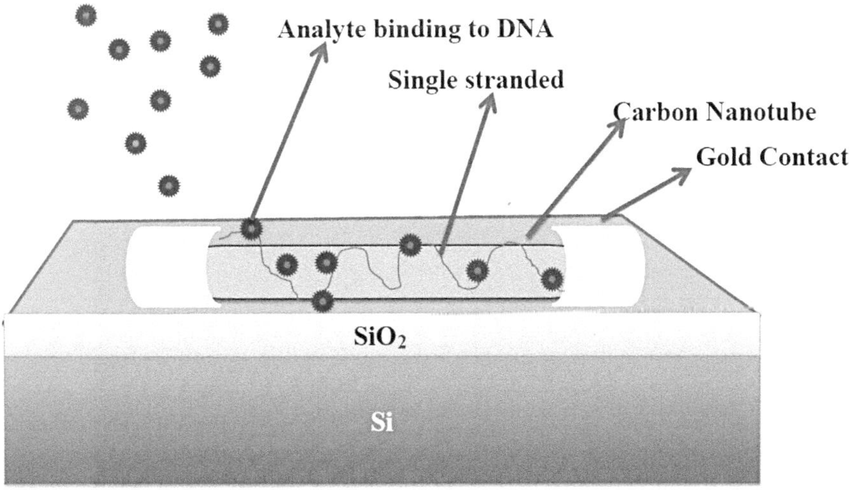

FIGURE 7.2 (See color insert.) Basic layout of CNTs based senor.

European Union-funded project called Bio-Finger, which could be used for the detection of pathogens in food and water samples as well as diagnosis of cancer.

7.4.4 ELECTRICAL NANOSENSORS

In the current scenario, electrical detection of biological species is an emerging concept with economically feasible fabrication techniques. The electrical detection is manifested by nano field-effect transistors (FETs), which have simple and direct online measurements with carrying capacities.[60] FET-based electrical nanosensors have been utilized for resistivity measurements by the binding analyte with sensor surface through nanotubes, nanoribbons, and nanowires. These devices laid superiority in terms of their cross-sectional flow of current from the nanoscale materials as compared with the planar surface. Silicon nanowires are most preferably used due to easy modification in chemical functional groups with high sensitivity.

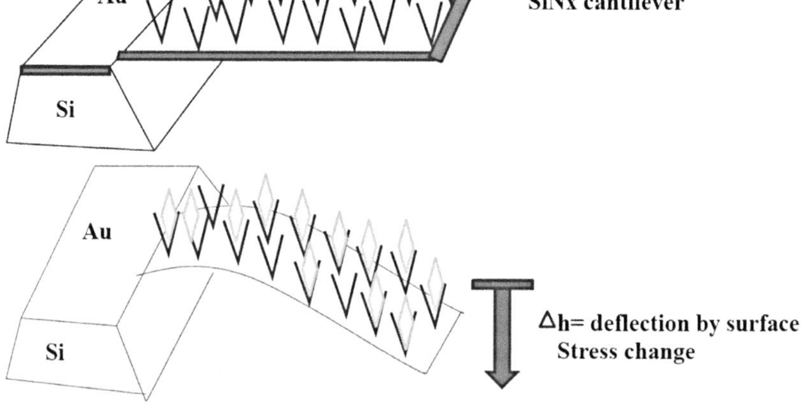

FIGURE 7.3 Design of nanocantilever-based biosensor.

7.4.4.1 NANOWIRE-BASED NANOSENSORS

Nanowires are considered as key nanomaterials because of their electrical controllability for accurate measurement, and chemical-friendly surface for various sensing applications. Silicon nanowires are quasi-one-dimensional (1D) structures with a diameter of less than 100 nm. Due to the high surface to volume silicon ratio and unique properties, these devices can outperform their traditional counterparts and confers fast and highly sensitive detection. The antibodies are attached with nanowires in a three-step process in which the introduction of terminal aldehyde groups has been done on the silicon nanowire surface followed by binding of aldehyde groups with monoclonal antibodies and thirdly the reacting with amines to block non-reacted aldehyde groups.[92] The detection is based on the conductance changes on the surface by binding of protein with their receptors. The elimination of erroneous signals, as well as differentiation of target-binding signals from the noise, have been done through differential doping of (p- and n-type) silicon nanowires.[64] These sensing devices has been explicitly utilized for the multiplexing assays (barcode method) in which the binding of selective antibodies to specific regions of magnetic (and non-magnetic) multi-metal nanowires, has been utilized for simultaneous detection of foodborne pathogens and intoxicants.

7.4.5 MAGNETIC NANOSENSORS

One of the main biological parameter known as spin–spin relaxation time is used to distinguish between tissue types in magnetic resonance imaging (MRIs) in which the alignment of unpaired nuclear spins (hydrogen atoms in water and organic compounds) occurs in the presence of magnetic field. The same principle is postulated for detection of biological interactions at molecular level with incorporation of magnetic nanoparticles as magnetic nanosensors. They work on the basis of measurement of the corresponding decrease in transverse relaxation time (T_2) of the surrounding molecules after binding with their molecular target.[74] The remarkable property of magnetic nanoparticle sensor technology permits the rapid detection of targets without extensive purification of the sample or signal amplification. The various contrast agents are available that can be used for enhancing the sensitivity as well as reduces the proton relaxation time. The examples of various contrasting agents include paramagnetic small molecule agents and super-paramagnetic iron oxide (SPION) nanoparticles. Among these, SPION are most preferred nanomaterials that can be utilized for applications in cell labeling, imaging, and food analysis due to their excellent physicochemical properties.[26,67]

7.5 CURRENT APPLICATIONS OF NANOSENSORS

7.5.1 FOR DETECTION OF FOODBORNE PATHOGENS

The detection of foodborne pathogens is an important aspect of food safety that prevents the microbiological hazards associated during the production, processing, storage, transportation, and distribution of food products as well as the economic loss of the entire nation. The conventional methods include traditional culture plating (gold standard method), polymerase chain reaction (PCR), enzyme-linked immunosorbent assay (ELISA) and immuno-chromatographic lateral flow assay (strip). The methods require more processing time and need complex nucleic acid extraction procedures and expensive infrastructure or have relatively high false positives and lack sensitivity. Therefore, there is dire need for the fast, inexpensive detection methods that can be used as "point-of-care" devices in the area of food safety.[42] It has been visualized that microbial sensors fabricated with nano-scale materials possess excellent optoelectrical properties, highly functionalized surfaces resulting in the marvelous modifications in selectivity, sensitivity and response time as compared to the sensors based on macroscale materials. In the past decade, research efforts are directed toward to develop new methods for rapid detection of foodborne pathogens. The literature reported the utilization of nanosensors for the detection of various foodborne pathogens. Nanoparticles such as metal NPs, NRs, NWs, CNTs[76,86] and so on modified with antibodies having immunological interactions have been preferentially utilized for the foodborne pathogens detection. The various nanoparticles for the detection of foodborne pathogens have been well documented in literature.

7.5.1.1 DETECTION OF SALMONELLA

Salmonella is a major foodborne pathogen and causes Salmonellosis that is major foodborne bacterial disease caused by *Salmonella enteritidis* and *Salmonella typhimurium* serotypes. The pathogen is associated with raw or undercooked eggs, poultry, beef, and unwashed fruit. The detection of this bacterium is an important milestone in the foodborne pathogen analysis due to the serious health implications such as loss of appetite, high fever, gastrointestinal pain, vomiting, diarrhea, nausea, and weakness. Bio-sensing methods, such as optical, impedimetric, electrochemical and piezoelectric methods, have shown immense potential for the fast detection of salmonella but current research has highlighted the role of nanotechnology for the sensing of this deadly foodborne pathogen.

Dungchai et al.[21] reported electrochemical immunosensor fabricated with gold nanoparticles (GNPs) based on anodic stripping voltammetric (ASV) technique for the detection of *S. typhimurium*. The immunosensor was fabricated by the immobilization of monoclonal antibodies on polystyrene support for the bacterial adhesion and further followed by the addition of polyclonal antibody GNPs conjugate to bind the bacteria. This interaction was mediated by copper as catalyst with ascorbic acid proving acidic medium. The ascorbic acid leads to the deposition of complex onto AuNPs tags through the reduction of Cu^{2+} ions to Cu and the quantification of deposited copper was directly co-related with the quantification of AuNPs tag. The limit of detection was found to be 98.9 CFU/mL which depicted high sensitivity of this sensor.

Later on, a capacitive electrochemical immunosensor based on EIS was reported by Yang et al.[99] This sensor was utilized for the highly sensitive detection of *Salmonella* spp. in pork products by fabricated with ethylenediamine and GNPs on glassy carbon electrode. The limit of detection was reported to be 1×10^2 CFU/mL and the developed sensor revealed the working concentration range of 1×10^2–1×10^5 CFU/mL. This immunosensor displayed fast analysis time (40 min) with highly selective toward nonspecific interactions that showed higher sensitivity as compared to the PCR method. Though nanosensing exhibit remarkable sensitivity for the detection of target analyte but sensing in foods matrices has associated with various challenges such as color, opacity of samples, light scattering properties, and other numerous interferences. Therefore, a detection technique has been devised to isolate the desired organism from these interferences for alleviating the signal-to-noise ratios. This technique is known as immunomagnetic separation in which the target analyte is separated from the food matrix by the selective attachment of nanomagnetic particles with antibodies by applying magnetic field. The research studies suggested that this technique has been utilized for the detection of *Salmonella* in complex food samples such as milk. An antibody conjugated MNPs were utilized for capturing the bacteria and with the application of external magnetic field, the separation of the bacteria-adsorbed probe was accomplished. Further, this complex has been exposed for the absorption of UV light by the antibody immobilized TiO_2 nanocrystals followed by the magnetic separation of MNPs *Salmonellae* TiO_2complexes from solution of unbound TiO_2 nanocrystals with an UV visible spectrometer. The lowest detection limit (LOD) for *Salmonella* was found to be 100 CFU/mL in milk samples.[39] Furthermore, more sensitive electrochemical biosensor based on CNTs was reported by Jain et al.[38] The fabrication of this sensor was performed on glassy carbon electrode by the immobilization of CNTs functionalized monoclonal antibodies for *S. typhimurium* detection. The analysis was done by the EIS by correlating the change in impedance and resistance through charge transfer. The lowest limit of detection was found to be 1.6×10^4 CFU/mL with the linear response of 1×10^{-1} to 1×10^{-6} M. Furthermore, advancement has been seen in the fluorescent labeling of optical biosensors by incorporating the use of QDs. The detection of *S. typhimurium* was reported by Yang and Li[101] in which the bacterium was captured by magnetic beads coated with anti-*Salmonella* antibody. The biotin-labeled anti-*Salmonella* as secondary antibody reacted with streptavidin-coated QDs resulting in the measurement of alteration of fluorescence intensity. The lower detection limit was reported to be 10^3 CFU/mL with linear response in the range of 10^3–10^7 CFU/mL. Weeks et al.[93] reported a nano-cantilever embedded biosensor for the identification of *S. enterica*. The silicon nitride cantilever was utilized for binding the bacterium and detection was carried by monitoring the bending of the cantilever's surface that is directly correlated with the concentration of bacteria bound on cantilever. The no. of salmonella was found to be 25 that showed the highly sensitive detection of this deadly foodborne pathogen. More recently, single-walled carbon nanotubes (SWCNTs) gained popularity for the detection of various pathogenic microorganisms. The aptasensing coupled with CNTs has been viewed as a promising technique for the highly sensitive and selective detection. In an attempt for this, Yang et al. (2013)[104] reported the detection of *S. paratyphi* by self-assembled SWCNTs coupled with DNAzyme-labeled aptamer.

7.5.1.2 DETECTION OF Escherichia coli

Escherichia coli O157:H7 is considered as deadliest foodborne pathogen and classified as most lethal serotype among *E. coli* strains. The widespread presence of this pathogen including pigs, deer, dogs, cattle, and poultry, does not cause illness in animals, but these served as good carriers of this bacterium. The infection causes severe intestinal inflammation with bloody diarrhea[82] The various research efforts have been directed towards the development of nano-sensing devices for the pathogenic *E.coli* detection.

Lin et al.[48] reported an electrochemical immunosensor based on amperometric screen-printed carbon electrode (SPCE) for the detection of *E. coli* O157:H7 in milk samples. The immunosensor was developed with 13-nm AuNPs functionalized with double antibodies coupled with ferrocenedicarboxylic acid (FeDC). The detection worked on the basis of indirect sandwich ELISA. The 13.1-folds current enhancement was observed with functionalized GNPs and FeDC and lower limit of detection was found to be 50 CFU/strip in milk samples. In a similar way, Zhang et al.[105] reported an electrochemical immunosensor based on ASV for detection of *E. coli*. In this report, they have used core-shellCu@Au NPs as anti-*E. coli* antibody labels in spite of AuNPs. Furthermore, electrochemical immunosensor based on cyclic voltammetry has been reported by Cho et al.[12] In this peptide nanotubes (PNs) were adsorbed on the SPCE. The anti-bacterial antibody immobilized on PNSPCE captured *E. coli* O157:H7 by antigen-antibody interaction and the analysis was performed by the current measurement. Later on, Mao et al.[52] using used streptavidin-conjugated MNPs for the development of nucleic acid-based sensor utilizing quartz crystal microbalance (QCM) technique. The hybridization process was done by complementary target DNA with self-assembled thiolated single-stranded DNA probe onto the QCM sensor. The asymmetric PCR was carried out for gene amplification with biotin-labeled primers. The concomitant change in QCM frequency was correlated with detection by change in mass. The working concentration range of developed sensor was found to be 2.67×10^2–2.67×10^6 CFU/mL with detection limit of 2.67×10^2 colony forming units (CFU)/mL. The *E. coli* O157:H7 detection has been done by using GNPs-conjugated thiolated probe through circulating-flow piezoelectric (PEB) biosensor.[8] The assessment has been done by analyzing the shift in frequency that can be correlated with change in mass of PEB that are conjugated with target gene-specialized thiolated probe. The detection limit was found to be 1.2×10^2 CFU/mL with working range of 10^2–10^6 CFU/mL. Research studies also reported the development of nanosensors with epifluorescent microscopy for the detection of *E. coli*. El-Boubbou et al.[24] developed the biosensing platform with D-mannose functionalized MNPs followed by the incubation at 4°C for 12 h with fluorescein-labeled concavalin A (Con A) and *E. coli* cells. The aggregates were stained with a fluorescent dye (PicoGreen) after removing the supernatants and fluorescent microscopy was performed. The analysis of fluorescent images revealed limit of detection 10^4cells/mL for the *E. coli* cells. Later on, based on this concept metal nanoshells have been used as excellent materials for the fabrication of highly sensitive sensor for

E. coli detection. Kalele et al.[43] fabricated SPR biosensor based on rabbit immunoglobulin G(IgG) antibody bound with nanoshells made up of silver. The limit of detection of *E. coli* was found to be in the range of 5–10^9 cells. Furthermore, Maurer et al.[53] developed the nanobiosensing platform for the detection of pathogenic *E. coli* with RNA functionalized GNPs through interaction of CNTs. The results revealed excellent selectivity toward *E. coli* with respect to other Gram-negative microorganisms.

7.5.1.3 DETECTION OF LISTERIA MONOCYTOGENES

Listeria monocytogenes is the most virulent microorganism responsible for listeriosis, considered as a highly infectious disease. The innovative rapid detection method embracing nanotechnology for the detection of *L. mono cytogenes* has been reported by Grossman et al.[31] In this report, the sensing device was designed to monitor the binding rate between antibody-linked supermagnetic nanoparticles (ASMNPs) and target bacteria using a superconducting quantum interference technique (SQUID) at high-transition-temperature. The MNPs of 50 nm in size was coated with antibody and further, the bacterial sample has been added. The magnetic dipole moments were aligned by the application of pulsed magnetic field. The SQUID measures the magnetic flux dissipated in the form of Brownian rotation by the free nanoparticles and Neel relaxation by the bound nanoparticles to *L. monocytogenes*. The varied sample volume was taken 20 μL and 1 mL and the biomass concentration and detection limit were found to be 230 cells and 5.6×10^6 M in respective samples.

7.5.1.4 DETECTION OF Mycobacterium avium

The primary cause of Johne's disease is *Mycobacterium avium* subspecies paratuberculosis (MAP) affecting the animals including domestic livestock. The detection of MAP is quite tough because it is difficult to grow as well as not visualized under an ordinary microscope. The fast and sensitive detection of this bacterium was reported by Kaittanis et al.[41] They developed magnetic nanosensors for the bacterial detection based on SPIONs in milk and blood samples. The detection principle relies on the property of nanoparticles to undergo magnetic relaxation on their concomitant change in spin–spin relaxation time with target molecules through the switching ability between clustered and dispersed state. The sensing device was developed by conjugating the SPIONS with G protein-coupled anti-MAP antibodies and analysis was done by subsequent addition of MAP in a dose-dependent manner. At room temperature, the sensor displayed working concentration range of 2 μg Fe/mL nanoparticles and the detection limit was found to be in the range of 15.5–775 CFUs/mL after a 30-min incubation. The high sensitivity for the detection of MAP could be achieved at 37°C within 2% milk. The developed sensor displayed excellent selectivity with respect to other foodborne pathogenic bacteria such as *Enterococcus faecalis, E. coli, Proteus vulgaris, Staphylococcus aureus*, *Pseudomonas aeruginosa*, and *Serratia marcescens*. In addition to that a sandwich immunoassay for fast and efficient detection of MAP based on SERS was reported by Yakes et al.[97] This immunosensor was based on the immobilization of MAb13E1 for capturing MAP2121c on MAP surface. This selective capturing leads to the production of readable SERS signals by the development of an extrinsic bi-specific Raman label. This labeling was done by adsorption of sulphur-containing compounds that further leads to the formation of 5,5′-dithiobis(succinimidyl-2-nitrobenzoate) (DSNB) layer on the surface of 60-nm GNPs that can tether antibodies. The developed immunoassay was analyzed through integration of 13E1 monoclonal antibody in SERS signals and displayed very good selectivity for the detection of MAP. The lower limit of detection was found to be 100 ng/mL in PBS and 200 ng/mL in pasteurized whole milk within 24 h.

7.5.1.5 DETECTION OF *Vibrio parahaemolyticus*

Vibrio parahaemolyticus is found in brackish saltwater and responsible for gastrointestinal illness in humans. The bacterial infection can occur by ingestion of raw or undercooked seafood, mainly oysters, which lead to the acute gastroenteritis. The recent progress in nanotechnology led to the development of disposable enzyme immunosensor for the detection of *V. parahaemolyticus*. Zhao et al.[106] developed a gorse coated GNPs with screen-printed electrode for direct electron transfer by amperometry. The incubation of the bacterium incorporated sensor for 30 min at 25°C, led to the formation of immunocomplex on the sensor surface by horseradish peroxidase (HRP) coated anti *V. parahaemolyticus* antibody. The analysis was performed by the measurement of lowering in cathodic peak current mediated by enzyme inhibition due to the oxidation of thionine by H_2O_2.

The developed sensor showed selectivity over various foodborne pathogenic microorganisms such as *E. coli* O157:H7, *S. aureus* and *S. pullorum* in the spiked samples.

7.5.1.6 DETECTION OF *Pseudomonas aeruginosa*

This organism is considered as opportunistic and recognized for its ubiquity as advanced antibiotic resistance mechanisms. This bacterium is responsible for nosocomial infections including urinary tract infections (UTIs), pneumonia and bacteremia that leads to the inflammation and sepsis. This organism is very well thrived on moist surfaces, therefore, causing cross infections in hospitals and clinics. Recently, advancements in nanotechnology led to detect microorganisms by selective destruction of pathogenic bacteria through the use of nanomaterials. This concept was further explored by Norman et al.[62] with selective destruction of *P. aeruginosa* cells with gold nanorods. The *P. aeruginosa* cells bound to conjugated nanorods (NRs) that were further treated with near-infrared (NIR) radiation followed by incubation with purified antibodies. The analysis was performed by live (green)/dead (red) cell staining and followed by counting. The results showed that cell viability was reduced up to 75% for the radiation-exposed nanorod-coated *P. aeruginosa* cells as compared with cell viability of approximately 80% revealed by both NIR-exposed cells with and without nanorods.

7.5.1.7 DETECTION OF MULTIPLE FOODBORNE BACTERIAL PATHOGENS

The recent technological interventions in nanotechnology have envisaged the development of methodologies for the multiple detections of pathogenic microorganisms. The concepts were found to be quite fascinating and facilitated high-throughput detection of single bacterial cell. Based on this concept, an immunosensor doped with silica nanoparticles has been utilized for the simultaneous detection of *S. typhimurium, B. cereus,* and *E. coli* O157:H7. Zhao et al.[107] fabricated immunosensor by conjugating the antibody with dye-adsorbed silica nanoparticles for incubation of bacteria. The unbound bacteria were removed by centrifugation and detection was carried by both fluorescence methods and counting the bacteria by gold-standard plate method. The similar results were obtained from both the methods but the response time of the fluorescence method was found faster than conventional approach. Furthermore, this method could be applied for the high-throughput detection of single bacterial cell as well as more foodborne pathogens. In addition to that, more comprehensive detection of twelve bacteria was detected through the nanosensor based on GNPs without the use of costly antibodies. In this study, GNPs were conjugated with p-conjugated polymer poly (para-phenyleneethynylene) followed by bacterial incubation. The interaction of bacteria with nanoparticles led to quenching in GNPs that emit fluorescence was reported by Philips et al.[68] The authors used three different nanoparticle preparations and differential fluorescence intensity has been observed for various investigated bacterium including *Bacillus licheniformis*, *Amycolatopsis azurea*, *Bacillus subtilis*, *Amycolatopsis orientalis*, different *E. coli* strains, *Streptomyces coelicolor*, *Lactobacillus plantarum*, *Pseudomonas putida*, *Lactobacillus lactis*, and *Streptomyces griseus*. The sensor displayed high sensitivity (1×10^9 CFU/mL) and proved to be cost-effective strategy for multiple detections.

Furthermore, the negative charge of bacterial surface was exploited for the pathogen detection. The fast and efficient detection of bacteria was done by amine-functionalized MNPs (AF-MNPs) based nanosensing device reported by Huang et al.[37] The high affinity electrostatic interactive force with cationic AF-MNPs and anionic bacterial surface was utilized for the binding of eight bacterial species including *E. coli, Sarcina lutea*, *B. subtilis, B. cereus, S. aureus, Salmonella, P. aeruginosa.* The developed sensor was able to differentiate biochemically between Gram-positive and gram-negative bacterium. However, the developed sensing device showed less selectivity as compared to their antibody conjugated counterparts.

More recently, nanosensors embracing aptasensing have been preferentially utilized for highly sensitive detection of three bacteria including *Staphylococcus aureus, Vibrio parahemolyticus, and Salmonella typhimurium*. Wu et al.[95] reported the nanosensor coupled with aptamers with multicolor upconversion nanoparticles (UCNPs) as luminescence labels. In this study, systematic evolution of ligands by exponential enrichment (SELEX) strategy was used for the aptamers synthesis. The luminescence labels were formed by the doping with various rare-earth ions and analyzed through well-differentiated emission peaks. The quantification was assessed on the basis of the distinct peaks and the linear correlation between the concentration of bacteria and the luminescence signal from 50 to 10(6) CFU/mL was observed. The performance was improved by the magnetic separation through incorporating the MNPs (Fe_3O_4) and the limits of detection were found to be 25, 15, and 10 CFU/mL for *S. aureus, S. typhimurium*, *V. parahemolyticus* respectively. The method was validated with gold plate standard method and the results were found to be consistent. This device led to the development of more sensitive multiplex nanosensors.

7.5.2 DETECTION OF FOODBORNE INTOXICANTS

The food contamination is also prevailed by toxic compounds secreted by bacteria, fungi, and other organisms known as endo-exotoxins. These compounds are by-products of the organism's normal metabolic processes and they have remarkable properties of resistance toward heat and cold treatment. These toxins are not contagious in nature but impart significant effects on organs and tissues. The fast and highly sensitive detection of these toxins is an important avenue of the nano-biosensing research as these toxins can survive after the death of their viable pathogenic microorganisms. The wide panorama of technologies is being utilized having multiplexing capability includes SPR based biosensors, polystyrene microbeads embedded with antibodies, ELISA, antibody microarrays, and western blots. However, these methods require tedious sample preparation methodologies and work on purified samples therefore, require significant time consumption and workforce. Thus, the development of nano-sensing devices with nanomaterials as a biological recognition element is very much crucial.

7.5.2.1 DETECTION OF SHIGA TOXIN

Shiga toxin (ST) is the metabolic product of *E. coli* 0157:H7 and disrupts the translational machinery and activate apoptosis and severe tissue damage. Till date, the sensing method utilizes the interaction between β subunit of Shiga toxin and trisaccharide αGal(1→4)βGal(1→4)βGlc of globotriose (P^k) blood group antigen. To materialize this concept, GNPs based SPR chip competition assay was reported by Chien et al.[7] In this method different size of GNPs was modified by the glycosylation through self-assembling of two derivatives of P^k. The binding affinity of P^k moiety was enhanced through the optimization of length of GNPs. The longer chain length has been observed with high affinity toward toxin surface. Furthermore, a colorimetric sensor was developed by utilizing Gal-α1,4-Gal glyco-polydiacetylene nanoparticle for quantification of *E. coli* O157:H7.[59] The microwells are coated with Shiga toxin-producing bacterial cells followed by the addition of glycopolydiacetylene nanoparticles. In less than 5 min, the color of the well plates changed from purple color to brown color if they contain toxin-producing cells whereas as control well plates where non-toxin producing cells are present, the solution remained purple. The sensitivity of this detection was found to be (LOD 1200 U/μL) and linear dynamic range (1200–7200 U/μL), respectively.

7.5.2.2 DETECTION OF CHOLERA TOXIN

Cholera toxin (CT), a hexameric protein complex produced through *Vibrio cholera*. The infection causes massive, watery diarrhea in humans. It consists of single, enzymatically active A subunit (CTA, 27 kDa), linked non-covalently with pentameric core of five identical receptor-binding B subunits (CT-B, 58 kDa). The activation of adenylatecyclase is responsible for their biological activity. Nowadays based on their toxicity the development of highly selective and sensitive methods for the detection of cholera toxin is urgently required. Viswanathan et al.[89] fabricated an electrochemical immunosensor based on Nafion-supported multiwalled CNTs. The surface of CNTs has been coated with an anti-CT-B subunit monoclonal antibody with thiol compounds that are encapsulated in liposomes. The sandwich-type assay was utilized for the detection through the binding of functionalized ganglioside liposome with anti-CT antibody-toxin complex. An adsorptive square-wave stripping voltammetry method was used for measurement of potassium ferrocyanide molecules released from liposomes bound onto working electrode. In another study, ganglioside lipid bilayer was used for detection of CT by tethered GNPs. This method was compared with other fluorescent immunoassays and results showed 100 times more sensitive detection. Furthermore, Schofield et al.[77] developed a colorimetric method for the CT detection. They used GNPs (16 nm) on which the thiolated-lactose derivative self-assembled and aggregated followed by the binding of the CT-B subunit. The analysis was performed on the basis of colorimetric shift from red to purple. This colorimetric nanosensing device displayed limit of detection of 3 μg/mL.

7.5.2.3 DETECTION OF RICIN

Ricin is classified as a type 2 ribosome-inactivating protein (RIP) and a toxic lectin and considered as a bioterrorism threat. The binding of β-D-galactopyranose or β-D-N-acetylgalactosamine to glycol reactive sites to the surface of the host cell mediates the toxicity in living cells. Moreover, the colorimetric bioassay reported by Uzawa et al.[87] utilizing GNPs modified with a thiolated glycosylated ligand for the sensitive detection of ricin. The detection works on the principle of colorimetric change from red color to purple color by binding of ricin with functionalized nanoparticles. The sensitivity of the sensor was evaluated at 10 and 30 minute intervals and the values obtained < 3.3 μg/mL and 1.7 μg/mL, respectively. The sensor displayed very high sensitivity with respect to other plant lectins peanut, clary sage, jack bean, soybean, osage orange, and complex proteins such as BSA.

7.5.2.4 DETECTION OF STAPHYLOCOCCAL ENTEROTOXIN

This is an enteric toxin and the leading cause of major gastroenteritis due to the consumption of contaminated foods. *Staphylococcal* enterotoxin s (SEs) are highly resistant to heating temperatures. From the current studies, it was confirmed that *S. aureus* produces 21 heat-stable toxins and associated with other foodborne diseases. *Staphylococcal* enterotoxin A (SEA), B (SEB), C (SEC), and D (SED) are the most common is foodborne intoxicants because of its remarkable toxicity and stability, it is considered as a promising biological weapon for mass destruction. The current immunological assays including ELISA do not confer appreciable sensitivity for the selective detection of these toxins. Therefore, the sensing devices based on nanomaterials enhanced the detection of *Staphylococcal* enterotoxins (SEs). In continuation of these, an optical immunosensor relies on CNTs reported by Yang et al. for detection of SEs[100] The detection protocol is based on immobilization of anti-SE primary antibody on CNTs with further binding with HRP-labeled secondary antibody followed by measurement of HRP fluorescence. The fluorescence intensity of this sandwich method was found to be far much higher than standard immunoassay with detection limit of 0.1 ng/ml and working concentration range of 0.1–100 ng/ml. However, the developed method could not be applied for the endotoxin detection in real samples as an additional sample purification step by carboxymethyl cellulose chromatography was needed. Owing to the toxicity of CNTs as nanomaterials as well as the requirement of the acid functionalization in the food matrices, Yang et al.[102] again reported the GNPs-based chemiluminescence (ECL) immunosensor for SEs detection in food matrices. The sensing device was based on the adsorbed anti-sera onto a GNP surface and followed by immobilization of the antibody-GNPs conjugated on polycarbonate platform. The sandwich ELISA technique was used for the quantification using secondary antibody (HRP-conjugated anti-rabbit IgG). The results revealed high sensitivity (approximately 10 times) as compared with other analytical methods such as ELISA and CNT-based immunosensor. The lowest limit of detection was observed as 0.01 ng/mL in the samples.

7.5.2.5 DETECTION OF BREVETOXIN

BreveToxin (BTXs) are lipid-soluble polyether marine neurotoxins produced by the dinoflagellate *Karenia brevis* (Shellfish). Their effect is excitatory, mediated by the enhancement of cellular Na^+ influx leading to muscle pain, loss of coordination, abdominal pain, respiratory irritation, slow heart rate, headache, dilated pupils. The fast detection is manifested by the development of electrochemical immunosensor by Tang et al.[85] The GNPs was used as a platform for nanosensing which was anchored with amine-terminated polyamidoamine dendrimers (GNP-PAADs) that served as a substrate for the immobilization BTX-B-BSA conjugate. The competitive immunoassay was performed and the detection limit, as well as working concentration range, was found to be 0.01 ng/mL and 0.03–80 ng/mL BTX-B, respectively.

7.5.2.6 MULTIPLEX DETECTION OF BACTERIAL TOXINS

Evidently, the multiple detection of foodborne toxins is an interesting alternative for the fast and selective detection to improve the safety of food materials. The high throughput fluoroimmuno assays were developed for detection of CT, ricin, Shiga-like toxin, and SEs was reported by Goldman et al.[29] The luminescent semiconductor nanocrystals (CdSe-ZnS core-shell QDs) were conjugated with antibodies and sandwich immunoassay was used to quantify four antigens in the single sample. However, the developed sensor suffers a limitation for cross-reactivity but has scope for overcoming this shortcoming by optimizing the assay conditions and selective usage of antibodies. Furthermore, Branen et al.[4] reported an assay based on quantification and identification of *E. coli* O157:H7, *S. entericaserovar. S. typhimurium*, and SEs simultaneously in same samples. In this research communication, the MNPs were conjugated with antibodies and quantification was performed by the DNA amplification of templates bound onto the antibody in both buffer and milk samples.

7.5.3 NANOSENSORS FOR FOOD PACKAGING

The food safety will be enhanced by the intervention of the novel food packaging technologies that will ensure the improvement of the characteristics of food, conservation, product freshness, flavor stability, longer shelf-life, monitoring, and traceability. The monitoring reveals the idea of governing factors that leads to the spoilage of food from external factors, such as heat, light, humidity control, oxygen level, relative pressure, enzymes, odors, microorganisms, insects, contaminants, dust, gaseous emissions, and lipid oxidation. Therefore, the packaging is a constantly evolving field and has been explicitly utilized in the food industry. In the current scenario, the scientific community, as well as industry analysts, predicted that nanotechnology has tremendous potential for the transformation of food from "farm to fork" concept with best health perspectives[80] The research on nano-packaging has been accelerated by the recent technological advances and current scientific research that is directed toward the field of printed electronics, microfluidics, and carbon and silicon photonics.[36] These scientific advances offer the possibility of developing a new generation of food packaging called "smart packaging" that incorporates nanomaterial's that has remarkable ability to respond with respect to environmental conditions and indicating about various food contaminations which alert the consumer for further consumption. The overview comprising the role of nanosensing devices in food safety has been shown in Figure 7.4.

The nanosensor interacts with intrinsic factors of food components and extrinsic factors for the generation of response (visual signal, electric signal) that correlates with the nature of the food product. The information retrieved has been useful for communication with users providing them with information on safety and quality of products but also can be used by the producers in for the effective production process and distribution channels for the food products. Based on this concept, currently "smart packaging" has two major avenues mentioned below.

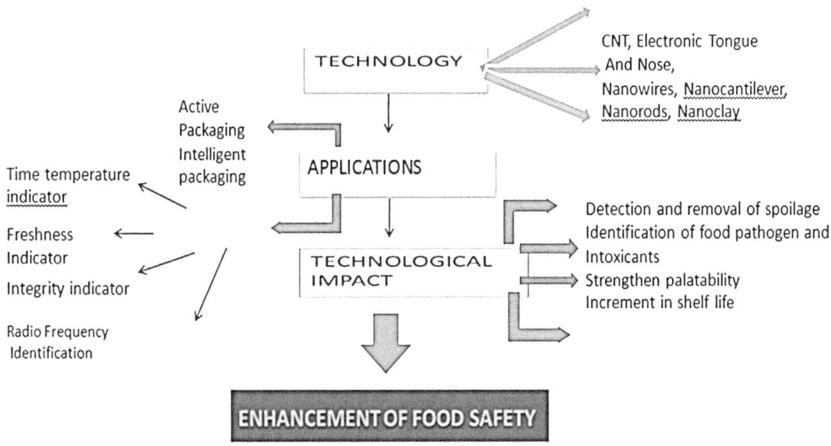

FIGURE 7.4 An overview of the role of nanosensors in food safety.

7.5.3.1 ACTIVE PACKAGING (AP)

This is the most innovative technique and considered as the main engine of nanomaterial applications in the food packaging industry. This technology ensures the food safety by alterations in biochemical reactions through the addition of specific components including salt, sugar, carbon dioxide, antioxidants, antifungal, antimicrobial, natural acids, and oxygen removing compounds by maintaining or improving the condition and quality of the packaged food.[71] Antimicrobials such as nanometals and nanometal oxides like nanozinc, nano silver, nano magnesium dioxide, and nano titanium dioxide can be used in food packaging for the prevention of spoilage.[30] Furthermore, the spoilage can also be alleviated by the incorporation of nano-clay into the food packaging to alleviate the gaseous components from the food materials. The successful and commercialized example has been viewed by the application of engineered plastic film called Durethan developed by Chemical giant Bayer Incorporation. This transparent film contains nanoclay based on polyamide 6 and polyamide 66 and confers excellent properties like high tensile strength, resistance to abrasion and chemicals as well as cracking. Therefore, such packaging corroborating the nano-particles greatly enhances the food safety.

7.5.3.2 INTELLIGENT PACKAGING (IP)

Nowadays, the demand for the automated systems has been tremendously increased thus the packaging system requires to perform smart and intelligent functions. Therefore, IP systems are considered as "next generation" packaging and are capable of performing intelligent functions such as detection, acquisition, recording, monitoring, communication, and application of scientific logic, to facilitate decision making, extending the life, and warning potential problems Siro et al.[81] These technologies utilized the quality indicators for the online monitoring and providing information on the status of the food storage as well as the internal and external environment without generally exerting any action on the food. The information generated by intelligent packaging is useful for both consumer's as well as manufacturer's perspectives, in their decision support systems to determine when and what measures should be taken over the entire production process and distribution channels for products.[19] The IP is further categorized based on the assessment of external and internal conditions as follows.

7.5.3.2.1 Time and Temperature Indicators (TTIS)

These are the controlling devices capable of maintaining the quality of food through a color response that matches or correlates with the quality of a food at a given temperature. The principle behind this detection is based on the different operating mechanism including physical systems such as diffusion, chemical systems such as polymerization reactions, and biological systems such as an enzymatic reaction. They are categorized as partial temperature history indicator, abuse indicators, and full-temperature history indicator. Abuse indicators are recognized as critical temperature evaluators as they assess the desired temperature of the packaged food.[27] Partial temperature history indicator works by correlating the time versus temperature changes by analyzing the temperature which could exceed a certain pre-set value. Full temperature history indicator measures a progressive timely change in temperature.

7.5.3.2.2 Radio Frequency Identification

Radio frequency identification (RFID) is technology of a collection of wireless data that uses electromagnetic waves (EW) to transfer between a transmitter and receiver. The food safety parameters have been greatly enhanced by the wireless data transfer from food sensors in the form of RFID tags. These are electronic chips that transfer high-frequency signals, which can be assessed from the distance of a few meters and nanosensors can installed at the packaging plant itself for providing the real-time data on the packaging by integrity control, authenticity, antitheft protection, anti-counterfeiting, quality, and traceability.[44] In the food industry, three types of RFID tags can be used with various combinations of nanosensors and their transducing platforms. *Active RFID tags*: These tags have their independent battery as capable of transmission of data independently as well as independent memory for the data storage. The active RFID tags have been utilized for the food packaging material with the incorporation of nanosensors. The anomalies or any error found in the food material can be determined by direct processing of the data during transport and storage conditions (package strapping, high temperatures, high bacteria concentration, etc).[2] Passive RFID tags: these sensing devices require an external power supply for the processing and transmission of the data. BAP (battery assisted passive) tags: these are broad range sensing devices which have a battery but require external power supply for data transfer. Food packaged with these RFID tags must be under constant supervision of RFID reader.

7.5.4 DETECTION OF GASES

The moisture and oxygen content present in the food are leading causes of food spoilage therefore that directly affect the palatability of the food products and more emphatically concerned with the consumer's health. The conventional approaches for the detection of gaseous constituents require the destruction of the packages. Therefore, a non-invasive method has been desired that can give continuous monitoring of the gas content of a package headspace that leads to ensure the safety evaluation of the food will be more sustainable and can be applied in the food industry. The thorough understanding of the development of nanosensors can be implied successfully for the detection of gases produced during food spoilage. The sensors detect gases such as hydrogen, nitrogen oxides, ammonia, hydrogen sulfide, and sulfur dioxide. These gas sensors are comprising metals such as gold, palladium, and platinum as well as conducting polymers. The electrical, optical, and magnetic properties are the prime factors for the development of conducting polymers. The assessment of the gas sensors has been done on the basis of resonance changes in the presence of gaseous components followed by changes in the response pattern on the conducting polymer. Based on this concept, Mill et al.[56] developed nanosensor for the detection of oxygen in the packaged food product. The sensor is based on the use of nanosized TiO_2 or SnO_2 particles and a redox-sensitive dye (methylene blue) adsorbed on photo-activated indicator ink. The presence of oxygen has been analyzed by the detection of the color change imparted by the redox dye. However, the developed sensor only determines the presence of oxygen qualitatively not quantitatively. In addition to that, photoluminescent sensor doped by fluorophore-encapsulated polymer nanobeads[55] were also reported for the non-invasive measurement of carbon dioxide content in MAPs. The developed CO_2 displayed detection limit of 0.8%–100%, a resolution of 1%, and 0.6% cross-sensitivity with molecular oxygen. Research studies also reported the detection of indicators of meat and fish spoilage (gaseous amines) by the nanosensors. Furthermore, Nopinyuwong et al.[61] reported the detection of gaseous amines molecules at parts-per-trillion (ppt) level by the development of an optical biosensor. The sensor was based on fluorescence quenching of fluorophores containing perylene that was decorated with nanofibrils. A parts-per-million (ppm) level conductometric sensor-based for the detection of gaseous amines was reported by Hernandez-Jover et al.[35] that is based on measurement of change in conductance in composites of SnO_2 nanoparticles and TiO_2 microrods. In another study, confirmation of food spoilage has been reported by the development of various electronic sensors comprising ZnO–TiO_2 nanocomposites or SnO "nanobelts" for the simultaneous detection of volatile organics that is, acetone, ethanol, and carbon monoxide. The timely detection of the presence of fruit ripening hormone, ethylene gas, was done by nanosensors comprising WO_3–SnO_2 nanocomposites.[69] Some of the nanosensing devices have also been commercialized at industrial scale. Mars Incorporation[90] developed an invisible edible nano wrapper that acts as a food envelope to block the exchange of gas and moisture from the external environment. The electronic barcodes developed by the nanosensors embedded into food products as tiny chips that are invisible to the human eye would also act Food tracking devices. Therefore, the food safety has been well maintained by the incorporation of nanosensors that can detect the chemicals released during food spoilage into the packaging material, where they performed the functioning as biomimetics such as 'electronic tongue' or 'noses.'

7.5.5 DETECTION OF MOISTURE

The moisture content is one of the predominant factors required for the growth of microorganisms. Therefore, the development of the nanosensing device will be highly appreciable for identifying the freshness of the food products. Therefore, Luechinger et al.[50] reported the non-invasive nanosensing strip based on tenside film decorated with carbon-coated copper nanoparticles. The moisture content was analyzed by the reflection or absorbance of different colors of light induced by the inter-nanoparticle separation due to swelling of polymeric matrices in a humid environment. These sensors paved the way for the rapid determination of moisture levels in the packaged food products thus enhancing the food security.

7.5.6 DETECTION OF FOOD CONTAMINANTS OR RELEVANT ANALYTES

In the last decade, nanosensors have tremendous potential to accelerate the rate of detection, identification, and quantification of foodborne pathogens, pesticides, microbial toxins, intrinsic and extrinsic factors for food decaying, and allergy-causing proteins. Therefore, these nanodevices have the potential to significantly impact food safety. The beauty of the nanosensors to be installed at the packaging material itself has very much apt for the detection of contaminants. Therefore, the burden on economy has been significantly reduced by the costly sampling analysis. The information and fast communication generated by these devices enhance the awareness of consumers regarding the safety of the food product. Though we have mentioned various nanosensors which have been reported for various pathogens and analytes still some more foodborne molecular contaminants or adulterants that were found to be toxic to the well beings of human life needs to be discussed. There are various nanosensors reported till date for the detection of various food contaminants. One of the adulterant melamine, used as artificially inflating protein content which has been added in infant formulation as well as in pet foods was quantified with nanosensors. The sensor was based on the selective binding of melamine with functionalized GNPs (AuNPs) by cyanuric acid. This aggregation of melamine on AuNPs leads to reproducible, and target concentration-dependent colorimetric shift from red color to blue color that can easily measure the concentration of melamine in raw milk and infant formula with the naked eye and concentrations was found to be 2.5 ppb.[1] The colorimetric quantification of melamine in raw milk samples was also reported with GNPs (AuNPs) functionalized with crown-ether-modified thiols. The developed sensor displayed limit of detection of 6 ppb.[51] More recently, a part per billion (ppb) levels, nanosensors decorated with patterned gold nanostructures has been explicitly used for the highly sensitive detection of melamine and their derivatives, as well as fungicides such as malachite green and crystal violet and showed the detection limit of 0.2 ppb.[83] Apart from that, various functionalized colorimetric metal nanoparticle based nano-sensing devices have also been constructed for the detection of various food contaminants,[91] proteins,[34] and metal ions.[32] Nanosensors with fluorescence as transducing platform were developed for the quantification of various analytes. An enhanced fluorescence linked immuno-sorbent assay (EFLISA) has been developed for the detection of gliadin, major inflammatory protein present patients suffering from Celiac disease.[13] The silver nanoparticle film coated with rhodamine-labeled anti-gliadin antibodies have been used for the sensor construction. The binding of gliadin leads to changes in the fluorescence pattern in the developed sensor. The fluorescence quenching of gold nanoclusters has been successfully applied by Qiao et al.[72] for the detection of cyanide in drinking water. The sensitivity of the developed sensor was found to be 2 nM. The detection of pesticides in drinking water was also reported with nanoscale liposome-based nanosensor. The electrochemical nanosensors based on CNTs has been developed for quantification of food colorants 8 (Ponceau 4R and Allura Red in soft drinks[79] and Sudan 1 in ketchup or chili powder.[98] Therefore, the outstanding optoelectrical and chemical credentials of nanoscale particles have leveraged the growth of nanosensors that can be applied for the different avenues for enhancement in the food safety.

7.6 CURRENT PROGRESS ON ITS APPLICATION

The dynamic and evolving growth of food industry is due to the continuous change in the food habits as well as the variation in the consumers' demand. Therefore, food manufacturers are continuously innovating and developing products within the local and global markets to meet this need and face fierce competition. Due to their low-profit margins, there is less affordability in modern and expensive analytical methods as compared with other high-technology sectors and therefore faces technological inhibition for the entry of a major new analytical device. However, various legislative and regulatory regimes have been developed for enhancing the utility of biosensors in the food industry but biosensors face a great rivalry with other analytical methods in terms of reliability, cost, and analytical performance. The most common affordable systems like gas chromatography, HPLC and capillary electrophoresis have been preferentially utilized for simultaneous detection of various food relevant analytes. The great demand for the fast and sensitive detection of microorganisms or foodborne toxins is one of the quite appealing opportunities for biosensors in this industry but it is of worthy consideration that whether a biosensor can compete with other compatible technologies or not. In recent years the use of nanomaterials confers exceptional electromagnetic and optical properties that increase the surface area of the transducers of the different biosensing platforms that increase its analytical performance. The marvelous ability of analysis in very low sample volume is due to their functionalized surface resulting large surface-to-volume ratio, controlled morphology, and miniaturized structure. Therefore, the potential of "detect-to-protect devices" by sensor miniaturization, multiplexing opportunities, and higher sensitivities, nano-bio-sensing has profiled itself in a short time span as an interesting alternative to conventional biosensors. Presently, the most commercial nanosensors are RFIDs, that has been widely used for enhancing the quality of chain supply of various food products and for monitoring the extrinsic environmental parameters (temperature, humidity) affecting the properties of food. The food analysis sector has also been benefited with the development of NEMS that can be portable and miniaturized from millimeter to nano-scale and leveraged with various frequency levels that lead to the smart communication at low cost. The food preservation has been enhanced by the interventions of NEMS and they have remarkable ability to control the storage environment by detecting and monitoring any adulteration[6] in packaging as well. In this way, they act as active 'sell by' devices. Recently, the United States based company Polychromix (Wilmington, MA, USA) has commercialized microelectromechanical systems (MEMS) technology based digital transform spectrometer (DTS) for detection of trans-fat content in foods.[75] The long term focused scientific efforts and financial supports nano-biosensors are expected now to fulfill their promises such as being able to view product traceability, food safety of products, and potential foodborne epidemics. The increasing prevalence of food safety and security strongly indicate the need for further development of new disposable nano-biosensors suitable to be integrated in point-of-care-technologies (POCT) devices.

7.7 POTENTIAL RISKS

The advent of nanotechnology has been visualized by the dramatic improvements in the quality as well as safety of food products. As with any new technology, public perception plays key role in the acceptance of nanotechnology in food industry. In recent years, the safety issue of nanotechnological interventions in food industry has aroused controversy at global scale due to the potential risks involved for human health, associated environment, and the ecosystem. The research studies give supportive evidence that some nanoparticles possess the ability to cause cellular damage to biological systems as well as disrupt the normal functioning of cellular metabolism by direct attachment of cellular receptors of the immune system.[9] Some of nanomaterials implied in the fabrication of nanosenosrs have capability of attaching the cellular proteins[45] that lead to the degradation of the protein. Silver nanoparticles have been reported to affect the human lung fibroblast by altering the ATP concentration,[54] damaging mitochondrial DNA, increasing ROS production[94] as well as chromosomal aberration. The reduced size of the nanoparticles generates free radicals in the tissues by crossing the biological barrier and eventually leads to oxidative damage.[33] CNTs migrate into the food matrix and can lead to toxic effects on the dermal and alveolar tissues of human. Apart from that, nanomaterials disposed of from the food industries can highly affect the ecosystem of that particular region by migrating and accumulating in the water and soil and subsequently reduce the normal flora and fauna of that region. It was well proven that these significant changes are due to their inherent properties such as concentration, particle size distribution, electrical charge, and interfacial characteristics that lead to the serious outcomes for the biological systems.

Keeping this view, undesirable transfer of nanosensing particles may significantly affect the food safety and there should be proper way to monitor these discrepancies. Although some toxicological studies have been performed with nanosensing devices comprising nanomaterials but it was found that there should be more consumer acceptance toward nanotechnology derived packaging as compared with nano-foods.[78] The above-mentioned studies strongly supported the hypothesis that nanotechnology inside a food is less perceived and acceptable than being on the outside (i.e., in the food packaging). From the current state-of-the-art, it is clearly indicated that the applications of nanotechnology have immensely beneficial for the food sector with paradigmatic enhancements in better quality and safety. However, the exposure to certain nano-materials leads to the serious health concerns obviously necessitates the need for thorough investigations in order to unravel their biological outcomes. Therefore, there is a huge challenge regarding the utility and safety of nanosensors among government bodies and food industry.

7.8 REGULATIONS ON NANOSENSORS DEVELOPMENT

The stupendous stride has been envisaged for enhancement of nanosensors research and applications in food sector, however, it is still an infancy as compared to their medical applications. Owing to their complexity and toxicity, there is a dire need for the regulation of nanomaterials that are being utilized for food manufacturing and processing.[16] The previous reports confirmed that various national and international bodies have been involved for providing different opinions or recommendations on the nanomaterial safety assessment. European Food Safety Authorities (EFSA) published the first practical guidance of risk assessment related to the exposure time as well as their related toxicity of nanomaterials [EFSA Scientific opinion 22, 23]. The recommendations on the official definition of "nanomaterial" were published by the European Commission (EC). Then after, realistic scenario of safety evaluation of nanomaterials has been put forward by rigorous activities of European Union (EU). The United States Food and Drug Administration (USFDA) made a draft protocol for utilization of nanomaterials at industrial purpose in 2011,[25] and then published draft documents for both "Cosmetic Products" and "Food Ingredients and Food Contact Substances" in April 2012. These drafts constitute the recommendations on the manufacturing changes in food and cosmetics incorporating the nanomaterials and defined that any alterations in properties, dimensions can be treated as new product proposed by FDA. European regulatory body led to the Framework 1935/2004 regulation which states that there shall be no change in native and organoleptic properties of food by the incorporation of nanomaterials. The Regulation also stated that the assessment of dose-response and the toxic effect of nanocomponents can be done prior to their incorporation in the food materials. Directive 89/107/EEC states about the incorporation of nanocomponents as packaging material has to be allowed after proper assessment as direct food additive. Apart from that, Environmental Protection Agency (EPA), US Patent and Trademark Office (USPTO), Occupational Safety and Health Administration

(OSHA), US Department of Agriculture (USDA), National Institute for Occupational Safety and Health (NIOSH) and Consumer Product Safety Commission (CPSC), are also governing the use and application of nanosystems in food. These regulations drafted by the regulatory bodies have to be publicly accepted and implemented by one and all who are actively involved in the development of nanomaterials used in the food industry. However, the regulatory norms are sometimes not followed properly that creates the dilemmatic situation with regards to food insecurity. Therefore, such scenario should be avoided an on a global scale, and these regulations should be followed by further development, production, and marketing of nanotechnological products with precautionary research and regulation and also requiring periodic risk assessments.

7.9 PROSPECTIVE FUTURE AND RESEARCH PRIORITIES

The prevailing incidences of foodborne disease outbreaks as well as the rising awareness of quality and safety of the food products have metamorphosized the development of technologies capable of fast detection of pathogens and also create the awareness about the quality of the consumed products among consumers. The potential risks and challenges mentioned above are the prime concerns of the working areas for the nanosensors researchers. The complex nature of the variety of food matrices poses major technological challenge for the analysis, therefore, it is expected that advancements in multiplexed hybrid systems with different sensing elements and transducing platforms will be developed in the future. The diverse contaminants present in food matrix explored the development of receptor-less physical sensors as viable methods. In the present scenario, the personalized and individualistic consumer approach also leads to the development of electronic noses and electronic tongue to detect smell and flavor related to the food quality that can be integrated with other electronic devices such as infant products, refrigerated products, temperature sensors, and cellular phones. Apart from the RFIDs, printed nanosensor labeling technology with feasible cost are also expected to be developed in near future. In the more advanced stage, sensing devices embracing the nanomaterials will also be developed to detect the nutritional properties of foods according to person's dietary needs, their allergic sensitivity and their taste preferences that increase the satisfaction of the consumers.

7.10 SUMMARY

This chapter has highlighted the potential role of nanosensors in the field of enhancing the food safety. The food safety has been ensured by the prompt preventive action manifested by the rapid detection of food contaminants by the nanosensing devices. This is accomplished due to the remarkable properties of nanosensors in which the selective binding with the biological components with target analytes and their assessment through the transformation of detectable signals by the suitable transducers. This magical phenomenon confers various advantages of fast, sensitive and portable detection to the nanosensors. The complexity of the food matrix as well the toxicity of nanomaterials brought a serious attention among the scientific community at global level. The various technological and scientific obstacles must be addressed properly before the nanosensors' application used in food industry. The recommendations and guidelines of various regulatory bodies must be followed in evaluating the safety of nanomaterials in food manufacturing, food packaging, food analysis, and preservation. Therefore, novel approaches and standardized test methodologies have been formulated to study the impact of nanoparticles on living cells for the complete alleviation of potential hazards associated with them. Therefore, based on their marvelous utility there is a strong prediction that nanosensing devices will be available increasingly to reduce the burden of food insecurity for the global scale in the coming years. In brief, nanosensors are going to transform the entire food industry with respect to precision and personalized approach. Still, much-needed efforts are required to do the comprehensive research and investigations to learn about the role of nanosensors in food safety, a promising future is certainly on the horizon.

KEYWORDS

- **biosensors**
- **electrochemical nanosensors**
- **immunosensor**
- **nanoelectromechanical systems**
- **nanotechnology engineered foods**
- **smart packaging**

REFERENCES

1. Ai, K.; Liu, Y.; Lu, L. Hydrogen-bonding Recognition-induced Color Change of Gold Nanoparticles for Visual Detection of Melamine in Raw Milk and Infant Formula. *J. Am. Chem. Soc.* **2009,** *131* (27), 9496–9497.
2. Barge, P.; Gay, P.; Merlino, V.; Tortia, C. Item-level Radiofrequency Identification for the Traceability of Food Products: Application on a Dairy Product. *J. Food Eng.* **2014,** *125* (1), 119–130.
3. Bergwerff, A. A.; Van Knapen, F. Surface Plasmon Resonance Biosensors for Detection of Pathogenic Microorganisms: Strategies to Secure Food and Environmental Safety. *Int. J. Anal. Sci.* **2006,** *89* (3), 826–831.
4. Branen, J. R.; Hass, M. J.; Douthit, E. R.; Maki, W. C.; Branen, A. L. Detection of Escherichia Coli O157, Salmonella Entericaserovar Typhimurium, and Staphylococcal Enterotoxin B in a Single Sample Using Enzymatic Bio-Nanotransduction. *J. Food Protect.* **2007,** *70* (4), 841–850.
5. Bultzingslowen, C. V.; McEvoy, A. K.; McDonagh, C.; Mac Craith, B. D.; Klimant, I.; Krause, C.; Wolfbeis, O. S. Sol–gel Based Optical Carbon Dioxide Sensor Employing Dual Luminophore Referencing for Application in Food Packaging Technology. *Analyst* **2002,** *127* (11), 1478–1483.
6. Canel, C. et al. In *Micro and Nanotechnologies for Food Safety and Quality Applications*. Micro-and Nano-Engineering, Microsystems and their fabrication Proceedings, Barcelona, Spain, 2006.
7. Chein, Y. Y.; Jan, M. D.; Adak, A. K.; Tzeng, H. C.; Lin, Y. P.; Chen, Y. J.; Wang, K. T.; Chen, C. C.; Lin, C. C. Globotriose-functionalized Gold Nanoparticles as Multivalent Probes for Shiga-like Toxin. *J. Chem. Biol.* **2008,** *9* (7), 1100–1109.
8. Chen, S. H.; Wu, V. C.; Chuang, Y. C; Lin, C. S. Using Ologonucleotide-functionalized Au Nanoparicles to Rapidly Detect Food Borne Pathogens on a Piezoelectric Biosensor. *J. Microbiol. Methods* **2008,** *73* (1), 7–17.
9. Chen, T.; Yan, J.; Li, Y. Genotoxicity of Titanium Dioxide Nanoparticles. *J. Food Drug Anal.* **2014,** *22* (1), 95–104.
10. Chen, Y. S.; Lee, C. H.; Hung, M. Y.; Pan, H. A.; Chiou, J. C.; Huang, G. S. DNA Sequencing Using Electrical Conductance Measurements of a DNA Polymerase. *Nature Nanotechnol.* **2013,** *8*, 452–458.
11. Cheng, I. F.; Chang, H. C.; Chen, T. Y.; Hu, C.; Yang, F. L. Rapid (<5 min) Identification of Pathogen in Human Blood by Electrokinetic Concentration and Surface-enhanced Raman Spectroscopy. *Sci. Rep.* **2013,** *3*, 2365–2377.
12. Cho, E. C.; Choi, J. W.; Lee, M.; Koo, K. K. Fabrication of an Electrochemical Immunosensor with Self-assembled Peptide Nanotubes. *Colloids Surf. A Physiochem. Eng. Aspects* **2008,** *313* (314), 95–99.
13. Chu, P. T.; Wen, H. W. Sensitive Detection and Quantification of Gliadin Contamination in Gluten-free Food with Immunomagnetic Beads Based Liposomal Fluorescence Immunoassay. *Analytica Chimica Acta* **2013,** *787*, 246–253.
14. Clark, H. A.; Kopelman, R.; Tjalkens, R.; Philbert, M. A. Optical Nanosensors for Chemical Analysis inside Single Living Cells. 2. Sensors for pH and Calcium and the Intracellular Application of PEBBLE Sensors. *Anal. Chem.* **1999,** *71* (21), 4837–4843.

15. Comparelli, R.; Curri, M. L.; Cozzoli, P. D.; Striccoli, M. Optical Biosensing Based on Metal and Semiconductor Colloidal Nanocrystals. *Nanotechnol. Life Sci.* (VerlagChemie, Weinheim) **2007,** 123–174.
16. Cushen, M.; Kerry, J.; Morris, M.; Cruz-Romero, M.; Cummins, E. Nanotechnologies in the Food Industry: Recent Developments, Risks and Regulation. *Trends Food Sci. Technol.* **2012,** *24* (1), 30–46.
17. Dahlin, A. B. Size Matters: Problems and Advantages Associated with Highly Miniaturized Sensors. *Sensors,* **2012,** *12* (3), 3018–3036.
18. DelaEscosura-Muniz, A.; Ambrosi, A.; Merkoci, A. Electrochemical Analysis with Nanoparticle-based Biosystems. *Trends Anal. Chem.* **2008,** *27* (7), 568–584.
19. Dobrucka, R.; Cierpiszewski, R.; Korzeniowski, A. Intelligent Food Packaging Research and Development. *Sci. J. Logist.* **2015,** *11* (1), 7–14.
20. Duncan, T. V. Applications of Nanotechnology in Food Packaging and Food Safety: Barriermaterials, Antimicrobials and Sensors. *J. Colloid Interface Sci.* **2011,** *363* (1), 1–24.
21. Dungchai, W.; Siangproh, W.; Chaicumpa, W.; Tongtawe, P.; Chailapakul, O. Salmonella typhi Determination Using Voltammetric Amplification of Nanoparticles: A Highly Sensitive Strategy for Metal Loimmunoassay Based on a Copper-enhanced Gold Label. *Talanta* **2008,** *77* (2), 727–732.
22. EFSA Opinion of the Scientiac Panel on Food additives, Flavourings, Processing Aids and Materials in Contact with Food (AFC) on a Request Related to a 14th List of Substances for Food Contact Materials. *Eur. Food Safety Auth. J.* **2007,** *5* (2), 452–454.
23. EFSA Scientific Opinion of the Panel on Food Contact Materials, Enzymes, Flavorings and Processing Aids (CEF). *Eur. Food Safety Auth. J.* **2008,** *6* (12), 888–890.
24. El Boubbou, K.; Gruden, C.; Huang, X. Magnetic Glyco-nanoparticle: A Unique Tool for Rapid Pathogen Detection, Decontamination and Strain Differentiation. *J. Am. Chem. Soc.* **2007,** *129* (44), 13392–13393.
25. FDA, A Report of the U.S. Food and Drug Administration, Nanotechnology Task Force. *Nanotechnology*; FDA: Washington DC, 2011, 113.
26. Fornara, A.; Johansson, P.; Petersson, K.; Gustafsson, S.; Qin, J.; Olsson, E.; Ilver, D.; Krozer, A.; Muhammed, M.; Johansson, C. Tailored Magnetic Nanoparticles for Direct and Sensitive Detection of Biomolecules in Biological Samples. *Nano Lett.* **2008,** *8*, 3423–3428.
27. Fuertes, G.; Soto, I.; Vargas, M.; Valencia, A.; Sabattin, J.; Carrasco, R. Nanosensors for a Monitoring System in Intelligent and Active Packaging. *J. Sens.* **2016,** *8*.
28. Gao, X.; Nie, S. Molecular Profiling of Single Cells and Tissue Specimens with Quantum Dots. *Trends Biotechnol.* **2003,** *21* (9), 371–373.
29. Goldman, E. R.; Clapp, A. R.; Anderson, G. P.; Uyeda, H. T.; Mauro, J. M.; Medintz, I. L.; Mattoussi, H. Multiplexed Toxin Analysis Using Four Colors of Quantum Dot Fluororeagents. *Anal. Chem.* **2004,** *76* (3), 684–688.
30. Gonzalez, A.; Alvarez Igarzabal C. I. Nanocrystal-reinforced Soy Protein Films and Their Application as Active Packaging. *Food Hydrocoll.* **2015,** *43*, 777–784.
31. Grossman, H. L.; Myers, W. R.; Vreeland, V. J.; Bruehl, R.; Alper, M. D.; Bertozzi, C. R.; Clarke, J. Detetection of Bacteria in Suspension by Using a Super Conducting Quantum Interference Device. *Proc. Natl. Acad. Sci. USAm.* **2004,** *101* (1), 129–134.
32. He, L.; Kim, N. J.; Li, H.; Hu, Z.; Lin, M. Use of a Fractal-like Gold Nanostructure in Surface-enhanced Raman Spectroscopy for Detection of Selected Food Contaminants. *J. Agric. Food Chem.* **2008,** *56* (21), 9843–9847.

33. He, W.; Liu, Y.; Wamer, Y. G.; Yin J. J. Electron Spin Resonance Spectroscopy for the Study of Nanomaterial-mediated Generation of Reactive Oxygen Species. *J. Food Drug Anal.* **2014,** *22* (1), 49–63.
34. Hecht, A.; Commiskey, P.; Shah, Nicolaas.; Kopelman, R. Bead Assembly Magnetorotation as a Signal Transduction Method for Protein Detection. B*iosens. Bioelectron.* **2013,** *48*, 26–32.
35. Hernandez-Jover, T.; Izquierdo-Pulido, M.; Veciana-Nogues, M. T.; Mariné-Font, A.; Vidal-Carou, M. C. Biogenic Amine and Polyamine Contents in Meat and Meat Products. *J. Agric. Food Chem.* **1997**, *45* (6), 2098–2102.
36. Honarvar, Z.; Hadian, Z.; Mashayekh, M. Nanocomposites in Food Packaging Applications and their Risk Assessment for Health. *Electron. Phys.* **2016,** *8* (6), 2531–2538.
37. Huang, Y. F.; Wang, Y. F.; Yan, X. P. Amine-functionalized Magnetic Nanoparticles for Rapid Capture and Removal of Bacterial Pathogens. *Environ. Sci. Technol.* **2010,** *44* (20), 7908–7913.
38. Jain, S.; Singh, S. R.; Horn, D. W.; Davis, V. A.; Ram, M. J.; Pillai, S. R. Development of an Antibody Functionalized CarbonNanotube Biosensor for Foodborne Bacterial Pathogens. *J. Biosens. Bioelectron.* **2012,** S11:002.
39. Joo, J.; Yim, C.; Kwon, D.; Lee, J.; Shin, H. H.; Cha, H. J.; Jeon, S. A Facile and Sensitive Detection of Bacteria Using Magnetic Nanoparicle and Optical Nanocrystal Probes. *Analyst* **2012,** *137* (16), 3609–12.
40. Joyner, J. J. Nanosensors and Their Applications in Food Analysis: A Review. *Intern. J. Sci. Technol.* **2015,** *3* (4), 435–452.
41. Kaittanis, C.; Naser, S. A.; Perej, J. M. One-step, Nanoparticle-mediated Bacterial Detection with Magnetic Relexation. *Nano Lett.* **2007,** *7* (2), 380–383.
42. Kaittanis, C.; Santra, S.; Perej, J. M. Emerging Nanotechnology Based Strategies for the Identification of Microbial Pathogenesis. *Adv. Drug Del. Rev.* **2010,** *62* (4–5), 408–23.
43. Kalele, S. A.; Kundu, A. A.; Gosavi, S. W.; Deobagkar, D. N.; Deobagkar, D. D.; Kulkarni, S. K. Rapid Detection of Escherichia coli by Using Antibody-conjugate Silver Nanoshells *Small* **2006,** *2* (3), 335–358.
44. Kolarovszki, P. Research of Readability and Identification of the Items in the Postal and Logistics Environment. *Trans. Telecommun. J.* **2014,** *15* (3), 198–208.
45. Li, M.; Yin, J. J.; Wamer, W. G.; Lo, M. Mechanistic Characterization of Titanium Dioxide Nanoparticle Induced Toxicity Using Electron Spin Resonance. *J. Food Drug Anal.* **2014,** *22* (1), 76–85.
46. Li, N.; Brahmendra, A.; Veloso, A. J.; Prashar, A.; Cheng, X. R.; Hung, V. W. S.; Guyard, C.; Terebiznik, M.; Kerman, K. Disposable Immunochips for the Detection of *Legionella* Pneumophilausing Electrochemical Impedance Spectroscopy. *Anal. Chem.* **2012,** *84* (8), 3485–3488.
47. Li, Z.; Sheng, C. Nanosensors for Food Safety. *J. Nanosci. Nanotechnol.* **2014,** *14* (1), 905–912.
48. Lin, Y.; Lu, F.; Tu, Y.; Ren, Z. Glucose Biosensor Based on Carbon Nanotube Nanoelectrode Ensembles. *Nano Lett.* **2004,** *4* (2), 191–195.
49. Lin, Y. H.; Chen, S. H.; Chuang, Y. C.; Lu, Y. C.; Shen, T. Y.; Chang, C. A. Disposable Amperometricimmunosensing Strips Fabricated by Au Nanoparticles-modified Screen-printed Carbon Electrodes for the Detection of Foodborne Pathogen Escherichia coli O157:H7. *Biosens. Bioelectron.* **2008,** *23* (12), 1832–1837.

50. Lopez, B. P.; Merkoci, A. Nanomaterial Based Biosensors for Food Analysis Applications. *Trends Food Sci. Technol.* **2011,** *2* (11), 652–739.
51. Luechinger, N. A.; Loher, S.; Athanassiou, E. K.; Grass, R. N.; Stark, W. J. Highly Sensitive Optical Detection of Humidity on Polymer/metal Nanoparticle Hybrid Films. *Langmuir* **2007,** *23* (6), 3473–3477.
52. Mansooreh, R. M.; Rakhshanipour M.; Application of Nanoparticle Modified with Crown Ether in Colorimetric Determinations. *Arab. J. Chem.* **2015,** *42*, 215–219.
53. Mao, X.; Yang, L.; Su, X. L.; Li, Y. A Nanoparticle Amplification Based Quartz Crystal Microbalance DNA Sensor for Detection of Escherichia coli O157:H7. *Biosens. Bioelectron.* **2006,** *21* (7), 1178–1185.
54. Maurer, E. I; Comfort, K. K; Hussain, S. M; Schlager, J. J; Mukhopadhyay, M. M. Novel Platform Development Using an Assembly of Carbon Nanotube, Nanogold and Immobilized RNA Capture Element Towards Rapid, Selective Sensing of Bacteria. *Sensors* **2012,** *12* (16), 8135–8144.
55. McShan, D.; Ray, P. C.; Yu, H. Molecular Toxicity Mechanism of Nanosilver *J. Food Drug Anal.* **2014,** *22* (1), 116–27.
56. Mills, A. Oxygen Indicators and Intelligent Inks for Packaging Food. *Chem. Soc. Rev.* **2005,** *34* (12), 1003–1011.
57. Mills, A. and Hazafy, D. Nanocrystalline SnO_2-based, UVB-activated, Colourimetric Oxygen Indicator. *Sens. Actuat. B Chem.* **2009,** *136* (2), 344–349.
58. Minke, W. E.; Roach, C.; Hol, W. G.; Verlinde, C. L. Structure-based Exploration of Ganglioside GM1 Binding Sites of Escherichia coli Heat-labile Enterotoxin and Cholera Toxin for the Discovery of Receptor Antagonists. *Biochemistry* **1999,** *38* (18), 5684–92.
59. Murphy, C. J. Optical Sensing with Quantum Dots. *Anal. Chem.* **2002,** *74* (19), 520–526.
60. Nagy, J. O.; Zhang, Y.; Yi, W.; Motari, E.; Song, J. C.; Lejeune, J. T.; Wang, P. G. Glycopolydiacetylene Nanoparticles as a Chromatic Biosensor to Detect Shiga-like Toxin Producing Escherichia coli O157:H7. *Bioorg. Med. Chem. Lett.* **2008,** *18* (2), 700–703.
61. Nehra, A.; Singh, K. P. Current Trends in Nanomaterial Embedded Field Effecttransistor-based Biosensor. *Biosens. Bioelectron.* **2015,** *74,* 731–743.
62. Nopwinyuwong, A.; Trevanich, S.; Suppakul, P. Developmentof a Novel Colorimetric Indicator Label for Monitoringfreshness of Intermediate-moisture Dessert Spoilage. *Talanta* **2010,** *81* (3), 1126–1132.
63. Norman, R. S.; Stone, J. W.; Gole, A.; Sabo-Attwood, T. L. Targeted Photothermal Lysis of the Pathogenic Bacteria, *Pseudomonas aeruginosa*, with Gold Nanorods. *Nano Lett.* **2008,** *8* (1), 302–306.
64. Ozdemir, C.; Yeni, F.; Odaci, D.; Timur, S. Electrochemical Glucose Biosensing by Pyranose Oxidase Immobilized in Gold Nanoparticle-polyaniline/AgCl/gelatin Nanocomposite Matrix. *Food Chem.* **2010,** *119* (1), 380–385.
65. Pal, S.; Alocilja, E. C.; Downes, F. P. Nanowire Labeled Direct-charge Transfer Biosensor for Detecting Bacillus Species. *Biosens. Bioelectron.* **2007,** *22* (9), 2329–2336.
66. Palchetti, I.; Mascini, M. Electroanalytical Biosensors and their Potential for Food Pathogen and Toxin Detection. *Anal. Bioanal. Chem.* **2008,** *391* (2), 455–471.
67. Patrycja, C.; Wojciech W. Potentiometric Electronic Tongues for Foodstuff and Biosample Recognition-An Overview. *Sensors* **2011,** *11* (5), 4688–4701.
68. Perez, J. M.; Josephson, L.; Weissleder, R. W.; Use of Magnetic Nanoparticles as Nanosensors to Probe for Molecular Interactions. *Chembiochemistry* **2004,** *5* (3), 261–264.

69. Phillips, R. L.; Miranda, O. R.; You, C. C.; Rottelo, V. M.; Bunz, U. H. Rapid and Efficient Identification of Bacteria Using Gold-nanoparticle-poly(para-phenyleneethynylene) Constructs. *AngewandteChemie Int.* **2008,** *47* (47), 2590–2594.
70. Pimtong-Ngam, Y.; Jiemsirilers, S.; Supothina, S. Preparation of Tungsten Oxide–tin Oxide Nanocomposites and Their Ethylene Sensing Characteristics. *Sens. Actuat. A Phys.* **2007,** *139* (1), 7–11.
71. Pradhan, N.; Singh, S.; Ojha, N.; Shrivastava, A.; Barla, A.; Rai, V.; Boasal, S. Facets of Nanotechnology as Seen in Food Processing, Packaging, and Preservation Industry. *BioMed. Res. Int.* **2015,** 17.
72. Prasad, P.; Kocchar, A. Active Packaging in Food Industry: A Review. *J. Environ. Sci. Toxicol. FoodTechnol.* **2014,** *8* (5), 1–7.
73. Qiao, G.; Xu, Y.; Zhang, T.; Zhang, C.; Shi, Y.; Shuang L.; Dong C. Highly Selective and Sensitive Nanoprobes for Cyanide Based on Gold Nanoclusters with Red Fluorescence Emission. *Nanoscale* **2015,** *7* (29), 12666–12672.
74. Ramirez Frometa, N. Cantilever Biosensors. *BiotecnologiaAplicada* **2006,** *23* (4), 320–323.
75. Ravindranath, S. P.; Mauer, L. J.; Deb-Roy, C.; Irudayaraj, J. Biofunctionalized Magnetic Nanoparticle Integrated Mid-infrared Pathogen Sensor for Food Matrixes. *Anal. Chem.* **2009,** *81* (8), 2840–2846.
76. Ritter, S. K.; An Eye on Food. *Chem. Eng. News* **2005,** *83*, 28–34.
77. Sanvicens, N.; Pastells, C.; Pascual, N.; Marco, M. P. Nanoparticle-based Biosensors for Detection of Pathogenic Bacteria. *Trends Anal. Chem.* **2009,** *28* (11), 1243–1252.
78. Schofield, C. L.; Field, R. A.; Russell, D. A. Glyconanoparticle for the Cholorimetric Detection of Cholera Toxin. *Anal. Chem.* **2007,** *79* (4), 1356–1361.
79. Siegrist, M.; Cousin, M. E.; Kastenholz, H.; Wiek, A.; Public Acceptance of Nanotechnology Foods and Food Packaging: The Influence of Affect and Trust. *Appetite* **2007,** *49* (2), 459–466.
80. Sierra-Rosalesa, P.; Toledo-Neirab, C.; Squella, J. A.; Electrochemical Determination of Food Colorants in Soft Drinks Using MWCNT-modified GCEsP. *Sens. Actuat. Bl* **2017,** *240*, 1257–1264.
81. Silvestre, C.; Duraccio, D.; Cimmino, S. Food Packaging Based on Polymer Nanomaterials. *Progr. Polym. Sci.* **2011,** *36* (12), 1766–1782.
82. Siro, I.; Bhatt, R.; Gómez-López; V. M. *Intelligent Packaging in Food Safety*; JohnWiley & Sons: Chichester, UK, 2014.
83. Sonawane, S. K.; Arya, S. S.; LeBlanc, J. G.; Jha, N. Use of Nanomaterials in the Detection of Food Contaminants. *Eur. J. Nutr. Food Saf.* **2014,** *4* (4), 301–317.
84. Song, D.; Yang, R.; Wang, C.; Xiao, R.; Long, F. Reusable Nanosilver-coated Magnetic Particles for Ultrasensitive SERS-based Detection of Malachite Green in Water Samples. *Sci. Rep.* **2016,** *6* (22870).
85. Sozer, N.; Kokini, J. Nanotechnology and its Applications in the Food Sector- Review. *Trends Biotechnol.* **2008,** *27* (2).
86. Tang, D.; Tang, J.; Su, B.; Chen, G. Gold Nanoparticle-decorated Amine-terminated Poly (amidoamine) Dendrimer for Sensitive Electrochemical Immunoassay of Brevetoxin in Food Samples. *Biosens. Bioelectron.* **2011,** *26* (5), 2090–2096.
87. Tok, J. B. H.; Chuang, F. Y. S.; Kao, M. C.; Rose, K. A.; Pannu, S. S.; Sha, M. Y. Metallic Striped Nanowires as Multiplexed Immunoassay Platforms for Pathogen Detection. *AngewandteChemie Int. Ed.* **2006,** *45* (41), 6900–6904.

88. Uzawa, H.; Ohga, K.; Shinozaki, Y.; Ohsawa, I.; Nagatsuka, T.; Seto, Y.; Nishida, Y. A Novel Sugar-probe Biosensor for the Deadly Plant Proteinous Toxin, Ricin. *Biosens. Bioelectron.* **2008,** *24* (4), 929–933.
89. Valdes, M. G.; Gonzalez, A. C. V.; Calzon, J. A. G.; Diaz-Garcia, M. E. Analytical Nanotechnology for Food Analysis. *Microchimica Acta* **2009,** *166* (1), 1–19.
90. Viswanathan, S.; Wu, L. C.; Huang, M. R.; Ho, J. A. Electrochemical Immunosensor for Cholera Toxin Using Liposomes and Poly (3,4-ethylenedioxythiophene)-coated Carbon Nanotubes. *Anal. Chem.* **2006,** *78* (4), 1115–1121.
91. Vo-Dinh, T.; Cullum, B. M.; Stokes, D. L. Nanosensors and Biochips: Frontiers in Biomolecular Diagnostics. *Sens. Actuat. B Chem.* **2001,** *74* (1–3), 2–11.
92. Wang, Y.; Ravindranath, S.; Irudayaraj, J. Separation and Detection of Multiple Pathogens in a Food Matrix by Magnetic SERS Nanoprobes. *Anal. Bioanal. Chem.* **2011,** *399* (3), 1271–1278.
93. Wang, Z; Lee, S; Koo, K; Kim, K. Nanowire-Based Sensors for Biological and Medical Applications. *IEEE Trans. Nanobiosci.* **2016,** *15* (3), 186–99.
94. Weeks, B. L.; Camarero, J.; Noy, A.; Miller, A. E.; Stanker, L.; De Yoreo, J. J. A Microcantilever-based Pathogen Detector. *Scanning* **2003,** *25* (6), 297–309.
95. Wu, H. H.; Yin J. J.; Wamer, W. G.; Zeng, M.; Lo, Y. M. Reactive Oxygen Species-related Activities of Nano-iron Metal and Nano-iron Oxides. *J. Food Drug Anal.* **2014,** *22* (1), 86–94.
96. Wu, S.; Duan, N.; Shi, Z.; Fang, C.; Wang, Z. Simultaneous Aptasensor for Multiplex Pathogenic Bacteria Detection Based on Multicolor Upconversion Nanoparticles Labels. *Anal. Chem.* **2014,** *86* (6), 3100–3117.
97. Yakes, B. J.; Lipert, R. J.; Bannantine, J. P.; Porter, M. D. Detection of *Mycobacterium avium* subsp. Paratuberculosisby a Sonicate Immunoassay Based on a Surface-enhanced Raman Scattering. *Chin. Vacc. Immunol.* **2008,** *15* (2), 227–234.
98. Yang, D.; Zhu, L.; Jiang, X.; Guo, L.; Sensitive Determination of Sudan I at an Ordered Mesoporous Carbon Modified Glassy Carbon Electrode. *Sens. Actuat. B* **2009,** *141* (1), 124–129.
99. Yang, G. J.; Huang, J. L.; Meng, W. J.; Shen, M.; Jiao, X. A. A Reusable Capacitive Immunosensor for Detection of Salmonella spp. Based on Grafted Ethylene diamine and Self-assembled Gold Nanoparticle Monolayers. *AnalyticaChimicaActa* **2009,** *647* (2), 159–66.
100. Yang, H.; Li, H. Jiang, X. Detection of Foodborne Pathogens Using Bioconjugated Nanomaterials. *Microfluid. Nanofluid.* **2008,** *5* (5), 571–583.
101. Yang, L.; Li, Y. Quantum Dots as Fluorescencent Labels for Quantitative Detection of Salmonella typhimuriumin Chicken Carcass Wash Water. *J. Food Protect.* **2005,** *68* (6), 1241–1245.
102. Yang, M.; Kostov, Y.; Bruck, H. A.; Rasooly A. A Gold Nanoparticle Based Enhanced Chemiluminescenceimmunosensor for Detection of Staphylococcal Enterotoxin B (SEB) in Food. *Int. J. Food Microbiol.* **2009,** *133* (3), 265–71.
103. Yang, M.; Kostov, Y.; Rasooly, A. Carbon Nanotubes Based Optical Immunodetection of Staphylococcal Enterotoxin B (SEB) in Food. *Int. J. Food Microbiol.* **2008,** *127* (1–2), 78–83.
104. Yang, M.; Peng, Z.; Ning, Y.; Chen, Y.; Zhou, Q.; Deng, L.; Highly Specific and Cost-efficient Detection of *Salmonella* Paratyphi A Combining Aptamers with Single-walled Carbon Nanotubes. *Sensors* **2013,** *13* (5), 6865–6881.

105. Zhang, X.; Geng, P.; Liu, H.; Teng, Y.; Liu, Y.; Wang, Q.; Zhang, W.; Jin, L.; Jiang, L. Devlopment of an Electrochemical Immunoassay for Rapid Detection of E. Coli using Anodic Stripping Voltammetry Based on Cu@Au Nanoparticles as Antibody Labels. *Biosens. Bioelectron.* **2009,** *24* (7), 2155–2159.
106. Zhao, G.; Xing, F.; Deng, S. A Disposable Amperometric Enzyme Immunosensor for Rapid Detection of Vibrio Parahaemolyticusin Food Based on Agrose/Nano-Au Membrane and Screen-printed Electrode. *Electrochem. Commun.* **2007,** *9* (6), 1263–1268.
107. Zhao, X.; Hilliard, L. R.; Mechery, S. J.; Wang, Y.; Bagwe, R. P.; Jin, S.; Tan, W.; A Rapid Bioassay for Single Bacterial Cell Quntitation Using Biocojugated Nanoparticles. *Proc. Natl. Acad. Sci. US A* **2004,** *101* (42), 15027–15032.

INDEX

E

F

G

I

L

M

O

P

S

T